AF595325

Low Carbon Water Treatment and Energy Recovery

Low Carbon Water Treatment and Energy Recovery

Editors

Xin Zhao
Lili Dong
Zhaoyang Wang

Basel • Beijing • Wuhan • Barcelona • Belgrade • Novi Sad • Cluj • Manchester

Editors
Xin Zhao
Northeastern University
Shenyang, China

Lili Dong
Hainan University
Haikou, China

Zhaoyang Wang
Lanzhou University
Lanzhou, China

Editorial Office
MDPI
St. Alban-Anlage 66
4052 Basel, Switzerland

This is a reprint of articles from the Special Issue published online in the open access journal *Applied Sciences* (ISSN 2076-3417) (available at: https://www.mdpi.com/journal/applsci/special_issues/Treatment_Recovery).

For citation purposes, cite each article independently as indicated on the article page online and as indicated below:

Lastname, A.A.; Lastname, B.B. Article Title. *Journal Name* **Year**, *Volume Number*, Page Range.

ISBN 978-3-0365-9266-4 (Hbk)
ISBN 978-3-0365-9267-1 (PDF)
doi.org/10.3390/books978-3-0365-9267-1

Contents

About the Editors

Xin Zhao

Zhao Xin is an Associate Professor of Environmental Engineering PhD Supervisor at Northeastern University, and one of the most experienced researchers in Shenyang. His main research interests include water pollution control, waste resource utilization, and the molecular ecology of environmental microorganisms. He has presided over and participated in more than 20 scientific research projects, including projects of the National Natural Science Foundation of China and the Key Research and Development Program. He has published more than 90 papers in academic journals, including more than 60 SCI papers, 14 authorized patents, and 2 monographs.

Lili Dong

Dr. Lili Dong is a Master's Supervisor in the Department of Environmental Science, College of Ecology and Environment, Hainan University. She has published 13 scientific papers in the past 5 years. Her research focuses on the study of lignocellulosic biomass biohydroxanes and the development of high-concentration organic wastewater treatment technologies for meeting and improving the current standards of wastewater treatment.

Zhaoyang Wang

Wang Zhaoyang graduated from Harbin Institute of Technology and is now an Associate Professor and Master's Supervisor at the College of Resources and Environment, Lanzhou University. Dr. Wang has published more than 20 papers in peer-reviewed journals and has led a number of provincial- and ministerial-level projects, such as projects of the National Natural Science Foundation of China and various horizontal projects. Dr. Wang's main research interests include the deep treatment and reuse of municipal wastewater and the theory and technology of the high-efficiency treatment of industrial wastewater.

Preface

Climate change caused by excessive carbon dioxide (CO_2) emissions poses a global challenge. Developing innovative wastewater treatments that minimize or eliminate CO_2 emissions is crucial to achieving carbon neutrality. To reduce carbon emissions from water treatment processes, technological and scientific innovation is required, such as biomass production, the use of bubble-less-gas mass-transfer bioreactors, reduced aeration with enhanced microbial processes, high-efficiency pumps and blowers, low-pressure self-cleaning free membranes, and the integration of solar power and bioelectrical systems. Considering the present technology for water and wastewater treatment, there is significant scope for improvement. By exploring low-carbon sewage treatment technologies, we expect to establish a theoretical basis for practical engineering applications and contribute to achieving the goal of carbon neutrality, which is central to the promotion of sustainable socio-economic development. This rationale underpins this Special Issue, which focuses on low-carbon water treatment and energy recovery.

Xin Zhao, Lili Dong, and Zhaoyang Wang

Editors

Editorial

Low-Carbon Water Treatment and Energy Recovery

Xin Zhao [1,*], Lili Dong [2] and Zhaoyang Wang [3]

1 Department of Environmental Engineering, School of Resources and Civil Engineering, Northeastern University, Shenyang 110819, China
2 Key Laboratory of Agro-Forestry Environmental Processes and Ecological Regulation, School of Ecology and Environment, Hainan University, Haikou 570228, China; donglili0569@126.com
3 Department of Environmental Science and Engineering, College of Earth and Environmental Sciences, Lanzhou University, Lanzhou 730000, China; wzy@lzu.edu.cn
* Correspondence: zhaoxin@mail.neu.edu.cn; Tel./Fax: +86-24-83679128

1. Introduction

Climate change led by excessive carbon dioxide (CO_2) emissions poses a global challenge. Developing innovative wastewater treatments that minimize or eliminate CO_2 emissions is crucial to achieve carbon neutrality. To reduce carbon emissions from the water treatment processes, technological and scientific advances are required, such as biomass production to lower CO_2 emissions, the use of bubble-less gas mass transfer bioreactors, reduced aeration with enhanced microbial processes, high-efficiency pumps and blowers, low-pressure self-cleaning free membranes, and the integration of solar power systems and bioelectrical systems. The present technology for water and wastewater treatment offers significant room for improvement. By exploring low-carbon sewage treatment technologies, a theoretical basis for practical engineering applications is expected to be established, contribute toward the goal of carbon neutrality, which is of great significance for promoting sustainable socio-economic development [1]. This rationale underpins the Special Issue focusing on low-carbon water treatment and energy recovery.

2. Application of Energy Saving and Low-Carbon Technologies in Water Treatment

In light of the above, this Special Issue aims to gather cutting-edge research on pertinent subjects, specifically focusing on advancements in water treatment technologies based on conventional water treatment technologies, to achieve the best possible utilization of resources and energy consumption, as well as low-carbon treatment technologies to minimize greenhouse gases emissions. Challenging issues relating wastewater treatment in the process of sustainable and resource-based utilization were addressed. A total of 14 papers were submitted, of which 11 (79% acceptance rate) were selected for publication. The Special Issue encompasses a wide range of topics, primarily centered on energy recovery, pollutant removal, and the assessment of removal efficacy.

Fuel cells have been attracting increasing attention owing to their ability to directly convert chemical energy in fuel into electrical energy. The paper authored by Han et al. demonstrated that compared with other types of fuel cells, aluminum–air fuel cells have the characteristics of high-power density, high current density, high fuel utilization rate (up to approximately 95% energy utilization), stable structure, and strong adaptability to fuel [2]. In addition, aluminum–air fuel cells offer a promising solution for achieving carbon neutrality in wastewater treatment, presenting an alternative to traditional energy-consuming electrocoagulation systems [2]. Hydrogen (H_2), known for its cleanliness, cost-effectiveness, and renewability, was considered an attractive alternative to traditional fossil fuels, and could greatly alleviate the global greenhouse effect and energy crisis at present. Amongst various H_2 production methods, an electrocatalytic H_2 evolution reaction (HER) via electrochemical water splitting, an important energy recovery technique, has attracted considerable attention in the field. In their forthcoming paper, Jiang et al.

Citation: Zhao, X.; Dong, L.; Wang, Z. Low-Carbon Water Treatment and Energy Recovery. *Appl. Sci.* **2023**, *13*, 9758. https://doi.org/10.3390/app13179758

Received: 25 August 2023
Accepted: 25 August 2023
Published: 29 August 2023

found that HER activity is inherently correlated with the density of the GBs in Au NPs, where active sites bind H_2 more strongly [3]. Their finding holds promise for enhancing the catalyst's HER performance. Advances in nanotechnology have paved the way for antimicrobial technology in nanomaterials. The increased bacterial and viral resistance has hindered traditional methods of water purification and disinfection. The review by Xin et al. was an exhaustive summary of recent research advances on antimicrobial ZnO-based nanomaterials [4]. By summarizing previous studies, ZnO-based nanomaterials with excellent antibacterial effects were obtained. In detail, various strategies to enhance the antimicrobial ability of ZnO-based nanomaterials were proposed. Finally, current limitations of the improvement strategies outline future directions for development and potential applications in the field. Here, two papers focus on the degradation of pollutants in water, specifically addressing typical organic pollutants and metal ions. In the first paper authored by Liu et al., the persistence, toxicity, mobility, and widespread pollution of short-chain PFASs in the water cycle are discussed. The study compares how various treatment techniques, such as the adsorption, electrochemical oxidation, and photocatalytic degradation have certain removal effects on short-chain PFASs [5]. Among these techniques, adsorption proves to be the most widely applied technique for the effective removal of short-chain PFASs, suitable for a wide concentration range of pollution and aligning with low-carbon policies. The findings provide a scientific basis for the effective treatment and regulation of short-chain PFAS contaminations in diverse water sources. For metal ions, the coexistence of iron and manganese was common in groundwater. Ways in which to remove iron and manganese efficiently and stably have become the focus of research. Kang et al., [6] developed a three-dimensional response surface to investigate the interaction of three variables (pH, manganese sand dosage, and the initial Fe/Mn concentration ratio) on the removal of iron and manganese ions. The study revealed that pH exerts the most significant influence on the adsorption process, and the optimal conditions for Fe and Mn ion adsorption by modified Mn sand were pH 7.2, 3.54 g/L of sand, and an initial Fe/Mn ratio of 3.8. In addition, the authors confirmed that the relative error between the model predictions and experimental values was close to 1%.

Reliable evaluation and accurate prediction of water pollution indicators are crucial for effective water resource management and pollution control. In studies on water quality evaluation and prediction, water quality evaluation and prediction models were established, combining the autoregressive integrated moving average (ARIMA) model and the wavelet neural network (WNN) mode with the bat algorithm to determine the optimal weight of each individual model [7]. The trained T–S fuzzy neural network was applied to the water quality evaluation, achieving total positive water quality grade evaluation rates of 90.38% and 88.46%, respectively. For the prediction of water quality, the combined model (ARIMA–WNN) produced a higher prediction accuracy, with an improvement of up to 68.06%. In the following article authored by Zhang et al., the focus was on nitrogen removal through partial nitrification coupled with ANAMMOX technology treating livestock wastewater. The study aimed to evaluate the influences of non-biodegradable organic matter and microbial communities on the performance [8]. Nitrite accumulation efficiencies of 78.4% and 64.7% were obtained in an intermittent aeration sequencing batch reactor and a continuous aeration sequencing batch reactor, respectively, at a loading rate of 0.93 kg ammonium/m^3/d. With ANAMMOX activity at low loading rates (118 ± 13 mg COD/L and 168 ± 9 mg COD/L) and an average nitrite removal rate of 87.4%, partial nitrification treating the livestock wastewater was successfully achieved.

To address the gap in computer-aided multi-class environmental microorganism (EM) detection, Li et al., proposed a novel squeeze-and-excitation-based mask region convolutional neural network (SEM–RCNN) [9]. Mask RCNN, one of the most applied object detection models, uses ResNet for feature extraction. Specifically, in terms of technology, an improved method combining Mask RCNN with SENet was proposed in this paper. In terms of applications, model training and testing were conducted in a small dataset of EMs and a large dataset of blood cells, respectively. Currently, the field of transparent image analysis has emerged as a prominent research topic due to advancements in computer vision, enabling image analysis through computer use. To address the low contrast between the foreground and background of transparent images making their segmentation difficult for computers, Li and Yang [10] aimed to address these problems in transparent images by cropping the image into patches and classifying their foreground and background. They used CNNs and ViT deep learning methods to compare the patch- and pixel-level performances of the transparent image segmentation. The study highlighted the advantages of CNN and ViT models in image classification. CNN excels at extracting the local features of images, whereas ViT effectively captures the global features of images combined with contextual information. To address the limitation of contour conglutination of dense objects while counting, the next paper [11] proposed a novel pixel interval down-sampling network (PID-Net), which was an end-to-end convolutional neural network (CNN) model with an encoder–decoder architecture for dense, tiny object (yeast cells) counting tasks with higher accuracy. By comparing with the proposed PID-Net and classical U-Net-based yeast counting results, the evaluation indexes of accuracy, dice, Jaccard, precision, counting accuracy, and Hausdorff distance of PID-Net improved by 0.04%, 0.15%, 0.26%, 0.4%, and 5.7%, respectively, and the Hausdorff distance decreased by 0.0394. These results highlighted the enhanced segmentation performance of PID-Net for the accurate counting of dense, tiny objects in a small dataset.

Author Contributions: All authors have contributed equally to this book. All authors have read and agreed to the published version of the manuscript.

Funding: This research received no external funding.

Acknowledgments: We express our gratitude to all the authors and peer reviewers for their valuable contributions to this Special Issue. Congratulations to all the authors, regardless of the final decisions of the submitted manuscripts. The feedback, comments, and suggestions from the reviewers and editors have significantly aided in improving the quality of the papers. Finally, we extend our congratulations on the successful launch of *Applied Sciences*, an esteemed, international, peer-reviewed journal that provides readers with unrestricted access to comprehensive information encompassing various aspects of applied natural sciences.

Conflicts of Interest: The authors declare no conflict of interest.

References

1. Li, D.; Wang, Z.; Yang, Y.; Liu, H.; Fang, S.; Liu, S. Research Status and Development Trend of Wastewater Treatment Technology and Its Low Carbonization. *Appl. Sci.* **2023**, *13*, 1400. [CrossRef]
2. Han, X.; Qi, H.; Qu, Y.; Feng, Y.; Zhao, X. Simultaneous Phosphate Removal and Power Generation by the Aluminum–Air Fuel Cell for Energy Self-Sufficient Electrocoagulation. *Appl. Sci.* **2023**, *13*, 4628. [CrossRef]
3. Jiang, R.; Fu, J.; Wang, Z.; Dong, C. Grain Boundary—A Route to Enhance Electrocatalytic Activity for Hydrogen Evolution Reaction. *Appl. Sci.* **2022**, *12*, 4290. [CrossRef]
4. Xin, Z.; He, Q.; Wang, S.; Han, X.; Fu, Z.; Xu, X.; Zhao, X. Recent Progress in ZnO-Based Nanostructures for Photocatalytic Antimicrobial in Water Treatment: A Review. *Appl. Sci.* **2022**, *12*, 7910. [CrossRef]
5. Liu, Y.; Li, T.; Bao, J.; Hu, X.; Zhao, X.; Shao, L.; Li, C.; Lu, M. A Review of Treatment Techniques for Short-Chain Perfluoroalkyl Substances. *Appl. Sci.* **2022**, *12*, 1941. [CrossRef]
6. Kang, H.; Liu, Y.; Li, D.; Xu, L. Study on the Removal of Iron and Manganese from Groundwater Using Modified Manganese Sand Based on Response Surface Methodology. *Appl. Sci.* **2022**, *12*, 11798. [CrossRef]
7. Jiao, G.; Chen, S.; Wang, F.; Wang, Z.; Wang, F.; Li, H.; Zhang, F.; Cai, J.; Jin, J. Water Quality Evaluation and Prediction Based on a Combined Model. *Appl. Sci.* **2023**, *13*, 1286. [CrossRef]
8. Zhang, M.; Chen, X.; Xu, X.; Fu, Z.; Zhao, X. Evaluation of Non-Biodegradable Organic Matter and Microbial Community's Effects on Achievement of Partial Nitrification Coupled with ANAMMOX for Treating Low-Carbon Livestock Wastewater. *Appl. Sci.* **2022**, *12*, 3626. [CrossRef]
9. Zhang, J.; Ma, P.; Jiang, T.; Zhao, X.; Tan, W.; Zhang, J.; Zou, S.; Huang, X.; Grzegorzek, M.; Li, C. SEM-RCNN: A Squeeze-and-Excitation-Based Mask Region Convolutional Neural Network for Multi-Class Environmental Microorganism Detection. *Appl. Sci.* **2022**, *12*, 9902. [CrossRef]

10. Yang, H.; Zhao, X.; Jiang, T.; Zhang, J.; Zhao, P.; Chen, A.; Grzegorzek, M.; Qi, S.; Teng, Y.; Li, C. Comparative Study for Patch-Level and Pixel-Level Segmentation of Deep Learning Methods on Transparent Images of Environmental Microorganisms: From Convolutional Neural Networks to Visual Transformers. *Appl. Sci.* **2022**, *12*, 9321. [CrossRef]
11. Zhang, J.; Zhao, X.; Jiang, T.; Rahaman, M.M.; Yao, Y.; Lin, Y.-H.; Zhang, J.; Pan, A.; Grzegorzek, M.; Li, C. An Application of Pixel Interval Down-Sampling (PID) for Dense Tiny Microorganism Counting on Environmental Microorganism Images. *Appl. Sci.* **2022**, *12*, 7314. [CrossRef]

Review

Research Status and Development Trend of Wastewater Treatment Technology and Its Low Carbonization

Demin Li [1], Zhaoyang Wang [1,*], Yixuan Yang [1], Hao Liu [1], Shuai Fang [1] and Shenglin Liu [2]

[1] College of Earth and Environmental Sciences, Lanzhou University, Lanzhou 730000, China
[2] Xinjiang Haomiao Environmental Protection Technology Co., Ltd., Alaer 843300, China
* Correspondence: wangzhaoyanghit@126.com; Tel.: +86-931-8912404

Abstract: With the rapid development of the social economy, the demand for water resources is gradually increasing, and the corresponding impact of water pollution is also becoming more severe. Therefore, the technology of sewage treatment is developing rapidly, but corresponding problems also arise. The requirements of energy conservation and emissions reduction under the goal of carbon neutrality and dual carbon pose a challenge to the traditional concept of sewage treatment, and there is an urgent need for low-carbon sewage treatment technology aiming at energy conservation, consumption reduction and resource reuse. This review briefly introduces conventional sewage treatment technology and low-carbon sewage treatment technology, and analyzes the research status and development trend of low-carbon sewage treatment technology in detail. The analysis and comparison of conventional and low-carbon sewage treatment technologies is expected to provide a theoretical basis for the practical engineering application of low-carbon sewage treatment technologyto achieve the goal of carbon neutrality. It is of great significance to promote the sustainable development of society and the economy.

Keywords: sewage treatment; low-carbon technology; research status; sustainable development

Citation: Li, D.; Wang, Z.; Yang, Y.; Liu, H.; Fang, S.; Liu, S. Research Status and Development Trend of Wastewater Treatment Technology and Its Low Carbonization. *Appl. Sci.* 2023, 13, 1400. https://doi.org/10.3390/app13031400

Academic Editor: José Carlos Magalhães Pires

Received: 8 December 2022
Revised: 11 January 2023
Accepted: 18 January 2023
Published: 20 January 2023

1. Introduction

In recent years, with the rapid development of urbanization and the rapid growth of population, the pressure on our living environment is also increasing day by day [1,2]. The development of industry and people's efforts to meet their quality-of-life requirements have caused a certain degree of impact on the environment. Problems such as the greenhouse effect and water pollution can be seen everywhere in our daily life [3,4]. In addition, the ecological and environmental problems brought by these problems are gradually feeding back into our lives, causing a series of predictable troubles. Global warming caused by greenhouse gases, rising sea level, frequent bad weather, malodorous water and toxic and harmful waste water are causes for alarm, requiring countermeasures [5]. It has become the consensus of all countries in the 21st century to reduce greenhouse gas emissions, reduce carbon emissions and slow down global warming. At the General Debate of the 75th Session of the United Nations General Assembly, China proposed to strive for the grand goal of carbon neutrality before 2060 [6]. This is China's solemn commitment to the world, demonstrating its responsibility as a major country and enhancing its international influence [7]. It also makes clearer the importance of greenhouse gas emissions, low-carbon living and achieving carbon neutrality.

It has become an important trend of social development in the modern era to ensure the quality of our ecological environment, implement energy savings and emissions reductions, improve energy utilization, promote environmental protection and high-efficiency and high quality development, and seek low-carbon technology as the new way forward for the development of energy-consuming industries [8–10]. Sewage treatment accounted for much energy consumption in in our country, and the proportion of energy consumption in society as a whole was increasing year by year [11]. At present, sewage treatment technology in our country consists mainly of conventional pretreatment and conventional advanced treatment, two treatment units, and the treatment methods used differ according to sewage quality and other factors. However, whether physical or biochemical treatment is adopted, a large amount of resources and energy is consumed in the process [12]. In addition, a large number of additional pollutants such as CO_2, CH_4 and N_2O are produced in the treatment process, which is the main contribution of the sewage treatment system to global warming. To some extent, this is a non-green means of changing water quality via energy consumption by energy dissipation and pollution transfer [13]. At the same time, a variety of pollutants contained in sewage are rarely recovered for use, resulting in a great waste of resources [14]. Effective exploitation and utilization of potential resources in sewage or sludge is a means of achieving low carbonization of sewage treatment.

It is very important to vigorously develop energy-saving and low-carbon sewage treatment technology for each treatment process and treatment unit. With the demand for water resource protection and recycling due to economic and social development, sewage treatment in all aspects will increase significantly in the future [15,16]. Therefore, in large sewage treatment systems, low-carbon treatment technology should be promoted in all aspects, and any available resources, whether pollutants themselves or products generated by the treatment process, should be fully utilized to carry out reasonable carbon conversion [17]. In the treatment technology, those processes with high energy consumption and low treatment efficiency should be eliminated, and the treatment process should be improved by means of co-construction or expansion. In addition, accurate assessment of sewage quality and targeted water treatment should be achieved through automated control, to replace tedious and unstable manual control. The operating parameters of each processing unit should be optimized on the basis of accurate assessment, and the best processing effect achieved, while minimizing resource and energy consumption [18,19]. Through the research and development of emerging sewage treatment technologies at home and abroad, new sewage treatment technologies, new materials or microorganisms can be developed to achieve lowcarbon, carbon reuse, carbon sequestration and other clean and green treatment technologies as much as possible.

This review briefly introduces conventional sewage treatment technology and low-carbon sewage treatment technology, and then analyzes the research status and development trend of low-carbon sewage treatment technology in detail. Through the analysis and comparison of conventional and low-carbon sewage treatment technologies, it is expected to provide a theoretical basis for the practical engineering application of low-carbon sewage treatment technologies and continue towards achieving the goal of carbon neutrality, which is of great significance to promoting the sustainable development of the social economy.

2. Conventional Wastewater Treatment Technology

This section examines conventional urban sewage treatment technology, and takes the activated sludge removal process as an example to discuss and analyze.

The traditional activated sludge removal process is to supply oxygen to activated sludge in sewage through external aeration equipment, and convert organic matter in sewage (40–50%) into CO_2 through activated sludge, and the remaining organic matter (50–60%) into residual sludge, which is difficult to be degraded by microorganisms [20]. From long-term engineering practice and various studies, it has been found that, depending on the change of water quality, the characteristics of microbial metabolic activity, operation management, technical economics, discharge requirements, etc., a variety of operation modes and pool types have been developed. The main types are as follows: pushed-flow activated sludge method, completely mixed activated sludge method, adsorption-regeneration activated sludge method, delayed aerated activated sludge method, pure-oxygen-aerated activated sludge method, sequential-batch-reactor activated sludge method (SBR), etc. Their removal rates of biological oxygen demand (BOD), chemical oxygen demand (COD), total suspended solids (TSS) and other pollution indicators in municipal sewage, as well as their advantages and disadvantages, are shown in Table 1 below:

Table 1. Removal rate of BOD, COD, TSS and other pollution indicators in municipal sewage by activated sludge methods and their advantages and disadvantages.

Operation Mode of Activated Sludge Method	BOD	COD	TSS	Advantages	Disadvantages	References
	Removal Rate (%)					
Pushed-flow activated sludge process	90–95	90–95	90–95	① The degradation efficiency of sewage is higher. ② The treatment of wastewater is more flexible.	The phenomenon of insufficient aeration at the head of the tank and excessive gas supply at the tail of the tank increases the power cost.	[21,22]
Completely mixed activated sludge process	85–90	85–90	90–95	① Strong ability to bear the impact load, to weaken the peak load. ② It can save power and facilitate operation management.	①Continuous water inflow and outflow may cause short circuits. ②Prone to sludge swelling.	[23,24]
Adsorption-regeneration activated sludge process	80–90	80–85	85–90	① The contact time is shorter and the adsorption pool volume is smaller. ② Bearing a certain impact load, the sludge in the regeneration tank is convenient to use.	①The treatment effect of wastewater is lower than that of the traditional activated sludge process. ②The treatment effect of wastewater with high dissolved organic matter is poor.	[21,25]
Delayed aerated activated sludge process	75–95	85–95	90–95	① The organic load is low, the residual sludge is less, and the sludge is stable and does not need to be digested. ② It has high stability of treatment water quality, strong adaptability to the impact load of wastewater and does not require a primary sedimentation tank.	The tank capacity is large, the aeration time is long, the construction cost and the operation cost are high, and it occupies a large area.	[26,27]
Pure-oxygen-aerated activated sludge process	90–95	85–90	90–95	① Greatly improves oxygen diffusion ability in the mixture ② The volume of gas required can be greatly reduced, the volume load can be greatly increased, it is not prone to sludge swelling, it has high treatment efficiency, the required aeration time is short, the amount of residual sludge generated is less.	The device is complex, management is troublesome, and the structure of the closed container is demanding.	[28–32]
Sequential-batch reactor activated sludge process (SBR)	85–95	85–90	90–99	The operation management is simple, the cost is reduced, the impact load is resistant, the effluent quality is good, the activated sludge filamentous bacteria can be inhibited, the nitrogen and phosphorus removal.	Automation control requirements are high. Operation, management and maintenance require high quality of operation and management personnel. High requirements for drainage equipment.	[33–37]

2.1. Pushed-Flow Activated Sludge Process

The pushed-flow activated sludge process is also known as the traditional activated sludge process. The surface of the push-flow aeration tank is rectangular. Under the push of aeration and hydraulic conditions, the water in the aeration tank is evenly pushed to flow. The wastewater enters from the head end of the tank and flows out from the tail end of the tank, and the liquid flow in the front section and the liquid flow in the back section do not mix. The process flow chart is shown in Figure 1. In the process of aeration, with the change in environment from the head of the tank to the end of the tank, , the biological reaction rate, the F/M value, the quantity and quality of microbial community, the adsorption, flocculation and stabilization of activated sludge, and the settlement-concentration performance are all constantly changing [21,22].

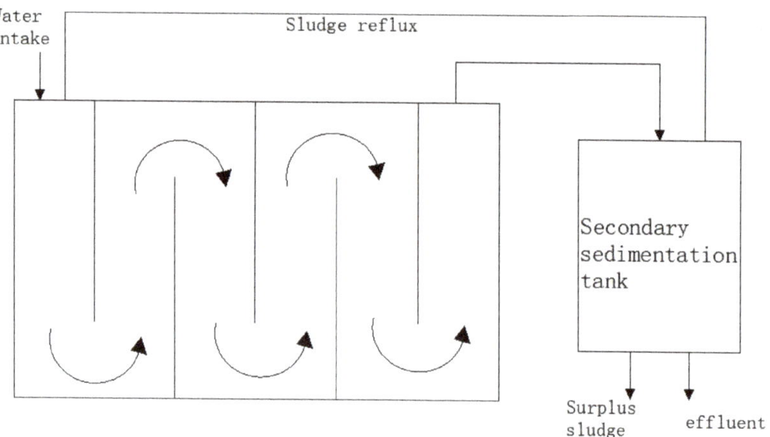

Figure 1. Process flow chart of pushed-flow activated sludge process.

The traditional activated sludge aeration tank is rectangular, the water flow is longitudinal mixed push flow, the aeration time of the mixed liquid in the aeration tank is usually 4–8 h, the sludge concentration is generally controlled within 2–3 g/ L, the amount of returned sludge requires 25–50% of the water intake, and the removal rate of BOD, COD and suspended matter can reach 90–95% [38]. The pushed-flow activated sludge method has the advantages of high treatment efficiency and flexible treatment modes, but it also leads to an energy surplus. The activated sludge at the head of the tank always absorbs gas from the head of the tank to the end of the tank, which increases the power cost to a certain extent and has the problem of energy waste, which is not conducive to the low-carbon emissions advocated by the present concept of carbon-neutral and sustainable development.

2.2. Completely Mixed Activated Sludge Process

A completely mixed aeration tank means that the waste water is fully mixed with the original mixture after entering the aeration tank. Therefore, the composition, F/M value and quantity and quality of microbial community of the mixture in the tank are completely uniform. The position of the whole process on the sludge growth curve is only one point. This means that the biological reaction is the same in all parts of the aeration tank, and the oxygen absorption rate is also the same. This process is characterized by a strong ability to withstand impact load, and the ability of the mixed liquid in the tank to dilute the wastewater and weaken the peak load. And because the aerobic requirement of the whole tank is the same, it can save power. The aeration tank and sedimentation tank can be built together for easy operation and management. However, the continuous inflow and outflow of water may cause short circuits, and the process is prone to sludge swelling and other problems [23]. The technological process is shown in Figure 2. For the treatment of municipal wastewater, the BOD load (Ns) is 0.2–0.6 kg BOD5 / (kg MLSS·d), the volume load (Nv) is 0.8–2.0 kg BOD5 / (m^3·d), the sludge age (mean residence time of biosolids) (θr) is 5–15 days, the concentration of suspended solids (MLSS) is 3000–6000 mg/L, the concentration of volatile suspended solids (MLVSS) is 2400–4800 mg/L, the sludge reflux ratio (R) is 25–100%, and the aeration time (t) is 3–5 h. The removal rate of BOD, COD and TSS reaches more than 85% [24]. Since the completely mixed aeration tank requires continuous water inflow and outflow, and the tank is fully and evenly mixed, there is the problem that the sludge resources cannot be better treated, and it is easy to cause sludge expansion. At the same time, it is not easy to adjust the treatment method after certain changes in sewage quality making the recycling of sludge resources is especially difficult. Compared with the traditional activated sludge process, it may be more efficient in energy utilization. However, the poor utilization of sludge is a shortcoming.

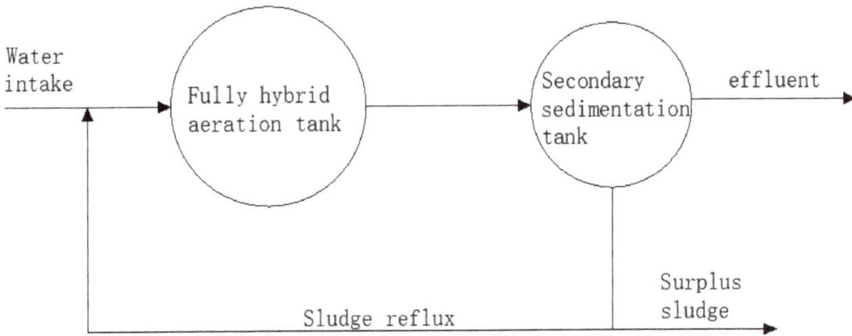

Figure 2. Process flow chart of completely mixed aeration tank.

2.3. Adsorption-Regeneration Activated Sludge Process

The adsorption-regeneration activated sludge method is also known as the biological adsorption method or contact stabilization method. The main feature of this operation mode is that the adsorption and metabolism of activated sludge for the degradation of organic pollutants are each carried out in their own reactors. The wastewater is fully recycled in the regeneration pool, and the activated sludge with strong activity enters the adsorption pool at the same time. The two contact fully in the adsorption pool, and most of the organic matter in the wastewater is absorbed by the activated sludge, and purified. The sludge separated from the secondary sedimentation tank enters the regeneration tank, where the activated sludge metabolizes and degrades the organic matter, and microorganisms proliferate. When the microorganisms enter the endogenous metabolic stage, the activity and adsorption function of the sludge are fully recovered, and it then enters the adsorption tank together with the wastewater [25,39]. The technological process is shown in Figure 3. For the treatment of municipal wastewater, the BOD load (Ns) is 0.2–0.6 kg BOD5/(kg MLSS· d), the volume load (Nv) is 1.0–1.2 kg BOD5/(m^3·d), the sludge age (mean residence time of biosolids) (θr) is 5–15 d, the mixed liquid suspended solids concentration (MLSS) is 1000–3000 mg/L, the mixed liquid volatile suspended solids concentration (MLVSS) is 3200–5200 mg/L, the adsorption pool concentration is 600–1200 mg/L, the regeneration pool concentration is 2400–7000 mg/L, and the adsorption pool reaction time is 0.5–1.0 h. The removal rates of BOD, COD and TSS can reach 80–90% under the conditions of 3–6 h of regeneration tank, 25–100% sludge reflux ratio (R) and 3–5 h aeration time (t). Its treatment effect on sewage is lower than that of the traditional activated sludge process, and it also has the disadvantage of poor treatment effect on wastewater with high dissolved organic matter [40]. Its advantages are that its energy consumption and sludge utilization are considerable, and it can more fully use the sludge, combined with the regeneration pool to carry out sludge conversion operation, reduce the discharge and disposal of sludge, and to a certain extent achieve the purpose of energy saving.

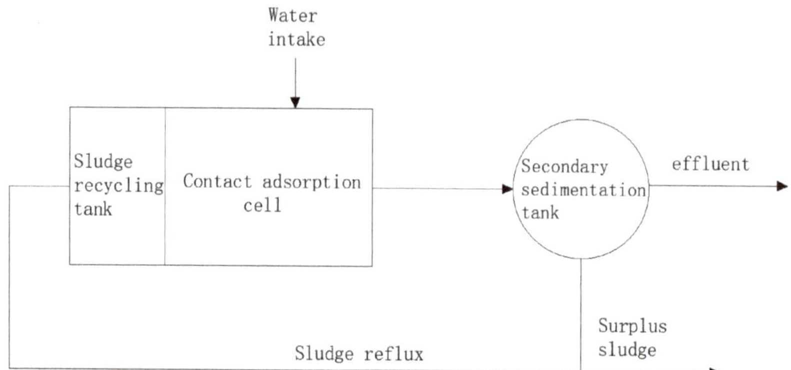

Figure 3. Process flow chart of adsorption-regeneration activated sludge method.

2.4. Delayed Aerated Activated Sludge Process

The delayed aerated activated sludge process is also called the completely oxidized activated sludge process. The main characteristics of the process are low organic load, continuous internal metabolism of sludge, less residual sludge, and stable sludge with no need for further digestion treatment. This process can be called the comprehensive wastewater and sludge treatment process. The process has the advantages of high stability of water quality treatment, strong adaptability to the impact load of wastewater, and no need to set up a primary sedimentation tank. The main disadvantages are large pool capacity, long aeration time, high construction and operation costs, and large size. This process is suitable for the sewage and industrial wastewater needs of a small town, which requires high water quality, and therefore should not be treated by sludge alone. The aeration tanks used in the process are completely mixed or push-flow. The technological process is as shown in Figure 2. When the reference values of the various design parameters used in the treatment of urban sewage are the same as 2.2, the removal rate of BOD, COD and TSS can only reach 85–90%. The removal rate is also lower than that of the previous activated sludge treatment processes [26,27].

This process has a very strong advantage in the operation and management of sludge, and can make full use of sludge resources, so as not to need much sludge disposal, and save part of the energy in the whole treatment system. However, its long aeration time, high operating cost and large area are great shortcomings of the process, and these shortcomings are difficult to make up objectively. Therefore, the process can only be applied if the site conditions are sufficient, and the water quality conditions are more suitable to better use the advantages of sludge treatment to better realize the purpose of energy saving, emissions reduction and low carbonization of sewage treatment.

2.5. Pure-Oxygen Activated Sludge Process

Compared with air aeration, pure-oxygen aeration, also known as enriched-oxygen aeration, has the following characteristics:

(1). The oxygen content of air is generally 21%; the oxygen content of pure oxygen is 90–95%, and the partial pressure of oxygen is 4.4–4.7 times higher than that of air, so pure oxygen aeration can greatly improve the diffusion capacity of oxygen in the mixed liquid.

(2). The oxygen utilization rate can be as high as 80–90%, while the air-aerated activated sludge method is only about 10%, so the volume of gas required to achieve the same oxygen concentration can be greatly reduced.

(3). Activated sludge concentration (MLSS) can reach 4000–7000 mg/L, so the volume load can be greatly increased at the same organic load.

(4). The sludge index is low, only about 100, which is not prone to sludge swelling;

(5). High treatment efficiency and short aeration time.
(6). The amount of residual sludge produced is small.

There are three types of pure oxygen aeration pool: (1) multistage- sealed, in which oxygen is introduced into the pool from the closed top cover, the sewage from the first stage is pushed forward step-by-step, oxygen flows from a centrifugal compressor through a hollow shaft into a rotary impeller, which mixes the sludge and oxygen in the pool to keep full contact, so that the sludge can greatly absorb unused oxygen and biochemical reaction metabolites from the previous level of discharge. (2) The old aeration tank is reformed, and a curtain is set on the pool. Not only pure oxygen enters, but also compressed air. Some tail gas is discharged, and can also be recycled. (3) Open pure oxygen aeration tank. The technological process is shown in Figure 4. When the reference values of the various design parameters used in the treatment of urban sewage are the same as 2.2, the removal rate of BOD, COD and TSS can reach 90% [28–32].

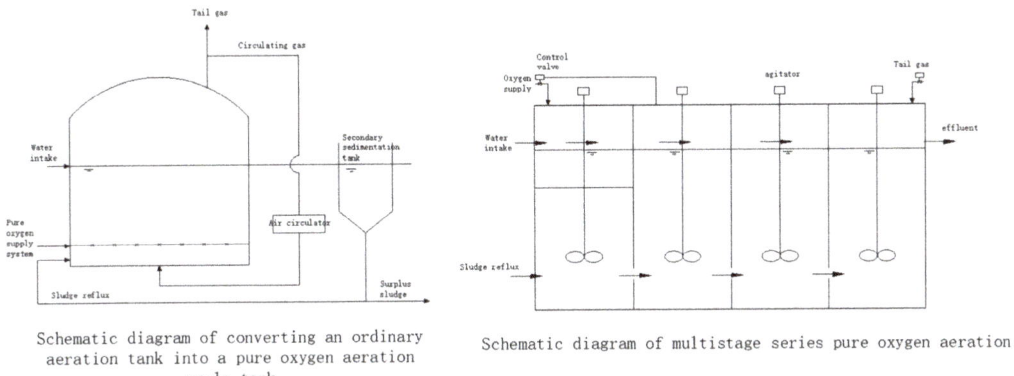

Figure 4. Process flow chart of pure-oxygen-aerated activated sludge method.

By supplying pure oxygen, the process can greatly improve the diffusion capacity of oxygen in solution and reduce the volume of gas, thus reducing aeration time, aeration energy consumption and carbon footprint. However, it still cannot achieve good sustainable development and utilization of resources in the disposal of sludge operation management. In addition, the structure and management requirements of the equipment are relatively high, which increase the operating costs of the treatment process.

2.6. Sequential-Batch Activated Sludge Process

The sequential-batch activated sludge process is also known as the SBR process because of the intermittent form of operation, so each reaction tank is a batch of sewage treatment; hence the name. Because of the high flexibility of the SBR operation, it can replace the continuous activated sludge process in most situations to achieve the same or similar results. By changing the operation mode of the SBR, you can simulate the operation mode of the full hybrid and push-flow processes. In the reaction stage, the organic matter in the reaction tank is degraded by microorganisms, and the wastewater concentration becomes lower and lower, which is very similar to the steady-state push flow, except that it is a kind of temporal push flow. If the influent period is long and the accumulation of organic matter in the wastewater of the reaction tank is very small during this period, then the situation is close to a complete mixture. The technological process is shown in Figure 5. When the reference values of the various design parameters used in the treatment of urban sewage are the same as 2.2, the removal rate of BOD, COD and TSS can reach 90% [33–37].

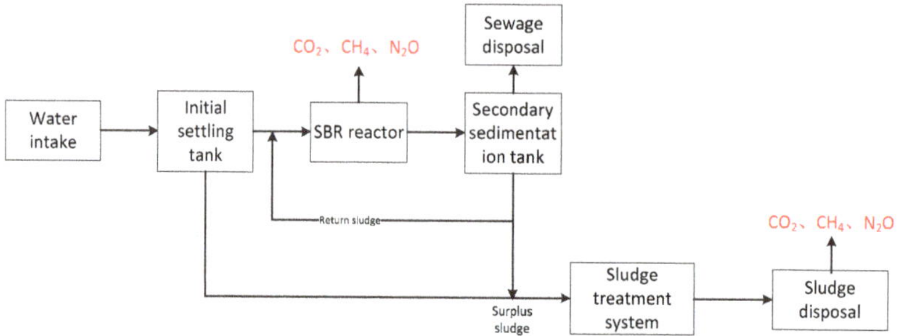

Figure 5. Process flow chart of sequential-batch activated sludge method.

The process improves on the traditional push-flow activated sludge method, and can simulate a complete mixture in certain cases, which can better make up for some problems, such as long aeration time and easy sludge swelling caused by sludge disposal. However, it still fails to solve the fundamental problems of aeration energy consumption and sludge resources, there are still problems of high operation, management and maintenance costs, and the process has relatively high requirements for drainage equipment. During the whole treatment process, when the sewage is biochemically treated by an activated sludge system, a large amount of external energy is consumed for the oxygen supply, and a large amount of greenhouse gases such as CO_2, CH_4 and N_2O are released during the treatment process. However, there are many ways to dispose the residual sludge. Simple treatment such as random accumulation and landfill will also result in the emission of CO_2, CH_4 and N_2O due to anaerobic fermentation and other factors.

2.7. Summary Discussion

Activated sludge is the main pollutant-removal process. In the cell composition of activated sludge microorganisms, carbon (C), hydrogen (H), oxygen (O) and nitrogen (N) account for about 90–97%, and the rest (3–10%) are inorganic nutrient elements, mainly phosphorus (P). The treatment of domestic sewage generally does not require additional nutrients, and some industrial wastewater treatment do require additional nutrients. In conventional aerobic treatment, N and P nutrients are added in accordance with BOD:N:p = 100:5:1. The removal rate of nutrients by the activated sludge method is more than 95%. A 5000 m^3/d municipal sewage treatment plant will produce a residual sludge volume of 7–19 g/(L·d) when treating municipal wastewater by the activated sludge processes described in 2.1, 2.2, 2.5 and 2.6. The 2.3 adsorption-regeneration activated sludge method can make full use of the regenerating tank to activate and reuse the residual sludge and reduce the sludge discharge. The amount of residual sludge produced is about 1–8 g / (L·d) [41–44]. The 2.4 organic load of the delayed aerated activated sludge method is low, the sludge is continuously in a state of endogenous metabolism, and the residual sludge is basically absent. The sludge produced by the activated sludge process needs to be transported to the dewatering plant and then treated. After the initial dewatering treatment of the sludge enrichment tank, it is further dehydrated through a plate-and-frame filter press and other measures. In this process, the main energy consumption is the electric energy of the plate-and-frame filter press, as well as the auxiliary energy consumption of the sludge transport conveyor belt. In the process of wastewater treatment, the main energy consumption in the biochemical reaction tank is air blast aeration and electricity consumption of some auxiliary equipment. An urban sewage treatment plant with a 15,000 m^3/d treatment capacity requires more than 3 blowers and at least 1 plate-and-frame frame filter press. Therefore, based on the calculation of 3 blowers and 1 plate-and-frame filter press, the daily energy consumption produced by full load operation would be 2582.68 Kw·h [45,46]. However, the wastewater produced in this process can be reused for greening, flushing, and so on. The dehydrated sludge can be used as plant fertilizer, building materials, and so on.

In the process of sewage treatment, which consumes much external energy, the form of pollution is actually changed from water pollution to air pollution and sludge pollution, which is obviously not in line with the concept of sustainable development. The COD in sewage can also be used for anaerobic digestion to produce CH_4 and fermentation to produce hydrogen. The CH_4 and H_2 produced can be reused as part of the plant thermal raw materials. The generated surplus sludge can also be used, and the best utilization is incineration capacity. The concept of resource reuse is also implemented to achieve sustainable development, realize low-carbon treatment of sewage and reduce carbon footprint [47].

In addition, in terms of resource utilization, traditional sewage treatment mainly recycles treated water for greening irrigation, washing or industrial cooling, but ignores the problem that sewage itself contains rich carbon resources [48]. Sewage is actually a carrier of resources and energy. It is estimated that the potential chemical energy of urban sewage with a COD of 400–500 mg/L is 1.5–1.9 kW·h/m^3 [49], and a metabolic heat of 0.14×10^8 J can be generated per kg of COD [50]. The heat generated by an increase or decrease of 5 °C of sewage is almost equal to the annual power generation of 332 large power plants, and about 4 times the metabolic heat of organic matter [51]. Sewage contains such a huge amount of energy that if the chemical energy or even heat energy of part of the COD were rationally utilized and converted into electric energy, it could theoretically achieve self-sufficiency in energy consumption, and even output energy (electric energy, heat energy) for use outside of the plant. There is much theoretical evidence to prove that new sewage treatment plants in the future could be not energy consumers, but energy suppliers [52]. However, the traditional sewage treatment process uses energy to supply oxygen to remove COD, which results in a large amount of chemical energy and heat energy in sewage not being extracted and utilized, which goes against the concept of sustainable development. Therefore, recycling and utilization of potential energy in sewage treatment has important practical significance, and plays an immeasurable role in practicing the concept of low-carbon development and promoting the low-carbon operation of sewage treatment.

3. Low-Carbon Sewage Treatment Technology

As one of the major sources of greenhouse gas emissions, the current sewage pollution removal technology mainly relies on energy dissipation and pollution transfer. The treatment process has high energy consumption and emits a large amount of greenhouse gases such as CO_2, CH_4 and N_2O. To effectively control the greenhouse gas emissions from sewage treatment plants and achieve the emission reduction requirements under the dual-carbon target, the traditional concept of sewage treatment must be challenged. It is required that the process of sewage treatment should be as low-carbon as possible, to realize self-sufficiency in energy and full utilization of resources in the process of sewage treatment, and make up for energy consumption and recycle resources by relying on the energy and resources contained in sewage treatment plants or the sewage itself [53].

The removal process of activated sludge in conventional municipal sewage treatment technology still has serious problems of high energy consumption. The removal rate of nutrients and pollutants in sewage can reach more than 95%, but the consumption of aeration energy and waste sludge cannot be ignored. In a hypothetical calculation, a municipal sewage treatment plant with a treatment capacity of 15,000 m^3/d consumes 2582.68 Kw·h of electricity in the biochemical treatment unit and sludge disposal alone, not including air pollution and greenhouse gas emissions produced in the treatment process. Thus it can be seen that the application of low-carbon wastewater treatment technology is very important.

The most serious problem of the push-flow activated sludge method is that the sludge at the head and tail of the tank receives the aeration time unevenly, which leads to a large amount of wasted aeration, resulting in unnecessary energy consumption. The characteristic problem of the push-flow reaction tank is that the BOD concentration is high where the water flows in, and low where the water flows out, so the number of aerators should decrease along the long direction of the tank, that is, to increase the spacing between aerators. In this way, the BOD concentration of the effluent can be met, one-time input can be reduced, and unnecessary aeration can be reduced [54]. The author believes that an intelligent aeration management system can be designed, through which the aeration capacity and aeration time of the push-flow gallery can be intelligently regulated, to make the oxygen demand of the sludge just enough to achieve the purpose of low carbonization and energy savings [55,56]. The completely mixed activated sludge method is a unified operation mode after the sludge and wastewater are completely mixed. Compared with the push-flow activated sludge method, it has no major problems in aeration energy consumption, but because it is a completely mixed operation, it is prone to sludge expansion and the production of excessive residual sludge, which affects effluent quality. In the face of the problems of the process system, it is necessary to monitor the process operation management in detail and continuously observe the water quality changes to prevent sludge buildup. In case of emergency, drugs can be added to enhance sludge settlement performance or directly kill filamentous bacteria through emergency measures [57]. Both the pure-oxygen-aerated activated sludge method and the SBR process are relatively mature. The biggest common problem is that the equipment is more complex and the requirements for management personnel are higher. In addition, these four process systems have two common problems that most need to be solved: aeration energy loss and sludge disposal energy loss. The essence of the activated sludge method is the adsorption and metabolism of activated sludge microorganisms to remove pollutants, and the metabolism needs to absorb oxygen, and the energy consumption of aeration is difficult to avoid. Therefore, in order to solve this problem, Tan Tiepeng [58] described the research process of oxygen demand in the activated sludge process at home and abroad, and the energy-saving technology of the microporous aeration system, and compared the microporous aerator with the perforated tube; it was proven that the former saves 4.4% more electricity compared with the latter. Wang Xian [59] et al. proposed a design concept of subsection control of dissolved oxygen concentration in an aeration tank by discussing oxygen demand and its distribution in the aeration tank, and applying oxygen transfer theory. If this design concept is applied to engineering practice, compared with the conventional design method, the energy consumption and equipment capacity of an aeration system can be reduced by about 15%. The author believes that energy consumption can be properly reduced through intelligent regulation of aeration equipment. At the same time, the metabolism mode of activated sludge microorganisms can be acclimated or changed to reduce the consumption of oxygen, or even operate without oxygen, or be combined with photosynthetic bacteria to achieve metabolism through photocooperation instead of oxygen energy consumption; these studies are very meaningful. As for the problem of sludge disposal, the author believes that the energy consumption of sludge disposal can be compensated by incineration power generation to achieve low carbonization as far as possible. In addition, the treated sludge can also be recycled as building materials and land fertilizer, to achieve the purpose of sustainable development [60–62]. The adsorption-regeneration activated sludge method and delayed aerated activated sludge method make good use of the residual sludge and reduce the sludge load. The sludge remains in a state of endogenous metabolism, thus greatly reducing the generation of residual sludge. Therefore, the problem of low carbonization in these two process systems is mainly the energy consumption of aeration. Similarly to the above four process systems, to solve the problem of aeration energy consumption, it is still necessary to change the microbial nature of activated sludge or reduce energy consumption as much as possible.

At present, Dai Xiaohu et al. [63], considering the development stage and international development trend of our country, put forward the development concept of "green, low-carbon, resource recycling, environmentally friendly, locally-adapted conditions" and corresponding key measures. Ruan Xiaoyang et al. [64] proposed that sludge is a pollutant with utilization value, and can become a resource as long as the sludge is treated and disposed by certain means. This is of great significance for improving the ecological environment, while also helping to reduce the consumption of natural resources and promote the sustainable development of cities in the process of recycling. For low-carbon sewage treatment technology, the main emission reduction measures are summarized as follows: resource recovery carbon conversion and reuse, comprehensive treatment of low carbon and carbon sequestration, operation parameter optimization and process improvement, reasonable carbon distribution, and other related upgrading and renovation treatment technologies. The discussion of the low-carbon operation modes in these sewage treatment processes is summarized in Table 2 below. All kinds of water treatment technologies are gradually upgraded and reformed. While high water treatment efficiency is required, clean treatment technologies such as low-carbon emission continue to emerge, and some sewage treatment plants have put them into use and achieved good results [65–67]. Therefore, this paper analyzes the current research status of low-carbon sewage treatment technology and the prospects for development in the future, so as to provide guidance for more understanding of low-carbon sewage treatment technology.

Table 2. Discussion and summary of low-carbon operation modes in the sewage treatment process.

Low-Carbon Operation Mode	Low Carbonization Pathway	Concrete Measure	Low-Carbon Achievement
Resource recycling carbon conversion and reuse	recycle water	After the sewage treatment is up to standard, it will be used for factory reuse.	"Turning waste into treasure" rationally utilizes all valuable and usable substances produced in the process of sewage treatment, maximizes the concept of sustainable development, reduces carbon emissions and carbon loss from all aspects, and reduces carbon footprint.
	Energy-carrying gas	Energy-carrying gases such as CH4 and H2 are used for fuel.	
	heat energy	Heat generated by microorganisms during sewage treatment is used for heating.	
	sludge	Recovery disposal sludge is used in a burning capacity.	
Comprehensive treatment of low carbon and carbon sequestration	energy-saving and cost-reducing	According to the nature of wastewater, the appropriate treatment process can be selected to reduce the energy consumption such as aeration.	To keep up with the pace of the times, the production and processing equipment is updated over time, and the low-carbon energy-saving equipment is used as much as possible to reduce carbon emissions. Actively develop and utilize new sewage treatment microorganisms to reduce energy consumption and carbon emissions as much as possible and reduce carbon footprint.
	equipment replacement	Upgrade equipment such as old blowers or mixers.	
	Development and utilization of carbon sequestration microorganisms	Reasonable development and utilization of photosynthetic bacteria and other microorganisms for low energy consumption and carbon sequestration methods.	
Operation parameter optimization and process improvement	Operation parameter optimization	Intelligent parameter control is carried out for each processing unit.	Under the premise of ensuring the standard of sewage treatment, the operation parameters of each treatment unit should be controlled, so as to achieve the standard of sewage treatment with the lowest energy consumption possible of each physical unit. The treatment process of each treatment unit should be updated in time, and the sewage treatment should be completed according to the concepts of sustainable development and low-carbon treatment.
	combined technology	According to the nature of sewage, can choose a lower carbonization treatment method combined treatment.	
	technology improvement	The low carbonization process was improved.	

3.1. Research Status

3.1.1. Resource Recovery Carbon Conversion and Reuse

In the process of sewage treatment, there are many resources that may be neglected and not fully utilized, for example, some of the energy and heat lost, the transfer of pollutants and the existing sludge. In fact, these can be converted into carbon to achieve recycling.

After treatment, those that can reach the standard of reuse can be reused in the factory for greening irrigation, washing or industrial cooling water and other methods of recycling, which is conducive to the dual carbon goal. The chemical and heat energy, and other energy resources generated in the sewage treatment process, may be directly consumed or discharged in the traditional treatment process without effective use, resulting in a great waste of resources. The heat energy may be relatively small, but it could still be put to effective use. Miricioiu et al. [68] prepared materials for CO_2 adsorption by activating the residual sludge generated during sewage treatment. After many tests, the adsorption capacity of the material was 11.87 cm 3/g, the separation efficiency was high, and the CO_2 recovery rates were 99.68 and 98.11%, respectively. Studies [69,70], have shown that the heat energy can be recycled through water source heat pump technology and used for heating the sewage treatment plant or surrounding facilities, which can more fully utilize energy resources, reduce carbon emissions, and realize the low-carbon potential of the sewage treatment plant. In addition, in sewage treatment plants with biological sewage treatment, the sludge discharged often contains many substances such as nitrogen and phosphorus converted from sewage. These resources can also be reused to reduce the use of new carbon, to achieve the purpose of relatively low carbonization. Studies [71] have shown that sludge containing nitrogen and phosphorus can be used as fertilizers for composting. And some concentrated sludge that does not contain harmless ingredients can be used as raw materials for incineration power generation after treatment, so that they are valuable for power generation, which can also effectively realize the purposes of energy saving, reuse and low carbonization.

3.1.2. Comprehensive Treatment of Low Carbon and Carbon Sequestration

The traditional activated sludge treatment process requires a large amount of carbon source input and aeration, both of which increase the utilization of carbon. The extensive use of aeration consumes a large amount of external energy for sewage treatment, which leads to the waste of resources, and the generated CO_2 and other gases increase air pollution. The traditional activated sludge method also produces a large amount of activated sludge, which represents a certain amount of sludge pollution, and does not conform to the concept of low carbonization. Compared with the activated sludge method, anammox technology can directly generate nitrogen by the reaction of nitrous nitrogen and ammonia nitrogen, which can greatly save the energy consumption of aeration and the input of carbon sources, but also reduce the production of sludge, greatly realizing the goal of low-carbon sewage treatment. It is a relatively clean biological nitrogen removal water treatment technology and has a good development prospect.

The annual energy consumption of the Sheboygan Wastewater Treatment Plant in the United States accounts for 3% of the total social energy consumption and is the most energy-consuming public facility [72]. In order to achieve the goals of energy self-sufficiency in wastewater treatment and sustainable utilization of resources, the United States Water Environment Research Foundation (WERF) has indicated that all wastewater treatment plants in the United States must achieve carbon-neutral operation by 2030 [73]. The Sheboygan Wastewater Treatment Plant is the first to start the practice of carbon neutral operation. Based on the operational goal and implementation plan of "zero energy consumption" of the "Wisconsin Focus Energy" project, the Sheboygan Wastewater Treatment Plant carried out a series of energy recovery plans from 2002 to 2011, adding 12 30 kW micro gas turbines and 4 heat recovery treatment plants. By 2012, the plant could generate 16,800 kW·h/d of electricity and 16,120 kW·h/d of heat by using cogeneration technology, which offset about 90% of the power consumption and 85% of the heat demand of the sewage plant, basically achieving energy self-sufficiency. In terms of sludge disposal, the equipment for sludge enrichment and dewatering disposal has been upgraded to achieve dehydration treatment with lower energy consumption. At the same time, the treated sludge can make up for the energy consumption of sludge disposal through incineration and power generation, and can also be used for farmland soil remediation, building materials, etc., to fully realize the sustainable development concept of low carbonization and resource recycling [74,75]. In the process of treating part of the industrial wastewater, additional nutrients need to be added, and underutilized nutrients can be recycled to reduce carbon loss. The wastewater discharged is to the standard of reuse for plant greening and equipment washing. In addition, the sewage plant self-funded nearly 1.1 million US dollars for a series of energy-saving upgrading and operation optimizations, updating the pump and blower and other mechanical equipment for energy savings of 20 and 13% respectively, installing a flow control valve (energy saving 17%), updating the digestion tank heating equipment, upgrading the programmable logic controller (PLC), supervisory control and data acquisition (SCADA) and so on, greatly reducing energy consumption. By 2013, the Sheboygan Sewage Treatment Plant had achieved a ratio of electricity production to electricity consumption of 90–115% and a ratio of heat production to heat consumption of 85–90%, basically close to the goal of carbon neutral operation.

The technology of the Sheboygan Wastewater Treatment Plant is worth studying for domestic sewage treatment plants. The energy treatment of sludge can not only effectively relieve the difficult situation of sludge disposal, but also relieve the pressure of fossil energy consumption, and greatly reduce the impact on the environment. Through the double measures of open source and cost reduction, a series of energy saving transformations has been carried out, which provides valuable experience for other sewage treatment plants.

As for low-carbon water treatment, it is still difficult to change only in the aspect of energy saving and carbon reduction. It still needs some technologies that can sequester carbon, directly use or otherwise not emit CO_2 and require low energy from the outside world. At present, it has been found that sewage treatment by photosynthetic bacteria and other organisms can absorb nitrogen and phosphorus and other substances in water for their own use without treatment, and CO_2 can also be absorbed and converted into organic carbon for internal use by photosynthesis [76–79].

3.1.3. Operation Parameter Optimization and Process Improvement

According to the water quality of different sewage, the treatment process selected by the sewage treatment plant will be different. When facing a specific sewage, the operation parameters of the different treatment process will also change slightly. Only by adjusting the most appropriate operation parameters can the treatment be most effective and require the lowest energy consumption. Duan Steel [80] envisioned the application prospect of energy-saving and low-carbon technology in sewage treatment through the construction of a precise aeration system. Liu Guitao [81] et al., based on fuzzy control of three-level inverter frequency conversion speed regulation energy-saving technology, studied the frequency conversion speed regulation system control of a high-powered pump in a sewage treatment plant, and finally realized effective control of a high-powered motor, which can significantly improve the energy-saving effect of a high-power motor in sewage treatment. Through the application of frequency converters in the energy-saving transformation of an aeration fan in a sewage treatment station, Mu Jian [82] adapted the frequency conversion to start slowly, effectively avoiding fan surge, effectively reducing starting current and running current, thus achieving the purpose of energy savings, and at the same time reducing the influence of an overly large or small opening of an intake valve on the load of the rear system. Ma Yong [83] et al. adjusted the sludge layer height of a secondary sedimentation tank to control the sludge return flow of the A/O process, and took the sludge layer height of secondary sedimentation tank as the control variable to establish the sludge return control strategy and controller. An A/O process pilot test device was used to treat real domestic sewage, and the established controller was verified. The results showed that the average effluent ammonia and total nitrogen concentrations were reduced by about 8 and 15%, respectively, compared with the traditional constant sludge reflux ratio control. With an increase in the returned sludge concentration, the sludge discharge and sludge return flow decreased correspondingly. The returned sludge concentration increased by 25.6% on average, and the sludge return flow decreased by 20% on average. The average SS concentration of effluent was reduced by 35.3%, which could avoid sludge loss and maintain the stable operation of the system. Compared with the traditional constant sludge reflux ratio control, the controller has great advantages in terms of system stability, operation cost and effluent quality by adjusting and improving the high- energy-consumption electrical equipment in the whole sewage treatment plant, especially the high-powered equipment such as the water pump and blower. Appropriate adjustment of operating parameters and process optimization can better realize the low-carbon sustainable development concept of sewage treatment. In the whole water treatment process, different water quality will lead to different treatment intensity of each treatment unit in the process. For example, with different concentrations of nitrogen and phosphorus, COD and BOD in the water, the treatment units are also different, so each unit needs to be adjusted to the most suitable operating parameters, to ensure the best treatment effect under the conditions of the lowest energy consumption and low carbon [84].

Traditional sewage treatment processes use physical or chemical methods to remove colloidal substances in sewage, and biological methods to treat organic pollutants in sewage. Existing studies [16] have shown that the improvement of the SBR process adds a pre-anoxic zone on the basis of the original, and the preliminary denitrification of the sewage is carried out through the pre-anoxic process, which provides a good reaction environment for the subsequent treatment, optimizes the selection of carbon sources in the distribution of raw water, and improves the efficiency of the sewage treatment overall and the recovery and utilization rate of water resources. With the improvement of the A2O process, the anaerobic phosphorus removal area and low oxygen aeration area are set next to the settlement area in order to form an integrated setting, which is conducive to improving work efficiency and shortening sewage treatment time. Making full use of the principle of air pressure, a low-oxygen aeration area in front of the air push zone is established to provide natural force and reduce energy consumption and impact load. The unique dissolved-oxygen control system can enhance the removal of COD, total nitrogen TN and total phosphorus TP. At the same time, the process has a wide range of applications, mainly in the efficient treatment of municipal sewage and various types of industrial waste water. It is the main process of urban sewage treatment with low-carbon sources.

For the first time, the coupling process of SHARON and anammox was successfully applied to the treatment of sludge digestion solution with high ammonia nitrogen in the Dokhaven Sewage Treatment Plant in Rotterdam, Netherlands. Compared with the traditional nitrification and denitrification process, the coupling process of SHARON and anammox can reduce the CO_2 emission by 88% and the operating cost by 90%. Domestic studies on this process are mostly in the pilot stage. Zhao Qing et al. [85] treated landfill leachate through the short-cut nitrification-coupled anammox process, and found that the average removal rate of ammonia nitrogen and nitrite nitrogen in this system was over 95%. Nimazelang et al. [86] started and acclimated the one-stage short-cut nitrification and anammox process in SBR. When the process reached stable operation, the SBR effluent was put into the filter column with different volume filling ratios of slow-release carbon sources to conduct in-depth nitrogen removal research. The results show that after 176 days of start-up and acclimation in SBR, stable operation of the one-stage short-cut nitrification and anammox process is successfully achieved, with the removal rate of ammonia nitrogen up to 98% and the total nitrogen removal rate up to 73%. Beijing Drainage Group adopted the anaerobic ammonia oxidation process to treat sludge digesters, with a total treatment scale of 15,900 m^3/d, which can reduce carbon emissions by 10,500 t per year [71].

In the context of rapid development, the value of development needs to be well considered. Successful development must be of good long-term value. Through the improvement of the industry, improvement of shortcomings, improvement of the quality and efficiency of the process, the process operation can produce better value. For the optimization of process operation parameters, the process and sewage quality change rapidly with various factors, so it is more necessary to monitor the operation parameters in detail, utilize the best reaction parameters, realize the most efficient process operation, and improve the process value.

3.2. Growing Trend

At present, the discharge standards of sewage treatment are very strict. All kinds of sewage treatment industries are following the discharge indicators, and vigorously developing and optimizing. At the same time, with the emergence of the development of carbon-neutral and low-carbon industry, the research on energy conservation and emission reduction has become an important task in industrial development. Therefore, although the sewage treatment industry is not a prominent carbon emission industry, its long-term water treatment projects and the accumulation of large amounts of high energy consumption will definitely contribute to carbon emissions [87]. Therefore, in the process of sewage treatment, it is very important to develop the low-carbon treatment technology of sewage through energy saving, consumption reduction and resource recovery as much as possible. However, at present, the new methods of low-carbon water treatment technology and related evaluation standards are still relatively lacking. When we are developing low-carbon water treatment technology, we also need to have an accounting of the carbon emissions in the whole treatment process, so as to promote the perfection of the certification and testing system for low-carbon sewage treatment technology. In addition, it would help us regulate put emissions requirements on CO_2 and other greenhouse gases from sewage treatment. According to the current situation, in the future we will need more advanced monitoring systems, and obtain the greenhouse gas pollutants produced in the process of sewage treatment to make the corresponding inventory descriptions, and develop a series of discharge standards. In terms of the carbon emission assessment of sewage treatment plants, more appropriate algorithms are also needed to establish a deep learning model, so as to better realize the accurate accounting of carbon emissions in the process of sewage treatment [88–90].

According to the above research status of low-carbon sewage treatment technology, we can find that it is possible to realize low-carbon sewage treatment. According to the current situation of each existing sewage treatment plant, reasonable improvements should be made under the supervision and technical support of relevant departments. Relevant departments cannot be one-size-fits-all, requiring the unified transformation of each sewage treatment plant, but specific rectifications must be carried out. The proposed resource recycling carbon conversion and reuse is the basic guarantee for the current low-carbon sustainable development. The concept of sustainable development can and must be realized under any conditions, which is the general trend of development. As for the implementation of low-carbon and carbon sequestration comprehensive treatment, optimization of operation parameters and process improvement, it is necessary for each sewage and water treatment plant to carry out the implementation according to its own regional water quality and plant conditions. Proper equipment upgrading and process optimization are beneficial, but it is also necessary to be guided by facts.

4. Conclusions

With the rapid development of urbanization and the rapid growth of population, the amount of sewage is also increasing sharply. Therefore, the problem of sewage treatment has become urgent. In addition, the waste of resources and energy consumption in the process of sewage treatment and the additional impact on the environment are also urgent matters. Since the conventional wastewater treatment activated sludge process is always achieved through activated sludge microorganisms to complete the degradation of pollutants, it necessitates oxygen energy consumption and sludge generation. No matter how the processing technology and parameters are changed, it cannot avoid energy consumption and the disposal of harmful substances. Therefore, it is difficult to achieve carbon neutrality and reduce carbon emissions. On the other hand, although some energy consumption is unavoidable, in low-carbon sewage treatment technology, low-carbon sewage treatment can be achieved through the resource recycling of carbon conversion and reuse, the generation of new methods of low-carbon carbon sequestration, and a certain degree of operational parameter optimization and process optimization, so as to implement the concept of low-carbon sustainable development as far as possible, and can be gradually improved through scientific research. Therefore, as described in this paper, on the basis of traditional conventional water treatment technology, it is also necessary to constantly study and update water treatment technology to realize the full utilization of resources and energy consumption as far as possible, as well as low-carbon water treatment technology, to reduce the production of CO_2 and CH_4 and other greenhouse gases. Promoting the innovation and improvement of sewage treatment technology and processes on the basis of saving energy and reducing consumption, high efficiency and low carbonization of sewage treatment can be realized, while the problem of sewage treatment is dealt with well.

Author Contributions: D.L.: Conceptualization, Methodology, Writing—Original Draft; Z.W.: Funding acquisition, Writing—Review & Editing; Y.Y., H.L. and S.F.: Investigation; S.L.: Data Curation. All authors have read and agreed to the published version of the manuscript.

Funding: This work is partially supported by the Natural Science Foundation of China (Grant no. 21906011), the Open Foundation of MOE Key Laboratory of Western China's Environmental System, Lanzhou University and the Fundamental Research Funds for the Central Universities (lzujbky-2021-kb01) and the finance science and technology planning project of Alaer City, Xinjiang Production and Construction Corps (Grant no. 2022GJJ04).

Institutional Review Board Statement: The study did not require ethical approval.

Informed Consent Statement: For studies not involving humans.

Data Availability Statement: No new data were created.

Conflicts of Interest: The authors declare no conflict of interest.

References

1. Zhang, L.; Gu, Q.; Li, C.; Huang, Y. Characteristics and Spatial–Temporal Differences of Urban "Production, Living and Ecological" Environmental Quality in China. *Int. J. Environ. Res. Public Health* **2022**, *19*, 15320. [CrossRef]
2. Cao, Y.; Kong, L.; Ouyang, Z. Characteristics and Driving Mechanism of Regional Ecosystem Assets Change in the Process of Rapid Urbanization—A Case Study of the Beijing–Tianjin–Hebei Urban Agglomeration. *Remote Sens.* **2022**, *14*, 5747. [CrossRef]
3. Wu, J.; Zhang, Q.; Guo, C.; Li, Q.; Hu, Y.; Jiang, X.; Zhao, Y.; Wang, J.; Zhao, Q. Effects of Aeration on Pollution Load and Greenhouse Gas Emissions from Agricultural Drainage Ditches. *Water* **2022**, *14*, 3783. [CrossRef]
4. Moiseenko, T.I. Surface Water under Growing Anthropogenic Loads: From Global Perspectives to Regional Implications. *Water* **2022**, *14*, 3730. [CrossRef]
5. Yoshida, H.; Mønster, J.; Scheutz, C. Plant-integrated measurement of greenhouse gas emissions from a municipal wastewater treatment plant. *Water Res* **2014**, *61*, 108–118. [CrossRef] [PubMed]
6. Ren, J.X.; Gao, Q.X.; Chen, H.T.; Meng, D.; Zhang, Y.; Ma, Z.-Y.; Liu, Q.; Tang, J.-J. Simulation research on greenhouse gas emissions fromwastewater treatment plants under the vision of carbon neutrality. *Clim. Chang. Res.* **2021**, *17*, 410–419.
7. Yu, J.; Zhao, R.Q.; Xiao, L.G.; Zhang, L.J.; Wang, S.; Chuai, X.R.; Han, Y.C.; Jiao, T.X. Study on carbon emission of municipal wastewater treatment System based on water-energy carbon correlation. *Resour. Sci.* **2020**, *42*, 1052–1062.
8. Imran, M.; Khan, S.; Zaman, K.; Khan, H.u.R.; Rashid, A. Assessing Green Solutions for Indoor and Outdoor Environmental Quality: Sustainable Development Needs Renewable Energy Technology. *Atmosphere* **2022**, *13*, 1904. [CrossRef]
9. Kim, J.; Kim, Y.-M.; Lebaka, V.R.; Wee, Y.-J. Lactic Acid for Green Chemical Industry: Recent Advances in and Future Prospects for Production Technology, Recovery, and Applications. *Fermentation* **2022**, *8*, 609. [CrossRef]
10. Androniceanu, A.; Sabie, O.M. Overview of Green Energy as a Real Strategic Option for Sustainable Development. *Energies* **2022**, *15*, 8573. [CrossRef]
11. Wang, X.; Dong, Y.; Yu, S.; Mu, G.; Qu, H.; Li, Z.; Bian, D. Analysis of the Electricity Consumption in Municipal Wastewater Treatment Plants in Northeast China in Terms of Wastewater Characteristics. *Int. J. Environ. Res. Public Health* **2022**, *19*, 14398. [CrossRef]
12. Yeoh, J.X.; Md. Jamil, S.N.A.; Syukri, F.; Koyama, M.; Nourouzi Mobarekeh, M. Comparison between Conventional Treatment Processes and Advanced Oxidation Processes in Treating Slaughterhouse Wastewater: A Review. *Water* **2022**, *14*, 3778. [CrossRef]
13. Cheng, G.F.; Zhang, J.; Yan, K.K. The technology of carbon neutralin the sewage treatment. *Jiangxi Chem. Ind.* **2017**, *2*, 225–227.

14. Jahan, N.; Tahmid, M.; Shoronika, A.Z.; Fariha, A.; Roy, H.; Pervez, M.N.; Cai, Y.; Naddeo, V.; Islam, M.S. A Comprehensive Review on the Sustainable Treatment of Textile Wastewater: Zero Liquid Discharge and Resource Recovery Perspectives. *Sustainability* **2022**, *14*, 15398. [CrossRef]
15. Song, X.X.; Lin, J.; Liu, J.; Gong, H.; Fan, H.T.; Zhang, L.; Wei, Y.S.; Sui, Q.W.; Peng, Y.Z. Foreword to the column on the Research and Development of key Technologies and Engineering Practice of wastewater treatment plants for the future. *J. Environ. Sci.* **2022**, *42*, 1–6.
16. Chang, J.W.; Jin, Y.Y.; Geng, Y.; Song, X.D. Promote the low-carbon transformation of municipal sewagetreatment industry and facilitate the realization of emission peak and carbon neutrality. *China Environ. Prot. Ind.* **2021**, *6*, 9–17.
17. Đurđević, D.; Žiković, S.; Čop, T. Socio-Economic, Technical and Environmental Indicators for Sustainable Sewage Sludge Management and LEAP Analysis of Emissions Reduction. *Energies* **2022**, *15*, 6050. [CrossRef]
18. Altowayti, W.A.H.; Shahir, S.; Eisa, T.A.E.; Nasser, M.; Babar, M.I.; Alshalif, A.F.; AL-Towayti, F.A.H. Smart Modelling of a Sustainable Biological Wastewater Treatment Technologies: A Critical Review. *Sustainability* **2022**, *14*, 15353. [CrossRef]
19. Hernández-Chover, V.; Castellet-Viciano, L.; Bellver-Domingo, Á.; Hernández-Sancho, F. The Potential of Digitalization to Promote a Circular Economy in the Water Sector. *Water* **2022**, *14*, 3722. [CrossRef]
20. Schaum, C.; Lensch, D.; Bolle, P.Y.; Cornel, P. Sewage sludge treatment: Evaluation of the energy potential and methane emissions with COD balancing. *J. Water Reuse Desalination* **2015**, *5*, 437–445. [CrossRef]
21. Zhang, Y.J.; Xing, X. Comparison between traditional activated sludge method and adsorption-regenerated activated sludge method. *Environ. Prot. Circ. Econ.* **2008**, *08*, 22–24.
22. Liu, Y.S. Review and prospect of activated sludge process technology development—And evaluation of SBR technology. *Chem. Water Supply Drain. Des.* **1988**, *1*, 1–8.
23. Sang, L.H.; Han, X.K.; Tang, J. Study on characteristics of SBR method and its development trend. *J. Jilin Inst. Civ. Eng. Archit.* **2002**, *1*, 34–38.
24. Zhou, F.C.; Sun, X.S. Experimental study on the treatment of domestic sewage by completely mixed activated sludge. *Water Treat. Technol.* **2009**, *35*, 50–52.
25. Sadegh, H.; Ali, G.A. The advantages and disadvantages of adsorption regenerated activated sludge for wastewater treatment. *Energy Energy Conserv.* **2016**, *2*, 87.
26. Ruan, S.C. Application of intermittent cycle delay aerated activated sludge process in the treatment of catering wastewater. *Jiangxi Chem. Ind.* **2009**, *4*, 179–181.
27. Li, S.G.; Zhang, L.Q.; Wu, X.W.; Zhang, K.F. Experimental study on the treatment of municipal sewage by intermittent cycle delay aerated activated sludge. *Ind. Water Wastewater.* **2004**, *6*, 57–62.
28. Wang, Z.C.; Li, X.D.; Wang, J.; Huang, Y.F. Treatment of rural domestic sewage by pure oxygen activated sludge. *Jiangsu Agric. Sci.* **2013**, *41*, 344–346.
29. Wen, G.Q.; Li, C.D.; Ji, C.W.; Zhang, Y.F.; Liu, Z.M. Application prospect of pure oxygen aeration in municipal wastewater treatment. *Guangdong Chem. Ind.* **2012**, *39*, 107–128.
30. Chen, S.; Tan, X.J.; Jiang, L.Y. Discussion on the treatment technology of pure oxygen aerated activated sludge. *Urban Roads Bridg. Flood Control* **2012**, *1*, 65–70.
31. Dong, W.H.; Yang, J.; Zhang, S.F.; Li, L. Progress in pure oxygen aeration. *China Resour. Compr. Util.* **2006**, *11*, 28–30.
32. Jiao, Z.W. Brief talk on pure oxygen aerated water treatment. *Ind. Water Treat.* **1992**, *6*, 3–5.
33. Yang, W.S.; Yang, Y.L. Discussion on SBR Process (Sequencing Batch Activated Sludge Process). *Sci. Tech. Inf. Dev. Econ.* **2006**, *10*, 159–160.
34. Wang, Z.X. The advantages and development status of SBR process. *Sci. Technol. Innov. Her.* **2008**, *26*, 95.
35. Chen, P.; Zhao, W.; Chen, D.; Huang, Z.; Zhang, C.; Zheng, X. Research Progress on Integrated Treatment Technologies of Rural Domestic Sewage: A Review. *Water* **2022**, *14*, 2439. [CrossRef]
36. Zainuddin, N.I.; Bilad, M.R.; Marbelia, L.; Budhijanto, W.; Arahman, N.; Fahrina, A.; Shamsuddin, N.; Zaki, Z.I.; El-Bahy, Z.M.; Nandiyanto, A.B.D.; et al. Sequencing Batch Integrated Fixed-Film Activated Sludge Membrane Process for Treatment of Tapioca Processing Wastewater. *Membranes* **2021**, *11*, 875. [CrossRef]
37. Jafarinejad, S. Simulation for the Performance and Economic Evaluation of Conventional Activated Sludge Process Replacing by Sequencing Batch Reactor Technology in a Petroleum Refinery Wastewater Treatment Plant. *ChemEngineering* **2019**, *3*, 45. [CrossRef]
38. Zhu, L.H.; Zhu, Z.B. Current situation and development of activated sludge process. *Environ. Exploit.* **1997**, *1*, 11–14.
39. Wei, H.T.; Liu, X.J.; Li, T. Review on the treatment of domestic sewage and wastewater by activated sludge. *Hebei Electr. Power Technol.* **2005**, *04*, 40–42.
40. Yu, Z.M. Development and application of activated sludge treatment technology. *Foreign Environ. Sci. Technol.* **1990**, *1*, 19–23.
41. Rong, F.; Zhou, R.L.; Zhu, L.S. Treatment of residual sewage by activated sludge. *Energy Environ. Prot.* **2007**, *21*, 41–42.
42. Wu, F.; Cheng, X.R. Calculation method of sludge volume in municipal sewage treatment plant. *J. Wuhan Univ.* **2009**, *42*, 244–247.
43. Yu, L.X.; Wang, H.C. Discussion on the calculation of residual sludge in activated sludge process. *Water Supply Drain.* **2003**, *29*, 3.
44. Zhou, B.L. Calculation of residual sludge by activated sludge method. *Water Supply Drain. China.* **1999**, *6*, 55–57.
45. Qi, X.R.; Xiong, Y.; Jin, W.J.; Liu, Y.; Cong, B.B. Calculation method of energy consumption of domestic urban sewage treatment. *J. Univ. Sci. Technol. Liaoning* **2015**, *38*, 155–160.

46. Cheng, D.D.; Pang, W.L.; Feng, L.X.; Wei, Z. Energy saving analysis of blast aeration system in urban sewage treatment plant. *Resour. Conserv. Environ. Prot.* **2015**, *5*, 19–20.
47. Yang, Q.; Wang, Y.X.; Cao, X.X.; Liu, X.H.; Zhang, S.Y. Research progress of carbon neutral operation technology for wastewater treatment. *J. Beijing Univ. Technol.* **2022**, *48*, 292–305.
48. Wang, L.Y.; Zhang, J.C. Study on carbon dioxide emission and carbon resource utilization in municipal sewage treatment. *Green Sci. Technol.* **2012**, *6*, 198–200.
49. Khiewwijit, R.; Temmink, H.; Rijnaarts, H.; Keesman, K.J. Energy and nutrient recovery for municipal wastewatertreatment : How to design a feasible plant layout? *Environ. Model. Softw.* **2015**, *68*, 156–165. [CrossRef]
50. Yue, Z.J. Analysis on low-carbon operation technology inbiological wastewater treatment process. *Environ. Sci. Manag.* **2013**, *38*, 8–11.
51. Hao, X.D.; Jin, M.; Hu, Y.S. Framework of futurewastewater treatment in the Netherlands: NEWs and theirpractices. *China Water Wastewater* **2014**, *30*, 715.
52. Liu, Z.X. Carbon capture and carbon redirection: New wayto optimize theenergy self-sufficient of wastewatertreatment. *China Water Wastewater* **2017**, *33*, 43–52.
53. Cai, J. Study on carbon neutraloperation of wastewater treatment. *Archit. Eng. Technol. Des.* **2017**, *24*, 4704.
54. Li, X.L. The aeration rate control of domestic sewage treated by activated sludge process. *Technol. Entrep.* **2013**, *3*, 211–256.
55. Wei, C.H.; Ru, X.; Yang, X.Z.; Feng, C.H.; Wei, Y.F.; Li, F.S. Energy saving strategy of biological wastewater treatment based on oxygen regulation. *Chem. Ind. Eng. Prog.* **2018**, *37*, 4121–4134.
56. Bai, S.Y. Selection and energy saving of Blower for sewage treatment by aerobic activated sludge method. *Shanxi Archit.* **2014**, *40*, 142–143.
57. Jian, J.F. Problems and measures of activated sludge process in wastewater treatment. *Sci. Technol. Innov. Her.* **2018**, *15*, 135–137.
58. Tan, T.P. Research status of oxygen demand in activated sludge and energy saving technology of oxygen supply system. *Environ. Dev.* **1996**, *11*, 16–20.
59. Wang, X.; Lv, Q.X. Control of dissolved oxygen concentration in aeration tank and energy saving. *Water Supply Drain.* **1996**, *22*, 22–24.
60. Zhang, Y.N.; Zhu., J.H. Review of municipal sludge disposal technology and resource utilization. *Environ. Prot. Circ. Econ.* **2019**, *39*, 5–7.
61. Tang, H.L.; Liu, K.; Ruan, W.Q. Review of municipal sludge disposal and recycling technologies. *Guangdong Chem. Ind.* **2020**, *47*, 166–167.
62. Wang, L.; He, R.; Lei, H.T. Summary of sludge treatment and disposal technology in urban sewage treatment plant. *Water Purif. Technol.* **2022**, *41*, 16–21.
63. Dai, X.H.; Hou, L.A.; Zhang, L.W.; Zhang, L.; Yang, D.H. Study on safe disposal and resource treatment of urban sludge in our country. *Chin. Eng. Sci.* **2022**, *24*, 145–153. [CrossRef]
64. Ruan, X.Y. Ways of sludge treatment, disposal and resource utilization. *Chem. Eng. Equip.* **2022**, *10*, 277–278.
65. Krahnstöver, T.; Santos, N.; Georges, K.; Campos, L.; Antizar-Ladislao, B. Low-Carbon Technologies to Remove Organic Micropollutants from Wastewater: A Focus on Pharmaceuticals. *Sustainability* **2022**, *14*, 11686. [CrossRef]
66. Xu, X.; Zhou, Q.; Chen, X.; Li, Y.; Jiang, Y. The Efficiency of Green Technology Innovation and Its Influencing Factors in Wastewater Treatment Companies. *Separations* **2022**, *9*, 263. [CrossRef]
67. Ali, S.A.; Mulk, W.U.; Ullah, Z.; Khan, H.; Zahid, A.; Shah, M.U.H.; Shah, S.N. Recent Advances in the Synthesis, Application and Economic Feasibility of Ionic Liquids and Deep Eutectic Solvents for CO_2 Capture: A Review. *Energies* **2022**, *15*, 9098. [CrossRef]
68. Miricioiu, M.G.; Zaharioiu, A.; Oancea, S.; Bucura, F.; Raboaca, M.S.; Filote, C.; Ionete, R.E.; Niculescu, V.C.; Constantinescu, M. Sewage Sludge Derived Materials for CO2 Adsorption. *Appl. Sci.* **2021**, *11*, 7139. [CrossRef]
69. He, X.D.; Ye, J.Z.; Li, J.; Jiang, H. Current situation and potential application of wastewater Thermal energy. *China Water Supply Drain.* **2019**, *35*, 15–22.
70. Chang, J.W.; Jin, Y.Y.; Geng, Y.; Song, X.D. Promote the low-carbon transformation of municipal wastewater treatment industry to help carbon peak and carbon neutrality. *China Environ. Prot. Ind.* **2021**, *6*, 9–17.
71. He, X.D.; Zhao, Z.C.; Li, J.; Li, S.; Jiang, H. Energy and resource recovery method and carbon emission calculation of wastewater treatment plant: A case study of Kakolanmaki wastewater treatment Plant in Finland. *Chin. J. Environ. Eng.* **2021**, *15*, 2849–2857.
72. Song, X.X.; Lin, J.; Liu, J. The current situation and engineering practice of sewage treatment technology facing the future. *Acta Scientiae Circumstantiae* **2021**, *41*, 39–53.
73. Hao, X.D.; Wei, J.; Cao, Y.L. A successful case of carbon-neutral operation in America: Sheboygan WWTP. *China Water Wastewater* **2014**, *30*, 1–6.
74. Gao, W.M.; Cheng, H.F. Progress in Research on sludge treatment and disposal technology in our country. *Chem. Miner. Process.* **2023**, 1–9. Available online: http://kns.cnki.net/kcms/detail/32.1492.tq.20230110.1252.001.html (accessed on 7 December 2022).
75. Wang, Q.X.; Zhang, Y. Carbon emission from sewage sludge treatment and its low carbonization strategy. *Leather Mak. Environ. Technol.* **2022**, *3*, 5–7.
76. You, X.; Yang, L.; Zhou, X.; Zhang, Y. Sustainability and carbon neutrality trends for microalgae-based wastewater treatment: A review. *Environ. Res.* **2022**, *209*, 112860. [CrossRef]

77. Liang, C.; Le, X.; Fang, W.; Zhao, J.; Fang, L.; Hou, S. The Utilization of Recycled Sewage Sludge Ash as a Supplementary Cementitious Material in Mortar: A Review. *Sustainability* **2022**, *14*, 4432. [CrossRef]
78. Sakiewicz, P.; Piotrowski, K.; Rajca, M.; Maj, I.; Kalisz, S.; Ober, J.; Karwot, J.; Pagilla, K.R. Innovative Technological Approach for the Cyclic Nutrients Adsorption by Post-Digestion Sewage Sludge-Based Ash Co-Formed with Some Nanostructural Additives under a Circular Economy Framework. *Int. J. Environ. Res. Public Health* **2022**, *19*, 11119. [CrossRef]
79. La Bella, E.; Baglieri, A.; Fragalà, F.; Puglisi, I. Multipurpose Agricultural Reuse of Microalgae Biomasses Employed for the Treatment of Urban Wastewater. *Agronomy* **2022**, *12*, 234. [CrossRef]
80. Duan, G. The application of energy saving and low carbon technology in sewage treatment is discussed. *Shanxi Archit.* **2017**, *43*, 186–187.
81. Liu, G.T.; Zhang, T.F.; Li, Z. Application of variable frequency speed regulating energy saving technology based on fuzzy control in wastewater treatment. *Mod. Electron. Technol.* **2017**, *40*, 135–138.
82. Mou, J. Application of frequency conversion technology in energy saving transformation of sewage treatment fan. *Mech. Electr. Prod. Dev. Innov.* **2016**, *29*, 53–55.
83. Ma, Y.; Peng, Y.Z.; Wang, S.Y. The sludge layer height of secondary sedimentation tank is used to control the sludge return flow of A/O process. *Chin. Environ. Sci.* **2008**, 121–125.
84. Hu, Y. Study on the Optimal operation of municipal wastewater treatment with low carbon source. *Eng. Constr. Des.* **2022**, *12*, 107–109.
85. Zhao, Q.; Liu, M.Y.; Lü, H.; Liang, J.Y.; Diao, X.X.; Zhang, X.; Meng, L. Setup and microbial community analysis of anammox system for landfill leachate treatment coupling partial nitrification denitrification process. *Environ. Sci.* **2019**, *40*, 4195–4201.
86. Ni, M.Z.L.; Mu, Y.J.; Xue, X.F.; Zhang, L.L.; Su, B.S.; Cao, Z.Q. Deep removal of total nitrogen by one-stage short-cut nitrification coupled with ANamMOx filter column for slow-release carbon source. *Chin. J. Environ. Eng.* **2021**, *15*, 2468–2479.
87. Guo, Q.; Liang, Z.; Bai, X.; Lv, M.; Zhang, A. The Analysis of Carbon Emission's Characteristics and Dynamic Evolution Based on the Strategy of Unbalanced Regional Economic Development in China. *Sustainability* **2022**, *14*, 8417. [CrossRef]
88. Wang, N.; Zhao, Y.; Song, T.; Zou, X.; Wang, E.; Du, S. Accounting for China's Net Carbon Emissions and Research on the Realization Path of Carbon Neutralization Based on Ecosystem Carbon Sinks. *Sustainability* **2022**, *14*, 14750. [CrossRef]
89. Zhao, Y.; Lin, G.; Jiang, D.; Fu, J.; Li, X. Low-Carbon Development from the Energy–Water Nexus Perspective in China's Resource-Based City. *Sustainability* **2022**, *14*, 11869. [CrossRef]
90. Guo, Y.; Zhang, Z.; Chen, Y.; Li, H.; Liu, C.; Lu, J.; Li, R. Sensor Fault Detection Combined Data Quality Optimization of Energy System for Energy Saving and Emission Reduction. *Processes* **2022**, *10*, 347. [CrossRef]

Article

Simultaneous Phosphate Removal and Power Generation by the Aluminum–Air Fuel Cell for Energy Self-Sufficient Electrocoagulation

Xiaoyu Han [1,2], Hanlin Qi [1], Youpeng Qu [3], Yujie Feng [2,*] and Xin Zhao [1,*]

1 Department of Environmental Engineering, School of Resources and Civil Engineering, Northeastern University, Shenyang 110819, China
2 State Key Laboratory of Urban Water Resource and Environment, School of Environment, Harbin Institute of Technology, Harbin 150090, China
3 School of Life Science and Technology, Harbin Institute of Technology, Harbin 150080, China
* Correspondence: yujief@hit.edu.cn (Y.F.); zhaoxin@mail.neu.edu.cn (X.Z.)

Featured Application: The Al–air fuel cell provides an energy self-sufficient electrocoagulation system for phosphate removal.

Abstract: A self-powered electrocoagulation system with a single-chamber aluminum–air fuel cell was employed for phosphate removal in this study. Electricity production and aluminum hydroxides in solution were also investigated. When the NaCl concentration increased from 2 mmol/L to 10 mmol/L, the phosphate removal increased from 86.9% to 97.8% in 60 min. An electrolyte composed of 10 mmol/L of NaCl was shown to obtain a maximum power density generation of 265.7 mW/m^2. When the initial solution pH ranged from 5.0 to 9.0, 98.5% phosphate removal and a maximum power density of 338.1 mW/m^2 were obtained at pH 6.0. Phosphate was mainly removed by aluminum hydroxide adsorption. These results demonstrate that the aluminum–air fuel cell can be applied as electricity-producing electrocoagulation equipment. Aluminum–air fuel cells provide an alternative method to meet the goal of carbon neutrality in wastewater treatment compared with traditional energy-consuming electrocoagulation systems.

Keywords: aluminum–air fuel cell; electrocoagulation; phosphate removal; electricity production

Citation: Han, X.; Qi, H.; Qu, Y.; Feng, Y.; Zhao, X. Simultaneous Phosphate Removal and Power Generation by the Aluminum–Air Fuel Cell for Energy Self-Sufficient Electrocoagulation. *Appl. Sci.* **2023**, *13*, 4628. https://doi.org/10.3390/app13074628

Academic Editor: Apostolos Giannis

Received: 8 February 2023
Revised: 27 March 2023
Accepted: 28 March 2023
Published: 6 April 2023

1. Introduction

The discharge of phosphorus into water bodies has caused a serious worldwide water eutrophication problem [1,2]. A concentration of 0.1–0.2 mg/L phosphate can induce incipient eutrophication in running water, while the critical concentration reduces to 0.005–0.01 mg/L of phosphate for still water [3]. The eutrophication of water threatens the health of aquatic creatures, livestock, and even humans. Therefore, the removal of phosphorus from municipal wastewater is essential to avoid water eutrophication. In China, the national limit for total phosphorus (TP) effluent of the Class 1A Discharge Standard is 0.5 mg/L [2]. The TP concentration is approximately 4–9 mg/L in China's municipal wastewater treatment plants [4]. Biological phosphorus removal processes are the conventional and most widely applied technology. However, the biological treatment system occupies a large area of land, requires high energy consumption, and needs to control the operation conditions to reach this standard [5]. For the high discharge standard of phosphate in municipal wastewater treatment, physicochemical treatments, such as chemical precipitation by adding aluminum or iron coagulants, are applied for the further removal of phosphate after biological treatment [6,7]. Chemical coagulation increases the operating costs, produces excessive amounts of chemical sludge, and leads to potential risks of secondary pollution.

Electrocoagulation (EC) is considered a promising wastewater treatment technology that comprises three aspects: coagulation, flotation, and electrochemistry [8]. In the EC process, a sacrificial metal anode, such as Al or Fe, dissolves and releases metal ions into the solution with an impressed current [8]. Various metal hydroxides are formed by the hydrolysis process. Hydrogen bubbles and OH^- ions are discharged at the cathode. Pollutants and impurities are mainly removed by multiple physicochemical reactions with metal hydroxides and generate precipitates simultaneously [9,10]. EC is an effective method for phosphate removal, and the advantages of the EC system compared with chemical coagulation are its easy operation, no additional chemical requirements, and less sludge production [10–12]. However, the electrical energy demands need to be reduced to make the EC process economically viable and eco-friendly. Recently, a new type of EC process using an air cathode has been investigated to reduce EC energy demands [13–15]. At the air cathode, an oxygen reduction reaction proceeds with a related catalyst, and oxygen gas is obtained through air diffusion without aeration. With a minor added operation voltage of 0.5 V, 98% of phosphate was removed in a shortened time of 15 min [16]. The reason for this decrease in the electrical energy requirement of the EC with the air cathode is that the metal anode and the oxygen reduction air cathode exist in a potential gradient.

Metal–air fuel cells, which consist of a metal anode and an air cathode, have been investigated for synchronous contaminant removal and energy generation [17]. Over the last few decades, metal–air fuel cells have been implemented to remove arsenate, chromium, p-arsanilic acid, and humic acid, and for algal recovery [18–24]. Some researchers have also applied metal–air fuel cells for phosphate removal and recovery. When wastewater contained nitrogen and phosphorus together, Mg–air fuel cells were employed to recycle the nitrogen and phosphorus by forming struvite precipitation [25,26]. Iron–air fuel cells are a new option for recovering phosphate from wastewater. Phosphate was recovered by the formation of vivianite precipitation. Within 3 months of continuous operation, the iron–air fuel cell output voltage achieved approximately 400 mV, and 97% of phosphate was removed from synthetic industrial wastewater [27]. Different configurations of iron–air fuel cells were constructed to treat wastewater containing anaerobically digested sludge. The phosphate removal rate of a two-chamber fuel cell obtained a rate of 11.60 mg-P/L/h [28]. Aluminum is also commonly used as an electrode in electrocoagulation for phosphate removal, and aluminum–air fuel cells have been utilized to remove arsenate. Thus, it is important to research the performance of phosphate removal and the influence of the phosphate removal rate on aluminum–air fuel cells.

In this study, an energy self-sufficient single-chamber aluminum–air fuel electrocoagulation system was constructed for phosphate removal. Different electrolyte concentrations, initial pH values of the solution, and phosphate concentrations were investigated to determine their influences on phosphate removal and power generation. The coagulated precipitation in an aluminum–air fuel cell was also selected and analyzed to demonstrate the phosphate removal process in the aluminum–air fuel cell. The self-powered aluminum–air fuel cell provides an energy-friendly electrocoagulation process for nutrient contaminant removal, which is significant in wastewater treatment for achieving the goal of carbon neutrality.

2. Materials and Methods

2.1. Reactor Construction

An aluminum–air fuel cell reactor was constructed using a rectangular plexiglass reactor of 80 mL (length: 4 cm, width: 2 cm, height: 10 cm, Figure 1). The area of the air cathode was 40 cm^2 (4 cm × 10 cm). The air cathode, including stainless steel mesh, activated carbon (Xinsen Carbon Co., Ltd., Nanping, Fujian, China), conductive carbon black (Jinqiushi Chemical Co. Ltd., Tianjin, China), and PTFE (60 wt%, Hesen Electrical Co., Ltd., Shanghai, China), was constructed using a rolling process, as described in a previous study [29]. For the diffusion layer, conductive carbon black and an appropriate amount of ethanol were mixed using mechanical and ultrasonic agitation for 15 min. The PTFE was then added and the solution was further stirred for 15 min. The carbon black was kneaded like dough. Then, it was compacted and shaped between the rollers, and the diffusion layer was pressed to a thickness of 0.4 mm. Finally, the diffusion layer sheet and the stainless-steel mesh were rolled together and calcined in a muffle furnace at 340 °C for 20 min. Activated carbon powder was used as the catalyst to create the catalytic layer. Similar to the operation of the diffusion layer, a certain amount of activated carbon powder, ethanol, and PTFE were mixed and rolled to create a catalytic layer sheet with a thickness of 0.3 mm. The catalytic layer was then rolled together on the other side of the stainless-steel mesh. The anode was a piece of aluminum mesh of 40 cm^2 (mesh size 1.0 mm × 1.5 mm), which was placed opposite to the cathode. Before operation, the aluminum mesh was washed with an acetone solution to remove the residual oil and stains on the anode surface. The anode was then immersed in 1.0 mol/L HCl solution for 10 min to remove the oxidation layer on the surface. The anode was used after it was washed with deionized water.

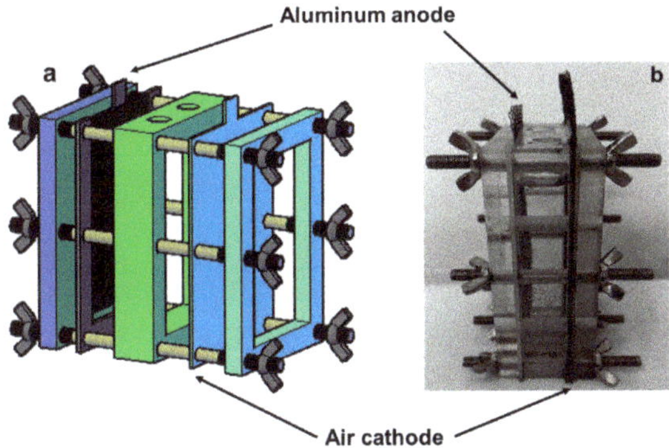

Figure 1. Schematic diagram (**a**) and photo (**b**) of the aluminum–air fuel cell for electrocoagulation.

2.2. Solutions and Experimental Procedure

The phosphate solution was dispensed with KH_2PO_4 (99.9%) and NaCl (99.5%), which were prepared to increase the solution's conductivity. The reactor was operated with a resistor of 10 Ω. The concentration of NaCl in the solution varied from 2 mmol/L to 10 mmol/L with 5 mg/L PO_4^{3-}-P to analyze the solution conductivity effect. The initial solution pH of 5.0–9.0 was adjusted by adding 1.0 mol/L HCl or 1.0 mol/L NaOH. The PO_4^{3-}-P concentrations ranged from 1 mg/L to 5 mg/L to compare the performances with different original PO_4^{3-}-P concentrations. Approximately 1.5 mL of solution was removed every 5 min for PO_4^{3-}-P analysis. The self-powered electrocoagulation experiments were conducted at room temperature.

2.3. Measurements and Calculations

The samples extracted at different times were tested after they were filtered through 0.45 μm pore diameter syringe filters. The phosphate concentration was measured using the ammonium molybdate spectrophotometric method (photoLab® 7600 UV, WTW, Munich, Germany) [30]. The pH was tested using a pH meter (pH 7110, WTW, Munich, Germany). The voltage across the resistor was monitored at 1 min intervals using a data acquisition system (PISO-813, ICPDAS Co., Ltd, Taiwan, China). The current density (I) was calculated based on Ohm's law (I = U/(RA)), and the power density (P) was calculated as P = IU/A, where U is the voltage (mV), R is the external resistor (Ω), and A is the surface area of the cathode. Polarization curves and electrode potential curves were obtained by recording the voltages under various external resistances (5–1000 Ω). A saturated calomel electrode (SCE, 0.242 V vs. a standard hydrogen electrode) was inserted into the reactor chamber as the reference electrode to test the electrode potential. The theoretical dissolved aluminum ion concentration was calculated, and the number of aluminum ions was calculated according to Faraday's law. The equation for Faraday's law is as follows:

$$m = \frac{I \times A \times t \times M}{Z \times F} \quad (1)$$

Here, m is the theoretical dissolved mass of aluminum ions (g), I is the current density (A/cm^2), t is the operation time (s), A is the surface area of the cathode (40 cm^2), M is the molecular weight of the aluminum material (M = 26.98 g/mol), Z is the number of electrons released from the aluminum anode (the Z of aluminum is 3 eq/mol), and F is Faraday's constant (96,485 C/mol). The volume of the electrolyte solution used for the aluminum ion concentration calculation was V = 80 cm^3. The P-precipitants obtained from the 10 mmol/L NaCl solution were collected and vacuum-dried at 40 °C for chemical phase analysis. The precipitates were tested by X-ray diffraction (D8 ADVANCE, Bruker, Karlsruhe, Germany) under operating conditions of 100 mA and 40 kV in the range of 2θ = 10–90°. The aluminum anode and precipitates were analyzed using scanning electron microscopy (Carl Zeiss AG, Sigma500, Oberkochen, Germany) and energy-dispersive spectroscopy.

3. Results and Discussion

3.1. Effects of Electrolyte Concentration

The electrolyte concentration of NaCl directly influenced the resistance between the electrodes and changed the output current [30]. In the EC process, increasing the electrolyte concentration could increase the phosphate removal efficiency and reduce the electrolysis operation time. The aluminum–air fuel cell system's phosphate removal performance was monitored with different concentrations of NaCl ranging from 10 mmol/L to 2 mmol/L with a resistor of 10 Ω. With the decrease in the electrolyte concentration, the phosphate removal rate in the aluminum–air fuel cell showed a downward trend, and the required reaction time was prolonged. The phosphate removal in the aluminum–air fuel cell containing 8 mmol/L of NaCl was slightly lower than that of the cell containing 10 mmol/L of NaCl. When the NaCl concentration decreased to 2 mmol/L, the phosphate removal reached 97.8%, 93.3%, 91.8%, and 86.9% at 60 min, respectively (Figure 2a). White flocculent precipitates were observed with different operating concentrations of electrolytes. At 10 mmol/L of NaCl, the phosphate removal was 98.0% at 50 min, and the rate of removal decreased with longer operation times, which demonstrated that the operation time could be reduced with a higher electrolyte concentration. When the NaCl concentration was decreased from 10 mmol/L to 2 mmol/L, the average output current density of the aluminum–air fuel cell decreased from 0.20 mA/cm^2 to 0.06 mA/cm^2 (Figure 2b). The low conductivity of the solution led to an increase in the internal resistance of the fuel cell, slowing down the mass transfer in solution and the dissolution rate of the aluminum anode [24]. When the NaCl concentration declined from 10 mmol/L to 2 mmol/L, the calculated dissolutions of aluminum ions were 33.7 mg/L, 32.2 mg/L, 20.3 mg/L, and 9.5 mg/L, respectively. A higher conductivity electrolyte increased the anode dissolution and produced larger amounts of aluminum hydroxides for complex precipitation reactions in the solution; thus, a higher electrolyte concentration was favorable for phosphate removal [20]. In a traditional EC system, an increase in the electrolyte concentration can improve energy consumption, leading to an increase in the EC operating costs. On the contrary, a higher electrolyte concentration will generate more electricity for the aluminum–air fuel cell.

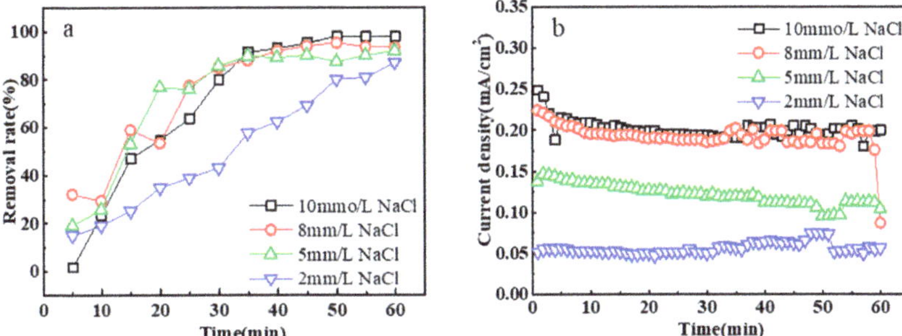

Figure 2. Effects of different electrolyte concentrations on (**a**) phosphate removal and (**b**) current generation in the aluminum–air fuel cell.

To analyze the influence of electrolyte concentrations on the electrochemical performance of the aluminum–air fuel cell, polarization curves and electrode potentials were tested with different electrolyte concentrations. As the electrolyte conductivity decreased, the maximum power density of the aluminum–air fuel cell reduced from 265.7 mW/m^2 to 76.6 mW/m^2 (Figure 3a). The electrode potentials of the aluminum anode and the air cathode both decreased with a decline in conductivity (Figure 3b). With a higher electrolyte concentration, more aluminum ions dissolved, generating a higher current output.

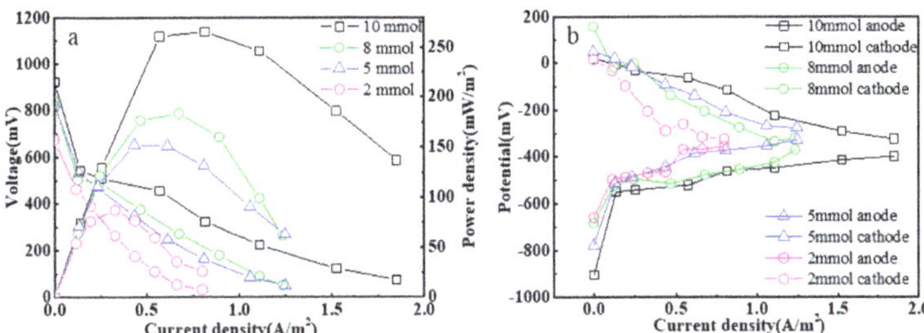

Figure 3. (**a**) Power density and polarization curves and (**b**) electrode potential curves for the aluminum−air fuel cell with different electrolyte concentrations.

3.2. Effects of Initial pH

The pH value can be a significant factor in the electrocoagulation process. In this work, the initial solution pH ranged from 5.0 to 9.0, the initial phosphate concentration was 5 mg/L, and the NaCl concentration was 10 mmol/L. As shown in Figure 4a, different initial pH values influenced the phosphate removal rate in the aluminum–air fuel cell system. When the solution pH was alkaline, the phosphate removal rate was faster than the rate under acidic conditions in the early stage. Within 50 min, all phosphate removal rates reached values higher than 90% with different initial pH values. At 60 min, as the initial pH 5.0 changed to pH 9.0, the phosphate removal reached 96.3%, 98.5%, 96.3%, 97.0%, and 95.6%. In contrast to the removal rate trend, the highest level of final phosphate removal was obtained at pH 6.0, which was likely due to the higher aluminum dissolution with a higher current output at pH 6.0 [31]. The average current density of the aluminum–air fuel cell ranged between 0.18 and 0.20 mA/cm^2, which changed slightly during electricity production. At pH values of 5.0 to 9.0, the calculated aluminum ion concentrations were 30.7 mg/L, 33.9 mg/L, 33.1 mg/L, 31.9 mg/L, and 32.2 mg/L. These results also indicated that the self-driven electrocoagulation could adapt to different pH conditions, as shown in previous research [20].

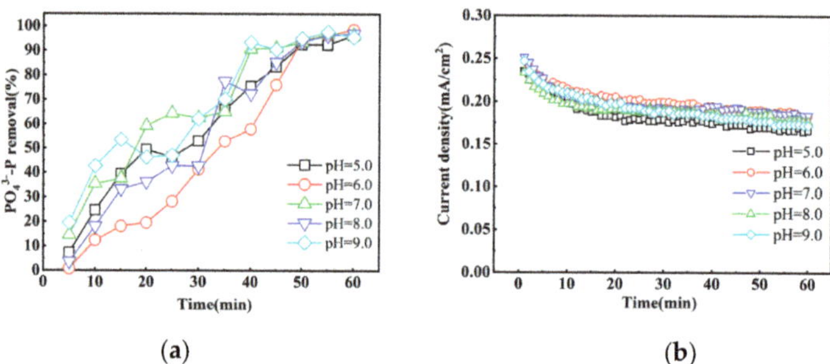

Figure 4. Effects of different initial pH values on (**a**) phosphate removal and (**b**) current generation in the aluminum−air fuel cell.

The power density of the aluminum–air fuel cell varied slightly under different pH conditions (Figure 5a,b). The maximum power densities of the aluminum–air fuel cell were higher at pH 6.0 and pH 7.0, being 338.1 mW/m^2 and 332.8 mW/m^2, respectively. Under alkaline conditions, the cell maximum power densities decreased; at pH 9.0, the maximum power density was 292.8 mW/m^2. The difference in the fuel cell electrode potential under different pH conditions was not obvious.

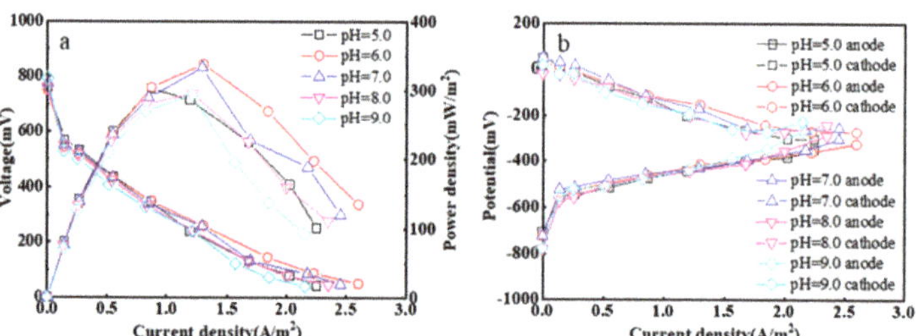

Figure 5. (**a**) Power density and polarization curves and (**b**) electrode potential curves for the aluminum−air fuel cell with different initial pH.

The electrocoagulation performance was influenced by the aluminum dissolution and the final pH. In this experiment, pH variation was also investigated with different initial pH values, as shown in Table 1. After 60 min of operation with the initial pH value ranging from 5.0 to 9.0, the final pH changed in different trends. For the acidic and neutral conditions, the pH value increased, while with initial pH values of 8.0 and 9.0, the final pH value decreased. The variation trends were similar to previously reported trends [20]. The dissolution rate of aluminum was also related to the current density and the pH value. The behaviors of the electrode and aluminum ions in solution also affected the solution pH. For the aluminum–air fuel cell, the electrode and overall reactions were as shown in Equations (2)–(4). At the aluminum anode, electrons were released from the aluminum, and then the aluminum ions reacted with water, as shown in Equation (5) [32,33]:

$$\text{Anode}: \mathrm{Al} \rightarrow \mathrm{Al}^{3+} + 3\mathrm{e}^{-} \mathrm{E}^{0} = 1.66\ \mathrm{V} \quad (2)$$

$$\text{Cathode}: \mathrm{O}_2 + 4\mathrm{H}^{+} + 4\mathrm{e}^{-} \rightarrow 2\mathrm{H}_2\mathrm{O} \mathrm{E}^{0} = 1.23\ \mathrm{V} \quad (3)$$

$$\text{Overall}: \mathrm{Al} + \mathrm{O}_2 + \mathrm{H}^{+} \rightarrow \mathrm{Al}^{3+} + \mathrm{H}_2\mathrm{O} \mathrm{E}^{0} = 2.89\ \mathrm{V} \quad (4)$$

$$\mathrm{Al}^{3+} + 3\mathrm{H}_2\mathrm{O} \rightarrow \mathrm{Al(OH)}_3 + 3\mathrm{H}^{+} \quad (5)$$

Table 1. The pH value changes of the aluminum–air fuel cell before and after operation for 60 min.

Initial pH	5.0	6.0	7.0	8.0	9.0
Final pH	5.6	6.4	7.3	7.9	8.8

At the air cathode, OH^- ions can also be generated through the four-electron oxygen reduction reaction with water, as shown in Equation (6):

$$O_2 + 2H_2O + 4e^- \rightarrow 4OH^- \tag{6}$$

At pH 5–6, for most of the polymeric species, the ratios of precipitation of Al^{3+} and OH^- were less than 3, which caused the OH^- ions to accumulate in the solution during the operation process and increased the pH [33]. Under alkaline conditions, more OH^- ions could be consumed by aluminum ions to form polymeric aluminum hydroxide, and this process decreased the pH value.

3.3. *Effects of Initial Phosphate Concentration*

Generally, the removal efficiency of phosphate was negatively correlated with the initial phosphate concentration in the EC process, and the phosphate removal was limited at the start of the operation, as the amount of coagulants produced was relatively small [14,34,35]. In this study, the phosphate concentration was set to approach the concentration of real domestic wastewater. A 10 mmol/L NaCl electrolyte and a 10 Ω operating condition were used, and the initial concentrations of phosphate in the simulated wastewater were set to 5 mg/L, 4 mg/L, 3 mg/L, 2 mg/L, and 1 mg/L, respectively. As the results in Section 3.1 show, the operation time could be reduced to 50 min with 10 mmol/L of NaCl. The maximum phosphate removal rate was achieved at 50 min, and the level of phosphate removal decreased after 60 min. At lower phosphate concentrations, a higher phosphate removal rate was obtained at the beginning of the operation. With the increase in the phosphate concentration, the phosphate removal rate increased at 50 min, achieving 92.2%, 93.0%, 92.6%, 96.9%, and 98.0%, respectively (Figure 6). In the actual sewage treatment process, the concentration of phosphate in sewage changes greatly. These results indicate that in a relatively low phosphate concentration range, the aluminum–air fuel cell system can achieve a high phosphate removal efficiency. Compared with the existing EC process, which consumes electricity, an electrical-power-generating electrocoagulation system was applied to this work, which provided an energy-friendly and promising electrocoagulation process for phosphate removal.

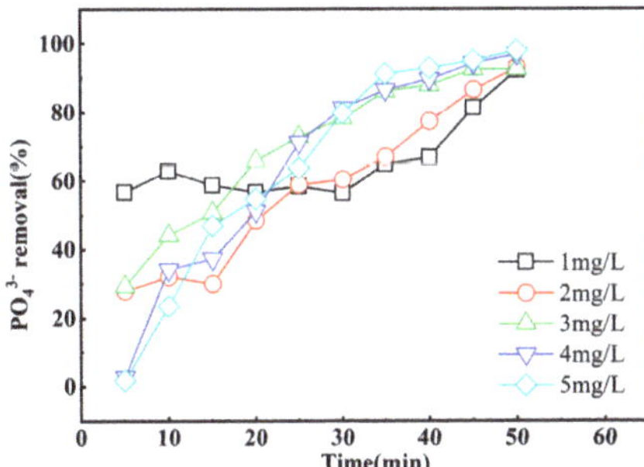

Figure 6. The relationship between the initial phosphate concentration and phosphate removal in the aluminum−air fuel cell.

3.4. *Comparison of Energy Demand and Implications*

In traditional EC systems for phosphorous removal using aluminum electrodes or a combination of aluminum and iron electrodes, the energy demand per cubic meter of P-wastewater was approximately 1.1–6.1 kWh, as shown in Table 2. When an air cathode was applied as the electrode, the energy consumption decreased compared with the traditional EC system. In the air cathode EC system which used a titanium inert electrode, only 0.009–0.06 kwh was added, and the operation time required was only 15 min [13]. Photovoltaic solar modules were also applied to power the EC process, and the energy from the solar energy modules was enough for the EC system, although the solar energy conversion efficiency was less than 13% [30]. For the aluminum–air fuel cell, with a 10 mmol/L NaCl electrolyte and a 10 Ω operation condition, the average power density at 60 min was approximately 160 mW/m^2. As the power density curves show, the aluminum–air fuel cell output power density was related to the output current. The power production was less than that of the solar cell; however, the aluminum–air fuel cell electrocoagulation system did not require an external electricity unit such as a solar cell module. This simple device is more practical in wastewater treatment, and solar energy is also highly dependent on weather. Compared with the traditional EC system, the operating cost for anode consumption is similar, but the cost of the air cathode in the aluminum–air fuel cell is higher than that of the conventional EC cathode [19]. The air cathode cost and lifetime should be one of the future research points considered for scale-up. Therefore, in future studies, the construction of a scale-up system, the electrode lifetime, and overall economic assessment including labor, sludge handling, maintenance, and depreciation costs will be key issues for practical application [18,36–38].

Table 2. Energy consumption of different electrocoagulation systems and the removal rates of phosphorus reported in different studies with Al electrodes.

Electrode Type	Wastewater Type	Energy (kWh/m^3)	Current Density (mA/cm^2)	Phosphate Removal (%)	Reference
Al–Al	urban wastewater	4.5	1	99	[39]
Al–Fe	synthetic solution	1.1–6.1	1–4	97–99	[3]
Al, air cathode	domestic wastewater	0.4–9.9	0.6–1.2	97–99	[14]
Al, air cathode, titanium sheet	domestic wastewater	0.009–0.06	1	71–98	[13]
Al, air cathode, graphite sheet,	domestic wastewater	0.05−0.14	1	85–98	[16]
Al–Al, photovoltaic solar	landscape water	——	1.3–3.8	96–99	[30]
Al–air cathode fuel cell	synthetic solution	——	0.06–0.2	86–98	This study

Another advantage of the metal–air fuel cell is that the air cathode is capable of generating hydrogen peroxide with a suitable oxygen reduction catalyst, which has the potential to treat wastewater by combining coagulation and oxidization processes. For domestic wastewater treatment, the aluminum–air fuel cell can be used at the pre-treatment segment and the advanced treatment segment to remove phosphate. Organic contaminants such as antibiotics, resistant genes, and persistent organic pollutants which are harmful to human health can coexist with phosphate in wastewater. The multiple physicochemical processes in aluminum–air fuel cells provide promising methods to treat the complex effluents in the advanced treatment stage.

3.5. Mechanisms of Phosphate Removal

The aluminum anode was analyzed before and after fuel cell operation using SEM to clarify the phosphate removal process (Figures 7 and 8). The anode surface formed a corrosion pit after the operation (Figure 7a,b), providing evidence that the anode spontaneously dissolved aluminum ions into the solution. Previous studies have shown that pitting corrosion exists on metal surfaces when they are covered by a compact oxide film [40]. Furthermore, chloride ions in the electrolyte solution are the aggressive ions on the metal surface. Chloride ions can trigger the pitting corrosion reaction, and the passivation layer on the electrode surface will decompose. For the Al anode, a corrosion-protective film consisting of aluminum oxide can easily form on the surface, and the protective film is stable within a pH range of 4–8.5. However, chloride ions can react with the protective film and expose the anode surface, which will increase the anode dissolution rate [34]. These processes are shown in Equations (7)–(9) [41]:

$$Al_2O_3 + Cl^- + 6H^+ \rightarrow 2AlCl_3 + 3H_2O \quad (7)$$

$$Al(OH)_3 + nCl^- \rightarrow Al(OH)_{3-n}Cl_n + nOH^-,\ n = 1, 2, 3 \quad (8)$$

$$AlCl_3 + Cl^- \rightarrow AlCl_4^+ \quad (9)$$

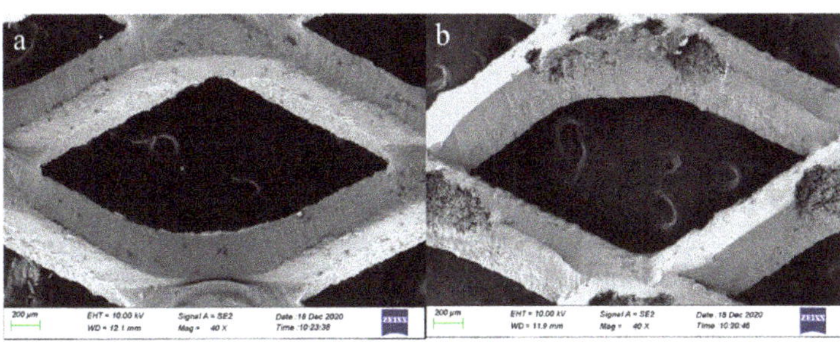

Figure 7. SEM results of the Al electrode in this study, (a) the original electrode, and (b) the electrode after the operation.

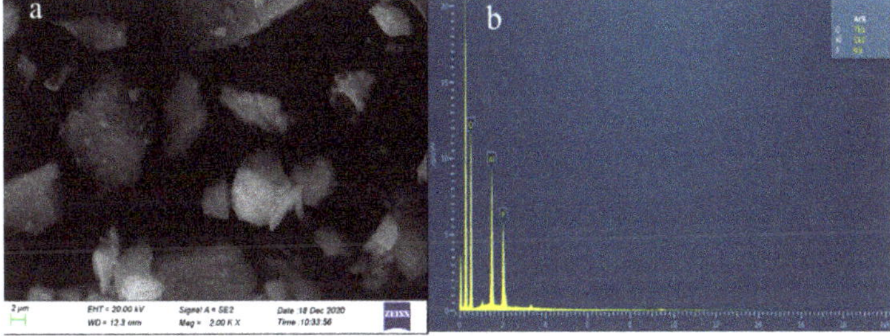

Figure 8. SEM images of phosphate precipitation in aluminum–air batteries (a) and EDS results (b).

The morphology of the precipitates from 10 mmol/L of NaCl and 5 mg/L of phosphate was observed by SEM, and the elemental compositions were analyzed by EDS, as shown in Figure 8a,b. The precipitates had a blocky structure, and the EDS results showed that the precipitates were mainly composed of P, Al, and O elements, which indicated that the phosphate was removed with the aluminum compounds.

There are two main mechanisms that exist in the aluminum coagulation phosphate removal process, namely co-precipitation by monomeric aluminum and adsorption by aluminum hydroxide [42]. In the EC process with aluminum electrodes, both direct precipitation and adsorption by aluminum hydroxide were observed, and direct precipitation for phosphate removal was more efficient than hydroxide adsorption removal [43,44]. The phosphate removal behavior was related to the water conditions and the aluminum species. $AlPO_4$ precipitate can be formed by Al^{3+} and phosphate, and the Al_{13} species can also induce formation of $AlPO_4$ by complex adsorption [45]. For the coagulation process, pH influences the speciation of aluminum and accordingly affects the phosphorus removal mechanism. When the pH value was below 3.5, the dominant aluminum species was Al^{3+}. As the pH value increased from 3.5 to 10.0, the dominant species were complex monomeric and polymeric aluminum hydroxides. $Al(OH)_4^-$ became the predominant species when the pH value was above 10.0 [46]. The precipitated products from different initial pH values were examined by XRD to reveal the crystalline structure (Figure 9). Three diffraction peaks were observed in the flocs. The second peak was in the range of 23–33°, and the third peak was at approximately 41°, which was likely within the spectrum of $Al(OH)_3$. The crystal plane of $AlPO_4$ was also found in the second peak range [45–47]. However, the absence of sharp peaks in the XRD results indicated that the flocs had a poor crystalline structure, and that many of the precipitates were in an amorphous state, which was similar to a previous report on aluminum coagulants [47]. Combining the precipitates, XRD results, and the aluminum species constructions with different pH values, it can be concluded aluminum hydroxide adsorption was the main mechanism for removing phosphate. A schematic diagram of the phosphate removal process is summarized in Figure 10.

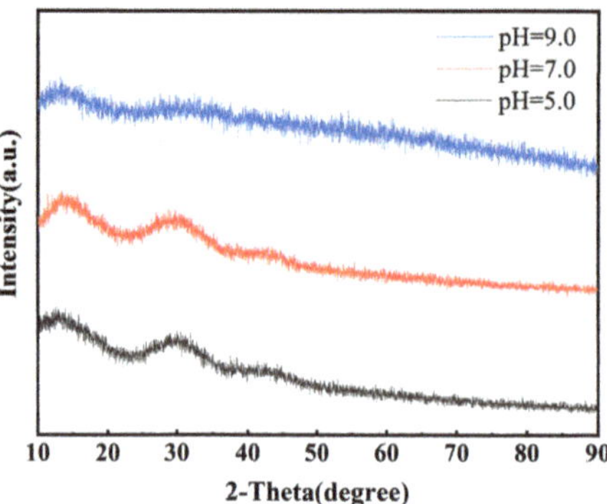

Figure 9. XRD patterns of floc precipitation in the aluminum–air fuel cell.

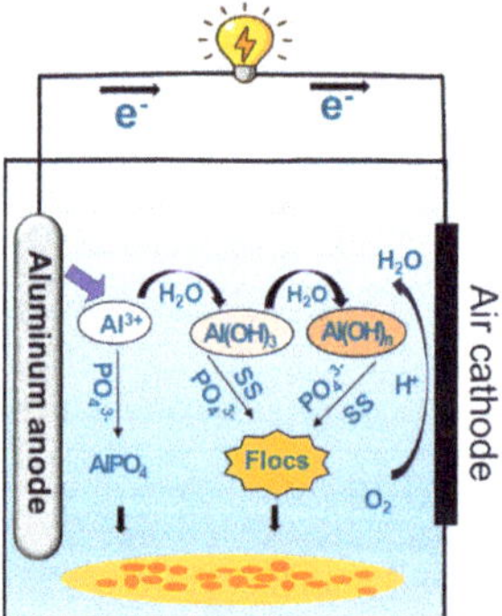

Figure 10. Schematic diagram of the phosphate removal process in the aluminum−air fuel cell.

4. Conclusions

A single-chamber aluminum–air fuel cell system was utilized to remove phosphate, and the effects of various factors were examined for phosphate removal. The phosphate removal mechanism was also discussed. An increase in the electrolyte concentration improved the level of phosphate removal and electricity production. With 10 mmol/L of NaCl, 97.8% phosphate removal was achieved, and the maximum power density reached 265.7 mW/m^2. For different initial pH values, the optimal condition for electricity production was pH 6.0, and 98.5% phosphate removal was achieved in 60 min. The average current density obtained was 0.20 mA/cm^2, and the maximum power density was 338.1 mW/m^2. Aluminum hydroxides were synthesized, and phosphate was primarily removed via the aluminum hydroxide adsorption process. The aluminum fuel cell can treat phosphate wastewater and generate energy simultaneously. Compared with conventional electrocoagulation systems, the aluminum–air fuel cell system can be considered an energy self-sufficient alternative, which is significant for achieving carbon neutrality in the wastewater treatment industry.

Author Contributions: X.H.: data curation and writing—original draft preparation; H.Q.: writing—original draft preparation; Y.Q.: methodology and supervision; Y.F.: conceptualization; X.Z.: writing—review and editing. All authors have read and agreed to the published version of the manuscript.

Funding: This research was funded by the National Natural Science Foundation of China (No. 52204184) and the Fundamental Research Funds for the Central Universities (No. N2201019).

Institutional Review Board Statement: Not applicable.

Informed Consent Statement: Not applicable.

Data Availability Statement: The data presented in this study are available on request from the corresponding authors.

Conflicts of Interest: The authors declare no conflict of interest.

References

1. Dall Agnol, P.; Libardi, N.; Muller, J.M.; Xavier, J.A.; Domingos, D.G.; Da Costa, R.H.R. A comparative study of phosphorus removal using biopolymer from aerobic granular sludge: A factorial experimental evaluation. *J. Environ. Chem. Eng.* **2020**, *8*, 103541. [CrossRef]
2. Xia, W.; Xu, L.; Yu, L.; Zhang, Q.; Zhao, Y.; Xiong, J.; Zhu, X.; Fan, N.; Huang, B.; Jin, R. Conversion of municipal wastewater-derived waste to an adsorbent for phosphorus recovery from secondary effluent. *Sci. Total Environ.* **2020**, *705*, 135959. [CrossRef]
3. Omwene, P.I.; Kobya, M.; Can, O.T. Phosphorus removal from domestic wastewater in electrocoagulation reactor using aluminium and iron plate hybrid anodes. *Ecol. Eng.* **2018**, *123*, 65–73. [CrossRef]
4. Jin, L.; Zhang, G.; Tian, H. Current state of sewage treatment in China. *Water Res.* **2014**, *66*, 85–98. [CrossRef] [PubMed]
5. Nguyen, D.D.; Ngo, H.H.; Guo, W.; Nguyen, T.T.; Chang, S.W.; Jang, A.; Yoon, Y.S. Can electrocoagulation process be an appropriate technology for phosphorus removal from municipal wastewater? *Sci. Total Environ.* **2016**, *563–564*, 549–556. [CrossRef] [PubMed]
6. Liu, B.; Liu, L.; Li, W. Effective removal of phosphorus from eutrophic water by using cement. *Environ. Res.* **2020**, *183*, 109218. [CrossRef] [PubMed]
7. Ren, J.; Li, N.; Wei, H.; Li, A.; Yang, H. Efficient removal of phosphorus from turbid water using chemical sedimentation by FeCl3 in conjunction with a starch-based flocculant. *Water Res.* **2020**, *170*, 115361. [CrossRef]
8. Moussa, D.T.; El-Naas, M.H.; Nasser, M.; Al-Marri, M.J. A comprehensive review of electrocoagulation for water treatment: Potentials and challenges. *J. Environ. Manag.* **2017**, *186*, 24–41. [CrossRef]
9. Bakshi, A.; Verma, A.K.; Dash, A.K. Electrocoagulation for removal of phosphate from aqueous solution: Statistical modeling and techno-economic study. *J. Clean. Prod.* **2020**, *246*, 118988. [CrossRef]
10. Dura, A.; Breslin, C.B. The removal of phosphates using electrocoagulation with Al−Mg anodes. *J. Electroanal. Chem.* **2019**, *846*, 113161. [CrossRef]
11. Hawari, A.H.; Alkhatib, A.M.; Hafiz, M.; Das, P. A novel electrocoagulation electrode configuration for the removal of total organic carbon from primary treated municipal wastewater. *Environ. Sci. Pollut. Res.* **2020**, *27*, 23888–23898. [CrossRef]
12. Ensano, B.M.B.; Borea, L.; Naddeo, V.; Belgiorno, V.; de Luna, M.D.G.; Balakrishnan, M.; Ballesteros, F.C. Applicability of the electrocoagulation process in treating real municipal wastewater containing pharmaceutical active compounds. *J. Hazard. Mater.* **2019**, *361*, 367–373. [CrossRef] [PubMed]
13. Tian, Y.; He, W.; Liang, D.; Yang, W.; Logan, B.E.; Ren, N. Effective phosphate removal for advanced water treatment using low energy, migration electric–field assisted electrocoagulation. *Water Res.* **2018**, *138*, 129–136. [CrossRef] [PubMed]
14. Tian, Y.; He, W.; Zhu, X.; Yang, W.; Ren, N.; Logan, B.E. Energy efficient electrocoagulation using an air-breathing cathode to remove nutrients from wastewater. *Chem. Eng. J.* **2016**, *292*, 308–314. [CrossRef]
15. An, J.; Li, N.; Wang, S.; Liao, C.; Zhou, L.; Li, T.; Wang, X.; Feng, Y. A novel electro-coagulation-Fenton for energy efficient cyanobacteria and cyanotoxins removal without chemical addition. *J. Hazard. Mater.* **2019**, *365*, 650–658. [CrossRef] [PubMed]
16. Tian, Y.; He, W.; Zhu, X.; Yang, W.; Ren, N.; Logan, B.E. Improved Electrocoagulation Reactor for Rapid Removal of Phosphate from Wastewater. *ACS Sustain. Chem. Eng.* **2017**, *5*, 67–71. [CrossRef]
17. Kim, J.H.; Park, I.S.; Park, J.Y. Electricity generation and recovery of iron hydroxides using a single chamber fuel cell with iron anode and air-cathode for electrocoagulation. *Appl. Energ.* **2015**, *160*, 18–27. [CrossRef]
18. Kim, J.H.; Maitlo, H.A.; Park, J.Y. Treatment of synthetic arsenate wastewater with iron-air fuel cell electrocoagulation to supply drinking water and electricity in remote areas. *Water Res.* **2017**, *115*, 278–286. [CrossRef]

19. Si, Y.; Li, G.; Wu, Y.; Zhang, H.; Yuan, Y.; Zhang, H.; Liu, B.; Zhang, F. Tradeoff between groundwater arsenite removal efficiency and current production in the self-powered air cathode electrocoagulation with different oxygen reduction pathways. *J. Hazard. Mater.* **2018**, *357*, 138–145. [CrossRef]
20. Maitlo, H.A.; Kim, J.H.; An, B.M.; Park, J.Y. Effects of supporting electrolytes in treatment of arsenate-containing wastewater with power generation by aluminumair fuel cell electrocoagulation. *J. Ind. Eng. Chem.* **2018**, *57*, 254–262. [CrossRef]
21. Liu, Y.; Yuan, J.; Ning, Y.; Tang, Y.; Luo, S.; Jiang, B. Efficient reduction of Cr(VI) and immobilization of Cr driven by an iron-air fuel cell: Reaction mechanisms and electricity generation. *Chemosphere* **2020**, *253*, 126730. [CrossRef] [PubMed]
22. Ni, C.; Wang, J.; Guan, Y.; Jiang, B.; Meng, X.; Luo, S.; Guo, S.; Wang, L. Self-powered peroxi-coagulation for the efficient removal of p-arsanilic acid: PH-dependent shift in the contributions of peroxidation and electrocoagulation. *Chem. Eng. J.* **2020**, *391*, 123495. [CrossRef]
23. Liu, Q.; Zhang, M.; Lv, T.; Chen, H.; Chika, A.O.; Xiang, C.; Guo, M.; Wu, M.; Li, J.; Jia, L. Energy-producing electro-flocculation for harvest of Dunaliella salina. *Bioresour. Technol.* **2017**, *241*, 1022–1026. [CrossRef]
24. Wei, W.; Gong, H.; Sheng, L.; Zhou, D.; Zhu, S. Optimum parameters for humic acid removal and power production by Al-air fuel cell electrocoagulation in synthetic wastewater. *Water Sci. Technol.* **2022**, *85*, 174–187. [CrossRef] [PubMed]
25. Kim, J.H.; An, B.M.; Lim, D.H.; Park, J.Y. Electricity production and phosphorous recovery as struvite from synthetic wastewater using magnesium-air fuel cell electrocoagulation. *Water Res.* **2018**, *132*, 200–210. [CrossRef]
26. Mahmood, A.; Kim, J.H.; Park, J. Development of an effective operation system in a magnesium-air desalination cell for electricity production with nitrogen and phosphorus removal. *Desalination* **2023**, *545*, 116164. [CrossRef]
27. Wang, R.; Liu, M.; Zhang, M.; Ghulam, A.; Yuan, L. An iron-air fuel cell system towards concurrent phosphorus removal and resource recovery in the form of vivianite and energy generation in wastewater treatment: A sustainable technology regarding phosphorus. *Sci. Total Environ.* **2021**, *791*, 148213. [CrossRef]
28. Wang, R.; Wan, S.; Lai, L.; Zhang, M.; Zeb, B.S.; Qaisar, M.; Tan, G.; Yuan, L. Recovering phosphate and energy from anaerobic sludge digested wastewater with iron-air fuel cells: Two-chamber cell versus one-chamber cell. *Sci. Total Environ.* **2022**, *825*, 154034. [CrossRef]
29. Dong, H.; Yu, H.; Wang, X.; Zhou, Q.; Feng, J. A novel structure of scalable air-cathode without Nafion and Pt by rolling activated carbon and PTFE as catalyst layer in microbial fuel cells. *Water Res.* **2012**, *46*, 5777–5787. [CrossRef]
30. Zhang, S.; Zhang, J.; Wang, W.; Li, F.; Cheng, X. Removal of phosphate from landscape water using an electrocoagulation process powered directly by photovoltaic solar modules. *Sol. Energy. Mater. Sol. Cells* **2013**, *117*, 73–80. [CrossRef]
31. Maitlo, H.A.; Lee, J.; Park, J.Y.; Kim, J.; Kim, K.; Kim, J.H. An energy-efficient air-breathing cathode electrocoagulation approach for the treatment of arsenite in aquatic systems. *J. Ind. Eng. Chem.* **2019**, *73*, 205–213. [CrossRef]
32. Picard, T.; Cathalifaud-Feuillade, G.; Mazet, M.; Vandensteendam, C. Cathodic dissolution in the electrocoagulation process using aluminium electrodes. *J. Environ. Monit.* **2000**, *2*, 77–80. [CrossRef]
33. Can, O.T.; Bayramoglu, M.; Kobya, M. Decolorization of Reactive Dye Solutions by Electrocoagulation Using Aluminum Electrodes. *Ind. Eng. Chem. Res.* **2003**, *42*, 3391–3396. [CrossRef]
34. Hu, Q.; He, L.; Lan, R.; Feng, C.; Pei, X. Recent advances in phosphate removal from municipal wastewater by electrocoagulation process: A review. *Sep. Purif. Technol.* **2023**, *308*, 122944. [CrossRef]
35. Irdemez, S.; Demircioglu, N.; Yıldız, Y.S.; Bingül, Z. The effects of current density and phosphate concentration on phosphate removal from wastewater by electrocoagulation using aluminum and iron plate electrodes. *Sep. Purif. Technol.* **2006**, *52*, 218–223. [CrossRef]
36. Elsheikh, A.H.; El-Said, E.M.S.; Abd Elaziz, M.; Fujii, M.; El-Tahan, H.R. Water distillation tower: Experimental investigation, economic assessment, and performance prediction using optimized machine-learning model. *J. Clean. Prod.* **2023**, *388*, 135896. [CrossRef]
37. Elsheikh, A.H.; Shanmugan, S.; Sathyamurthy, R.; Kumar Thakur, A.; Issa, M.; Panchal, H.; Muthuramalingam, T.; Kumar, R.; Sharifpur, M. Low-cost bilayered structure for improving the performance of solar stills: Performance/cost analysis and water yield prediction using machine learning. *Sustain. Energy Technol. Assess.* **2022**, *49*, 101783. [CrossRef]
38. Khoshaim, A.B.; Moustafa, E.B.; Bafakeeh, O.T.; Elsheikh, A.H. An Optimized Multilayer Perceptrons Model Using Grey Wolf Optimizer to Predict Mechanical and Microstructural Properties of Friction Stir Processed Aluminum Alloy Reinforced by Nanoparticles. *Coatings* **2021**, *11*, 1476. [CrossRef]
39. Rodrigo, M.A.; Cañizares, P.; Buitrón, C.; Sáez, C. Electrochemical technologies for the regeneration of urban wastewaters. *Electrochim. Acta* **2010**, *55*, 8160–8164. [CrossRef]
40. Guseva, O.; Schmutz, P.; Suter, T.; von Trzebiatowski, O. Modelling of anodic dissolution of pure aluminium in sodium chloride. *Electrochim. Acta* **2009**, *54*, 4514–4524. [CrossRef]
41. Mansouri, K.; Ibrik, K.; Bensalah, N.; Abdel-Wahab, A. Anodic Dissolution of Pure Aluminum during Electrocoagulation Process: Influence of Supporting Electrolyte, Initial pH, and Current Density. *Ind. Eng. Chem. Res.* **2011**, *50*, 13362–13372. [CrossRef]
42. Georgantas, D.A.; Grigoropoulou, H.P. Orthophosphate and metaphosphate ion removal from aqueous solution using alum and aluminum hydroxide. *J. Colloid Interf. Sci.* **2007**, *315*, 70–79. [CrossRef] [PubMed]
43. Lacasa, E.; Cañizares, P.; Sáez, C.; Fernández, F.J.; Rodrigo, M.A. Electrochemical phosphates removal using iron and aluminium electrodes. *Chem. Eng. J.* **2011**, *172*, 137–143. [CrossRef]

44. Pulkka, S.; Martikainen, M.; Bhatnagar, A.; Sillanpää, M. Electrochemical methods for the removal of anionic contaminants from water—A review. *Sep. Purif. Technol.* **2014**, *132*, 252–271. [CrossRef]
45. Hu, C.; Chen, G.; Liu, H.; Zhao, H.; Qu, J. Characterization of flocs generated by preformed and in situ formed Al13 polymer. *Chem. Eng. J.* **2012**, *197*, 10–15. [CrossRef]
46. Omwene, P.I.; Kobya, M. Treatment of domestic wastewater phosphate by electrocoagulation using Fe and Al electrodes: A comparative study. *Process Saf. Environ.* **2018**, *116*, 34–51. [CrossRef]
47. Ma, B.; Chen, G.; Hu, C.; Liu, Z.; Liu, H.; Qu, J. Speciation matching mechanisms between orthophosphate and aluminum species during advanced P removal process. *Sci. Total Environ.* **2018**, *642*, 1311–1319. [CrossRef]

 *applied sciences*

Article

Grain Boundary—A Route to Enhance Electrocatalytic Activity for Hydrogen Evolution Reaction

Ran Jiang [1], Jianyu Fu [1], Zhaoyang Wang [2] and Cunku Dong [1,*]

[1] Institute of New Energy Materials, School of Materials Science and Engineering, Tianjin University, Tianjin 300072, China; jiangran_2019@tju.edu.cn (R.J.); jianyufu@tju.edu.cn (J.F.)

[2] School of Earth and Environmental Sciences, Lanzhou University, Lanzhou 730000, China; wzy@lzu.edu.cn

* Correspondence: ckdong@tju.edu.cn

Abstract: The electrocatalytic hydrogen evolution reaction (HER) of a given metal catalyst is intrinsically related to its electronic structure, which is difficult to alter for further improvement. Recently, it was discovered that the density of grain boundaries (GBs) is mechanistically of great importance for catalytic activity, implying that GBs are quantitatively correlated with the active sites in the HER. Here, by modeling the atomistic structure of GBs on a Au(110) surface, we find that HER performance is greatly enhanced by Au GBs, suggesting the feasibility of the HER mediated by GBs. The promoted HER performance is due to an increase in the capability of binding adsorbed hydrogen on the sites around GBs. A Au catalyst with a dominantly exposed (110) plane is synthesized, where considerable GBs exist for experimental verification. It is found that HER activity is inherently correlated with the density of the GBs in Au NPs. The improvement in HER activity can be elucidated from the geometrical and electronic points of view; the broken local spatial symmetry near a GB causes a decrease in the coordination numbers of the surface sites and the shift up of the d–band center, thereby reducing the limiting potential for each proton−electron transfer step. Our finding represents a promising means to further improve the HER activity of a catalyst.

Keywords: grain boundaries; electrocatalytic hydrogen evolution; density functional calculations; gold

Citation: Jiang, R.; Fu, J.; Wang, Z.; Dong, C. Grain Boundary—A Route to Enhance Electrocatalytic Activity for Hydrogen Evolution Reaction. *Appl. Sci.* **2022**, *12*, 4290. https://doi.org/10.3390/app12094290

Academic Editor: Leonarda Francesca Liotta

Received: 30 March 2022
Accepted: 19 April 2022
Published: 24 April 2022

1. Introduction

In recent decades, human beings have relied on fossil fuels for over 80% of their total energy needs. Hydrogen, as a clean, economic and renewable energy carrier, is considered to be an attractive alternative to traditional fossil fuels, and it can greatly alleviate the global greenhouse effect and energy crisis at present [1]. Amongst various hydrogen production techniques, the electrocatalytic hydrogen evolution reaction (HER) via electrochemical water splitting, an important energy recovery technique, has received great attention as a hot research topic [2–4]. Consequently, ever–increasing efforts have been devoted to developing a variety of new catalysts with the aim of improving HER performance [5,6]. However, noble metal–based materials such as platinum (Pt) (e.g., Pt/C) are still the most efficient catalysts widely used to catalyze hydrogen evolution, owing to their outstanding thermodynamical and kinetic features for the HER [3]. Unfortunately, Pt is an extremely scarce and precious metal, which results in a high cost for the HER. Therefore, designing a new catalyst or boosting the activity of existing non–noble materials other than Pt has become urgent.

As an important descriptor, hydrogen bonding ability is often taken into account in the screening or designing of HER catalysts [7–9]. According to the Sabatier principle, either a too weak or too strong adsorption ability results in a drop in HER performance in that a too weak adsorption of H* leads to a pronounced decline in intermediate stabilization, while a too strong adsorption inhibits the desorption of gaseous hydrogen [10]. The superior HER of Pt is due to its optimal H* binding energy, or Gibbs free energy, which is close to zero. To date, many non–noble materials suffer from a weak H* binding ability. Identifying how to enhance their H* binding activity remains an enormous challenge. Therefore, various routes have been proposed to control the microstructure of catalysts in order to expose more active facets or sites [11–13]. However, further improving HER activity is still a great challenge due to the inherent electronic structure of the catalysts themselves, leading to a certain capacity for H* binding, which almost cannot be tuned. Inspired by a previous study on the relationship between grain boundaries (GBs) and catalytic activity in various reactions [14–16], GBs rich in the atomic arrangement disorder were found to alter the local electronic structure around GBs [17,18]. In addition, GBs feature under–coordinated sites and microstrains [16,19], which favor H binding. Kim [17], using extensive density functional theory (DFT) calculations to model the atomistic structure of GBs on a Au(111) surface, concluded that the grain boundary of Au is conducive to the adsorption of COOH by the active site, thus enhancing electrochemical CO_2 reduction. Dong [18] believes that the GB sites on a Au(110) surface lead to a high selectivity toward CH_3OH. Thus, GBs are expected to have a stronger H* binding capability so as to compensate for the unsaturated coordination of GB atoms resulting from such a disorder. However, these studies lack an investigation of the atomic and electronic structures of grain boundaries and an explanation of the enhanced catalytic performance of grain boundaries.

Herein, we first predict the impact of grain boundaries on a Au surface on the HER occurring at GBs via density functional theory (DFT) calculations, in light of the much weaker H* binding capacity of Au [20]. Then, a Au NP catalyst with GBs is prepared for the verification of GB–assisted HER activity. HER activity greatly increases with an increase in the density of GBs in Au NPs. Additionally, the atomic and electronic structures are analyzed to determine the role of the GBs. On the basis of the theoretical and experimental studies, it is found that the GBs can prodigiously promote the chemical bonding ability of H* intermediate, resulting in enhanced catalytic activity in hydrogen evolution.

2. Compactional and Experimental Methods

2.1. Computational Model and Method

Of many possible high–angle grain boundaries (HAGBs), an ∑6{2–21} HAGB atomistic model on a Au(110) surface was constructed using coincidence site lattice (CSL) theory, as shown in Figure 1. This HAGB surface model consists of three layers containing 52 Au atoms (Figure S1). Five atop catalytic active sites were selected, labeled as s1–s5 (Figure 1).

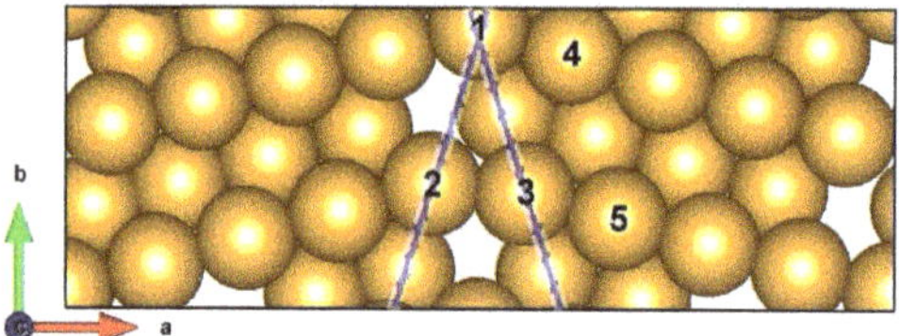

Figure 1. Top view of the Σ6{2–21} HAGB model on Au(110) surface with three atomic layers in an orthogonal supercell (25.29 × 8.94 × 17.80 Å^3). The GB area is highlighted by blue lines. The selected Au around GB is labeled as s1–s5.

Spin–polarized DFT calculations were performed using Vienna ab initio Simulation Package (VASP) [21]. The revised Perdew−Burke−Ernzerhof (rPBE) exchange–correlation functional within the generalized gradient approximation (GGA) was used with the projector augmented wave (PAW) pseudopotential [22]. A plane–wave cutoff energy of 450 eV was used for clean surface and adsorbate surface relaxation. The relaxation was complete when the residual force was less than 0.05 eV/Å. During the structure relaxation, two downmost layers were fixed to their original position of the slab without the adsorbate, while the uppermost layer and adsorbate were fully relaxed. Then, 2 × 1 × 1 Monkhorst−Pack mesh sampling was employed for the clean surface and adsorbate surface in the surface

Brillouin zone. A vacuum space of 15 Å perpendicular to the slab surface was used to avoid artificial interaction. Additional computational details about the calculation of the free energy diagram can be found in the Supporting Information.

2.2. Preparation of Au/CFP Electrodes

Au nanoparticles were deposited on carbon fiber paper by magnetron sputtering under a sputtering vacuum of 5×10^{-2} mbar and a sputtering current of 20 mA for 30 s using a tabletop DC magnetron sputtering coater (Leica EM SCD 500). The purity of the Au target used for deposition was 99.999%. The annealing of Au/CFP electrodes was performed using a tube furnace (Thermo Scientific) with a flowing Ar atmosphere at 100 sccm, heated at a given temperature (200~400 °C) for 2 h. The use of an Ar atmosphere prevented the formation of carbon coatings on the Au nanoparticles.

2.3. Electrochemical Characterization

All electrochemical measurements were performed on a CHI 1100C electrochemical workstation (Chenhua, China). The Pb underpotential deposition (upd) measurements of the Au/CFP electrodes were performed in 0.1 M NaOH solution containing 1 mM $Pb(OAc)_2$, and Pt and Ag/AgCl electrode (3.0 M KCl) were the counter electrode and reference electrode, respectively. The voltammogram scan rate was 50 mV s^{-1}.

Electrochemical hydrogen evolution measurements were performed in 0.5 M H_2SO_4 with continuous purging of N_2 (>99.999% purity) using a standard three–electrode cell, where the Au/CFP electrodes, the graphite rod (>99.999% purity) and the Hg/Hg_2Cl_2 electrode (saturated KCl solution) were used as the working, counter and reference electrodes, respectively. The potential was calibrated with respect to RHE in the high–purity hydrogen–saturated electrolyte, with a Pt plate as the working electrode. A flow of N_2 was maintained over the electrolyte (0.5 M H_2SO_4) during the electrochemical measurements in order to eliminate the possible effects of other gases. The polarization curves were recorded with a scan rate of 5 mV s^{-1} without *iR* correction.

3. Results and Discussion

3.1. Theoretical Prediction of GB–Assisted HER

3.1.1. Gibbs Free Energy

The HER is a multi–step electrochemical process occurring on the surface of an electrode [23]. In acid media, two possible routes have been elucidated for the HER, i.e., the Volmer–Heyrovsky and the Volmer–Tafel mechanisms, which can be described as follows [24,25]:

$$H_3O^+ + M + e^- \rightleftharpoons H*-M + H_2O \quad (1)$$

$$H*-M + H_3O^+ + e^- \rightleftharpoons M + H_2 + H_2O \quad (2)$$

$$2H*-M \rightleftharpoons 2M + H_2 \quad (3)$$

Step (1) refers to the Volmer reaction, while steps (2) and (3) refer to the Heyrovsky and Tafel reactions, respectively. We first explore the impact of the GBs on the Au(110) surface on the Volmer reaction, which dominantly determines the overall HER performance; for this, we chose various active sites at and near the Au GBs (Figure 1). Figure 2a shows the Gibbs free energy change (ΔG_{H^*}) profile of the HER in acid media. The ΔG_{H^*} values of the flat Au(110) and Pt(111) GB–free surfaces are also presented for comparison (Tables S1 and S2). It should be noted that ΔG_{H^*} dramatically reduces once the GBs are introduced. Additionally, ΔG_{H^*} exhibits site–dependent HER activity for sites in the vicinity of the GBs. To be specific, the ΔG_{H^*} for s1 is 0.18 eV, approximately 0.4 eV less than the flat Au(110) (0.58 eV), whereas s2 and s3 have a ΔG_{H^*} of about −0.07 eV; this is less than that of Pt (−0.09 eV), which is more close to zero. As is well known, the optimal value of ΔG_{H^*} for the HER is zero, as hydrogen binding is neither too strong nor too weak. Therefore, the activity for the Volmer reaction is greatly promoted by the GBs. In addition, the sites near the GBs (s4 and s5) reduce ΔG_{H^*} by ~0.13 eV with regard to the Au(110) surface, suggesting that GBs on the Au(110) surface not only greatly promote HER activity on their own (s1–s3) but also activate the sites nearby (s4–s5) to further lower the reaction barrier.

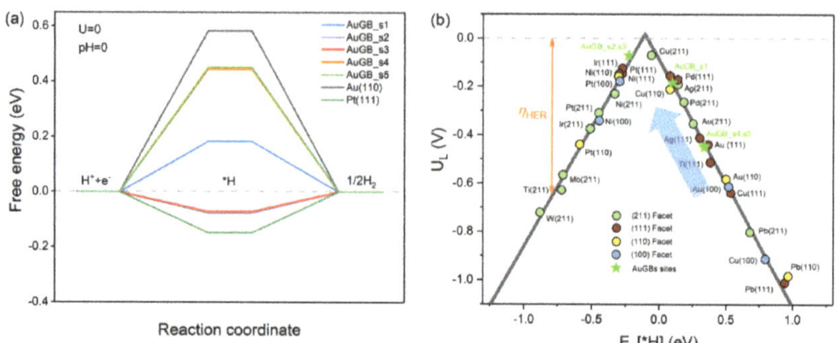

Figure 2. (**a**) The calculated free−energy diagram of HER at the equilibrium potential (U = 0 V) for Au(110) (black), Pt(111) (green) and different active sites on Au GBs. (**b**) The volcano relation of the limiting potential for HER as a function of $E_B(H^*)$. The arrow indicates the desired direction for catalyst design with higher activity, and the color pentagrams represent GB sites on Au surface for prodigious promotion of HER. Theoretical overpotentials (η_{HER}) are the vertical difference between the points and the equilibrium potential (red dashed line).

3.1.2. Limiting Potentials

To further evaluate the HER performance, we compared the limiting potential ($U_L = -\Delta G_{H^*}/e$) of various metal materials with different exposed facets (Tables S3 and S4). Figure 2b presents a volcano plot, which typically correlates $E_B(H^*)$ with U_L. GB–free Au(110) is located far from the top on the right side of the volcano plot, indicating that Au(110) binds H* weakly enough not to form a stable intermediate, which readily desorbs off the surface; thus, a much higher overpotential is required to yield gaseous hydrogen [10], which is in agreement with the previous discussion. When GBs are formed on Au(110), HER activity moves close to the top of the volcano. The active sites at the GBs (s1–3) possess superior HER performance to almost all the metal catalysts considered here. Additionally, the U_L required is also reduced for sites near the GBs (s4 and s5). In particular, s1, s2 and s3 lie close to the top of the volcano plot near Pt, although PDS is different, suggesting that their HER activity is similar to that of the Pt extended surface. The enhanced HER performance is mainly due to the increase in the ability to bind H* on the Au(110) surface with GBs. Therefore, the GBs on the Au(110) surface can dramatically improve H* binding strength and, thus, boost the HER process.

3.1.3. Exchange Current Density

For the purpose of the application of the GB concept in the HER experiment, we calculated the exchange current density (j_0), the most common experimental descriptor, to measure the catalysts' HER activity [10,26]. According to the micro–kinetic model, at equilibrium, j_0 can be theoretically computed as an indirect function of ΔG_{H^*} [10],

$$j_0 = Fk^0 C_{total}\left[(1-\theta)^{1-\alpha}\theta^{\alpha}\right] \tag{4}$$

$$\theta = \frac{exp\left(-\Delta G_{*H}/k_B T\right)}{1+exp\left(-\Delta G_{*H}/k_B T\right)} \tag{5}$$

where k^0 is the standard rate constant, α is the transfer coefficient (α was set to 0.5 in this work), C_{total} is the total number of HER active sites on the surface of the catalyst, and k_B is the Boltzman constant. The relationship of catalysts correlating $E_B(H^*)$ is presented with the logarithm of j_0 ($\log j_0$) (Figure 3a). Similar to the U_L–$E_B[H^*]$ plot, a typical volcano plot is also clearly observed, which is consistent with that in a previous study [10]. Pt and other HER–active materials are close to the top of the volcano curve, which denotes the scaling relationship between $\log j_0$ and $E_B[H^*]$. The flat Au(110) surface has $\log j_0 = -10.88$ A·cm^{-2} at $E_B[H^*] = 0.50$ eV, which implies that Au(110) is not a good HER catalyst because its j_0 is extremely small. Once GBs are formed on the Au(110) surface, the $\log j_0$ of the sites around the GBs is significantly increased (s1–s5), and they shift up toward the volcano top. s1–s3 are promoted several to a dozen million times in j_0 with respect to the Au(110) surface. It is more surprising that, for s1 and s2, $\log j_0$ jumps to -3.30 A·cm^{-2}, which is close to the volcano top, outperforming Pt(111) ($\log j_0 = -3.50$ A·cm^{-2}). This dramatic increase in j_0 is also ascribed to the strong H* binding capability caused by GBs on the Au surface.

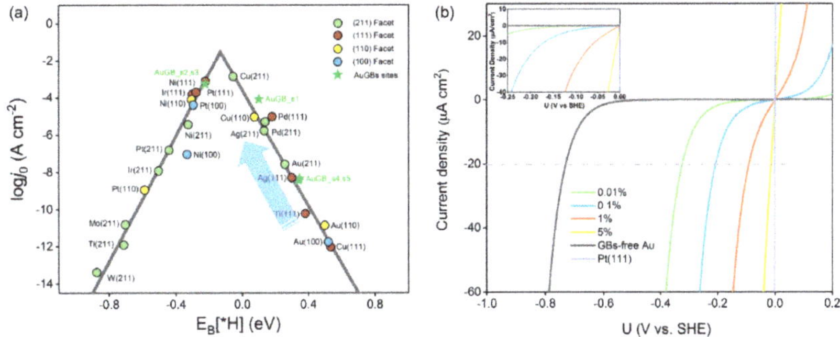

Figure 3. (**a**) Computational exchange current density ($\log(j_0)$) for hydrogen evolution over Au GB sites, and various metal surfaces plotted as a function of the hydrogen binding energy. (**b**) Current density of hydrogen evolution as a function of the applied potentials for Au(110) with varying concentrations of GBs on Au(110), and Pt(111) is presented for comparison. The inset presents a magnified version of the plot.

3.1.4. GB–Mediated Current Density

In fact, the active sites on the Au surface with GBs consist of GB sites and normal sites, which both contribute to the hydrogen evolution. Thus, the overall HER performance of the Au(110) surface with GBs should be described by current density (j) at a given applied potential, which results from the combination of Au GBs and a flat Au(110) surface. For this purpose, we plotted the current density of hydrogen evolution as a function of the applied potential (U) using the kinetics of electrode reactions [27], which can simulate a linear scan voltammetry (LSV) curve. Under the precondition of ruling out mass transfer effects, the LSV curve of the HER can be predicted using the Butler–Volmer Equation (Supporting Information),

$$j = j_{forward} + j_{backward} = j_0\left[e^{-\alpha f \eta} - e^{-(1-\alpha) f \eta}\right] \quad (6)$$

where α is the transfer coefficient, η is a certain overpotential, and f denotes F/RT. Figure 3b shows a j–U plot of Au(110) surfaces with varying amounts of GBs by controlling the ratio of GB sites (s1–s3)/normal sites. The GB–free Au(110) surface has the largest overpotential of -0.73 V (vs SHE) at a current density of 20 μA cm^{-2}. Surprisingly, GBs induce a significant decrease in the overpotentials required for the HER. Of particular note is that only 0.01% of the GBs on Au(110) lead to a dramatic decrease in the overpotential by almost one–half from -0.73 to 0.33 (V vs. SHE), suggesting that quite a low GB density on Au(110) can extremely boost the overall HER performance. As can be clearly seen in Figure 3, the higher the density of the Au GBs, the lower the potential that the catalyst will have. When the density of the GBs increases to 1%, the overpotential to achieve 20 μA cm^{-2} further reduces to -0.09 V (see inset). A total of 5% of GBs can further reduce the overpotential to 0.01 V.

However, we cannot expect even higher HER activity by continuously increasing the GB density, because there exist only a few GBs on a real catalyst surface.

3.2. Experimental Verification of GB–Assisted HER Activity

3.2.1. Exposed Surface Characterization

To determine the accessible exposed facet of the as–prepared Au electrodes, we used the lead underpotential deposition (Pb–upd) technique. The Pb–upd process is very sensitive to the surface structure of the gold electrode with a distinct voltammetric profile, where the main peaks appear at different characteristic potentials for different exposed facets of Au electrodes [28,29]. As a result, the main peaks can be used as indicators to characterize the surface structure of the electrode. In Figure 4, it can be clearly observed that the pristine Au/CFP electrode shows only one significant deposition peak in the Pb–upd voltammogram profile, which is located at ~−0.47 V (vs. Ag/AgCl) and corresponds to the (110) facet of Au. Upon scan reversal, the Pb–upd layer is oxidized to dissolved Pb^{2+} in the form of $Pb(OH)^-$ in alkaline solution in a sharp peak at ~−0.38 V (vs. Ag/AgCl), which is in good agreement with the results of a previous study [30]. Additionally, the CFP electrode exhibits no Pb–upd peak, suggesting that Pb–upd is attributed to the Au itself. Therefore, Au(110) is the dominant exposed surface faceting of the as–deposited Au/CFP electrodes, which provides the experimental basis for the verification of our GB–assisted HER on Au(110).

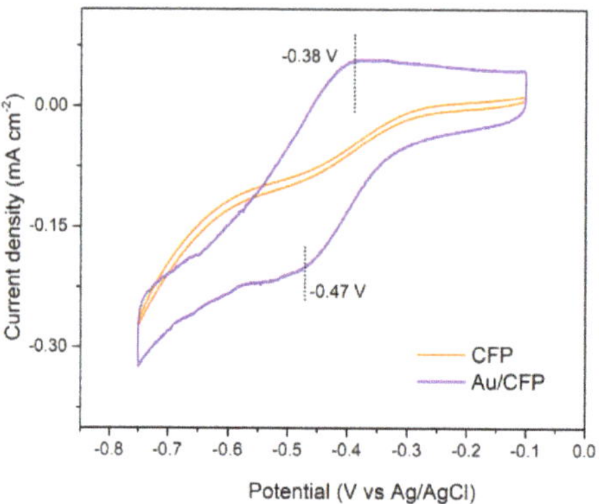

Figure 4. Pd–upd voltammetric profile of as–deposited Au/CFP and clean CFP electrodes (0.5×1 cm^2). Scan rate: 50 mV/s.

We performed a detailed characterization of the morphology and structure of the as–prepared and annealed Au/CFP. From SEM and compositional maps (Figure S2), it can be seen that Au is deposited on the CFP electrode with a uniform distribution and a small particle size. The as–deposited Au/CFP was characterized by X-ray diffraction (XRD) (Figure S3). When the sputtering time is short, such as 30 s or 2 min, the intensity of the Au diffraction peaks is weak, because the content of Au is small and the particle size is small.

The morphological structures of the as–deposited and annealed Au/CFP were investigated via transmission electron microscopy (TEM) observation (Figure 5). The high–resolution TEM (HR–TEM) images show a lattice fringe spacing of 2.35 Å corresponding to the (111) facet direction of the Au nanoparticles (NPs). It can be clearly seen that the as–deposited Au NPs feature a representative surface rich in GBs, showing a polycrystalline character to some extent (Figure 5a). Notably, the grain boundary angle in Figure 5a is similar to that in our calculated GBs surface model (Figure 1).

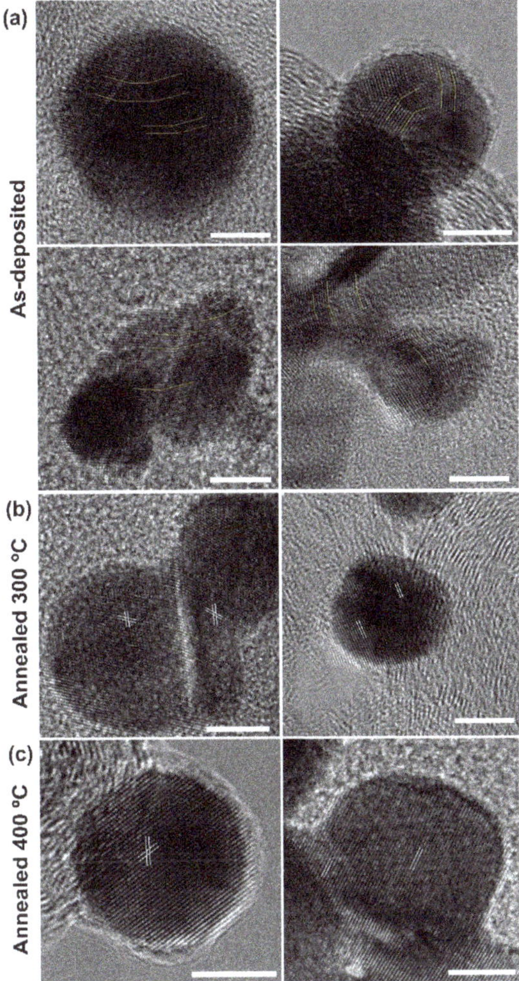

Figure 5. High–revolution TEM images of as–prepared (**a**) and annealed Au/CFP electrodes at 300 °C (**b**) and 400 °C (**c**). The light yellow lines indicate GBs in the Au NPs. The white lines indicate the lattice fringe spacing for Au NPs. The scale bar represents 5 nm.

To control the density of the GBs on Au(110), we annealed the pristine Au/CFP under an Ar atmosphere at different temperatures, namely, 300 °C and 400 (producing Au/CFP_{300} and Au/CFP_{400}, respectively). By comparing the change in the morphology before and after annealing, it is interesting to note that the (110) facet is still dominant on various Au/CFP electrodes. In addition, the density of the GBs in the Au NPs dramatically decreases as the annealing temperature is increased (Figure 5b,c). At 300 °C, only a few Au NPs with GBs are observed, and the GB density in the Au NPs becomes much lower, suggesting that most Au NPs are transformed into a monocrystalline form. Nevertheless, Au NPs are completely monocrystalline after annealing at 400 °C.

3.2.2. Electrochemical HER Characterization

The effect of GBs on the HER activity of the Au NPs was examined in acid aqueous solution (0.5 M H_2SO_4). Figure 6a shows representative LSV curves for the CFP substrate, pristine Au/CFP and annealed Au/CFP electrodes (Au/CFP_{200}, Au/CFP_{300} and

Au/CFP$_{400}$). The CFP substrate has negligible HER activity and, thus, is expected to exert no influence on the current density contribution by the GBs. The pristine as–deposited Au/CFP has the highest HER activity and shows an overpotential of 330 mV at a current density of 10 mA/cm^2. By contrast, the overpotentials required to achieve 10 mA/cm^2 HER current density for the annealed Au/CFP electrodes are 470 mV (Au/CFP$_{200}$), 530 mV (Au/CFP$_{300}$) and 550 mV (Au/CFP$_{400}$). In combination with the above TEM analysis, the HER activity significantly decreases with the decrease in the density of the GBs in the Au NPs. Au/CFP$_{300}$ has only a 20 mV lower overpotential than Au/CFP$_{400}$, as the GBs in the Au NPs almost diminish after annealing at 300 °C and 400 °C, which is consistent with the above TEM observation (Figure 5b,c). Obviously, the experimental LSV curve is in good agreement with our simulated current density curve (Figure 3b). Although it is quite difficult to measure the GB density accurately due to the irregular shape of Au NPs, the overpotential difference (220 mV) still indeed suggests that the GBs in Au NPs could significantly enhance HER activity.

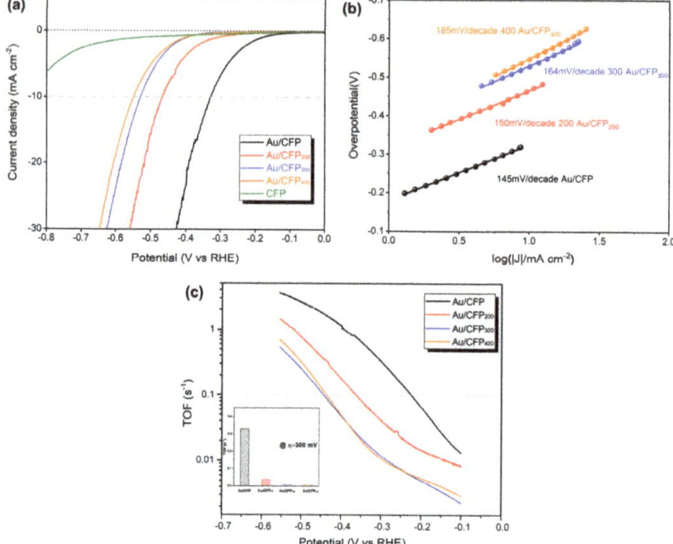

Figure 6. Effect of GBs on Au surface on the electrochemical HER activity of Au/CFP electrodes. (**a**,**b**) Electrochemically active (EA) surface area−calibrated linear sweep voltammetry (LSV) curves and Tafel plots for four Au/CFP electrodes with a size of 0.5 × 1 cm^2 (electrolyte: 0.5 M H_2SO_4, scan rate: 5 mVs^{-1}). (**c**) Turnover frequency per surface Au atom of various Au/CFP samples, and the inset is the TOF value obtained at an overpotential of 300 mV.

A comparison of Tafel plots for various Au/CFP electrodes is shown in Figure 6b. It can be seen that the introduction of GBs in Au NPs leads to a smaller Tafel slope compared with the GB–free counterparts, resulting in values of 145 and 185 mV dec^{-1} for Au/CFP and Au/CFP$_{400}$, respectively. As widely accepted, Tafel slopes of 120, 40 and 30 mV dec^{-1} have been observed for the Volmer, Heyrovsky and Tafel determining rate steps, respectively [31]. Regarding the HER mechanism, the rate–determining step for this Au catalyst is the Volmer reaction, i.e., the initial adsorption of protons from the acid solution to form adsorbed H [10]. Of note is that GBs may change the rate–determining step (RDS); that is, RDS may be changed from binding H* to desorbing hydrogen at Au atoms around the GBs by increasing the capability of binding H (Figure 2a, Table S2), which is in agreement with the theoretical ΔG_{H*} for the GB sites and normal sites (Figure 2a).

To exclude the improvement in HER activity caused by particle size and surface areas, we calibrated the electrochemically active (EA) surface area (Figure S4) of the as–

prepared Au/CFP electrodes. Based on the EA surface area, we then calculated the turnover frequency (TOF) per surface Au atom to compare the intrinsic activity of various Au/CFP electrodes. Figure 6c shows the experimental TOF as a function of applied overpotential. The TOF value increases with the overpotential following the Tafel behavior for all Au/CFP electrodes. The TOF value of the as–deposited Au/CFP is even higher than those of the other annealed Au/CFP electrodes, indicating that, after annealing treatment, the highly effective sites for the HER significantly decrease. Additionally, the TOF values of Au/CFP_{300} and Au/CFP_{400} are almost the same over the whole overpotential range. Specifically, at an overpotential of 300 mV, the as–deposited Au/CFP electrode exhibits a high TOF value of 0.33 s^{-1}, which is extremely higher than those of the annealed Au/CFP electrodes (0.04 s^{-1} for Au/CFP_{200} and 0.01 s^{-1} for Au/CFP_{300} and Au/CFP_{400}) (see inset, Figure 6c). The trend in the TOF value for various Au/CFP electrodes is consistent with LSV and Tafel, and it is inherently correlated with the decrease in the density of the GBs on the Au surface after annealing (Figure 5). Thus, creating GBs is a much more effective strategy to improve HER activity of the originally inactive metal electrocatalyst.

3.3. Origin of GB–Enhanced HER

Although Au(110) is not a traditional active catalyst for the hydrogen evolution reaction [20], it becomes a great activity catalyst for the HER by forming grain boundaries, as stated above. The improved HER activity is ascribed to the H* binding energy caused by the GBs in the Au electrocatalyst. As the strength to bind H* is inherently decided by the underlying electronic structure of the catalyst itself, we focus on the differences in atomic and electronic structures, which may provide good guidance. The density of states (DOS) is calculated to understand the interaction between the H atoms and Au GB sites and, thus, to gain an insight into the different HER activities of the GB sites. Only the 5*d*–projected DOS of Au atoms is presented because Au 5*d* orbitals decide the formation of H−Au bonds. Figure 7 shows the Au 5*d* orbital–projected DOS (PDOS) onto s1, s2 and s5 on Au(110) with GBs. The 5*d*–PDOS of the flat Au(110) with H* is also presented for comparison. Compared with the GB–free Au(110), the peak positions and shapes of Au(110) with GBs vary. The variation in the H* binding strength for different GBs sites can be explained by a *d*–band model, in which the *d*–band center (ε_d) is correlated with the adsorption energy of the adsorbate by considering either the adsorbed H* state by itself or the energy–level realignment in the HER process [32]. As previously mentioned, neither a too strong nor a too weak binding strength is suitable for HER, and an appropriate ε_d can ensure an appropriate strength of the H−Au bond and catalytic activity for the HER. The value of ε_d can be calculated by

$$\varepsilon_d = \frac{\int_{-\infty}^{+\infty} E \times DOS(E)dE}{\int_{-\infty}^{+\infty} DOS(E)dE} \tag{7}$$

Considering the dominant role of *d*–states under the Fermi level in the H* binding process, we integrated the domain of the Au 5*d* states below the Fermi energy to obtain an effective ε_d. As seen in Figure 7a, the ε_d values for s1, s2 and s5 around the GBs are calculated to be −3.38 eV, −3.08 eV and −3.47 eV, respectively, higher than the GB–free Au(110) surface (−4.05 eV). That is, GBs indeed allow ε_d to shift up toward the Fermi energy on the Au(110) surface. It is well known that a higher *d*–band center corresponds to stronger adsorption [32]. Consequently, the upper shift of ε_d suggests that the sites near GBs have a larger H* binding strength and, thus, a lower U_L in comparison with the clean Au surface. To quantitively evaluate the relationship between HER activity and the electronic structure of Au(110) GBs, we present the trend in U_L(H*) on different GB sites as a function of ε_d in Figure 7b. We can see that ε_d exhibits an approximately linear relationship with U_L with $r = 0.84$, suggesting that the sites just at the GBs are superior to the sites near the GBs.

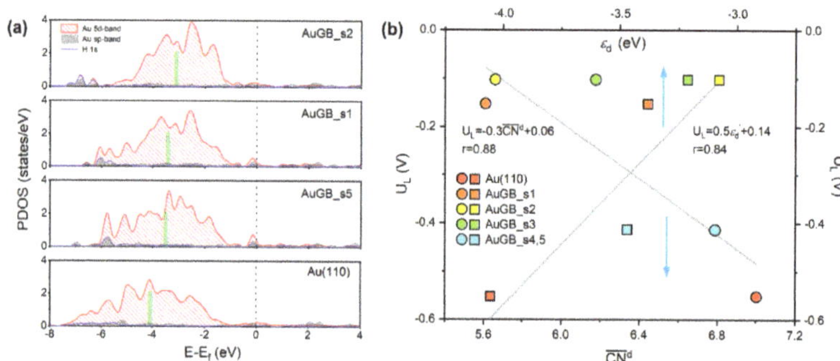

Figure 7. (**a**) Au 5*d* orbital−PDOS on active sites on Au(110) with and without GBs. The Fermi energy (E_f) is set to zero and is represented by a black vertical dashed line. The green, thick bar represents the ε_d of a specific atom on Au(110) surface in the absence or presence of GBs. (**b**) The trends in U_L for HER on different GBs sites as a function of $\overline{CN}^d$ and ε_d. The perfect Au(110) plane is also presented for comparison.

In fact, GBs feature disorder of the atomic arrangement, where under–coordinated atoms exist; we can also elucidate the origin of GB–assisted HER from a geometrical point of view. By considering the environment of their neighboring atoms, we use the concept of the bond–energy–integrated coordination number ($\overline{CN}^d$) proposed by us [33,34]. $\overline{CN}^d$ can differentiate the surface sites in HER activity. The $\overline{CN}^d$ of the five active sites mentioned above is calculated according to the following expression [34]:

$$\overline{CN}^d = \frac{\sum_{i \neq j}^{r_{ij}<r_c} \sqrt{\mu_{id}^{(2)}}}{\sqrt{\left(V_{nn}^{d,\infty}\right)^2}} \tag{8}$$

where $\mu_{id}^{(2)}$ is the second moment of the local density of states projected on the d orbital of site *i*, which is calculated by summing the square of the d–electron hopping integrals to the neighboring d orbitals by only counting the first nearest neighboring atom j within a cutoff distance of r_c = 3.2 Å. $V_{ds\sigma}$, $V_{dd\sigma}$, $V_{dd\pi}$ and $V_{dd\delta}$ are the d–electron hopping integrals to the s or d orbital of the first nearest–neighbor atom for the formation of σ, π and δ bonds. $V_{nn}^{d,\infty}$ is the corresponding d–electron hopping integral in bulk metal, which is a constant for the catalysts of the same metal.

In Figure 7b, we can see that $\overline{CN}^d$ exhibits an approximately linear relationship with U_L, with r = 0.88. The sites around the GBs (s1–s5) have a smaller $\overline{CN}$ than those on the flat Au(110) surface (=7.00). Of note is that a decrease in $\overline{CN}^d$ corresponds to a decrease in U_L. As stated above, U_L is strongly correlated with H* binding strength; therefore, a smaller $\overline{CN}^d$ implies a higher E_B[H*] for Au GBs and, thus, better HER performance. Therefore, HER activity (or H* binding) can be directly evaluated by the $\overline{CN}^d$ of the active sites around the GBs. Compared with the flat extended surface, atoms around GBs have no regular arrangement; that is, they are rich in atomic disorder, leading to the lack of complete coordination (unsaturated coordination). Thus, strong Au–H bonds have to form to compensate for the lack of coordination, following the bond–order conservation theory [35,36], leading to a higher H* binding capacity of the sites around GBs. By making a comparison with the U_L−$\overline{CN}^d$ relationship, it is evident that a smaller $\overline{CN}^d$ corresponds to a larger ε_d shift, indicating that the lack of a coordination number alters the ε_d of active sites around GBs. Hence, in the present case, the bond–energy–integrated coordination number or the *d*–band center theory explains the origin of the improved HER activity; both originated from the disorder of the atomic arrangement, which inherently exists at GBs.

4. Conclusions

In summary, via DFT simulations, we studied the HER performance of Au mediated by GBs. The active site around GBs exhibited superior HER performance with a remarkably low overpotential required, which arose from the significant enhanced capacity of binding H. The Au catalyst with GBs was experimentally prepared using the magnetron sputtering technique, and it exhibited superior HER activity compared to the Au catalyst without GBs. HER activity was inherently correlated with the density of the GBs in the Au catalyst, where active sites bind hydrogen more strongly. The possible reasons for this increase in H* binding ability were discussed, including geometrical and electronic features. The origin of the GB–assisted HER was the disorder of the atomic arrangement in the GB region on the Au surface, leading to the lack of complete coordination; thus, a strong H–Au bond formed to compensate for the lack of coordination. In addition, this disorder gave rise to the migration of the *d*–band center upward to the Fermi energy, leading to increased bonding. The GB–mediated HER performance emphasizes that the GB technique may be particularly suitable for applications in practical electrochemical hydrogen production.

Supplementary Materials: The following supporting information can be downloaded at: https://www.mdpi.com/article/10.3390/app12094290/s1, Figure S1: Front and side views of fully relaxed periodic supercells for $\sum 6\{2\text{–}21\}$ HAGB; Figure S2: SEM images of as–prepared (a) and annealed Au/CFP electrodes at 300 °C (c) and 400 °C (e). Compositional maps of as–prepared (b) and annealed Au/CFP electrodes at 300 °C (d) and 400 °C (f), with red indicating Au and green indicating carbon; Figure S3: X–ray diffraction patterns of Au/CFP for 30 s, 2 min and 5 min magnetron sputtering; Figure S4: Cyclic voltammograms under different scan rates for (a) Au/CFP, (b) Au/CFP_{300} and (c) Au/CFP_{400}. (d) Electrochemically active (EA) surface area derived from a change in charging current density related to scan rate; Table S1: Calculated values for conversion of electronic energies to free energies. Assumed fugacities for gaseous H_2 are also included; Table S2: Calculated values for conversion of electronic energies to free energies on the GBs on Au(110) surface. Assumed fugacities for gaseous species are also included; Table S3: Calculated values for conversion of electronic energies to free energies of metals on (111) and (211) surfaces. Assumed fugacities for gaseous species are also included; Table S4: Calculated values for conversion of electronic energies to free energies of metals on (100) and (110) surfaces. Assumed fugacities for gaseous species are also included; Table S5: The flat surface and grain boundary active sites on Au(110). The coordination numbers and band centers are given.

Author Contributions: Data curation, formal analysis and writing—original draft preparation, R.J.; methodology and software, J.F.; supervision, Z.W.; supervision, funding acquisition and writing—review and editing, C.D. All authors have read and agreed to the published version of the manuscript.

Funding: This research was funded by Natural Science Foundation of China (No.21403152) and Tianjin Natural Science Foundation (16JCQNJC05700).

Institutional Review Board Statement: Not applicable.

Informed Consent Statement: Not applicable.

Data Availability Statement: This study did not report any data.

Acknowledgments: This study was supported by the Natural Science Foundation of China (No. 21403152) and the Tianjin Natural Science Foundation (16JCQNJC05700).

Conflicts of Interest: The authors declare no conflict of interest.

References

1. Jacobson, M.Z.; Colella, W.G.; Golden, D.M. Cleaning the air and improving health with hydrogen fuel-cell vehicles. *Science* **2005**, *308*, 1901–1905. [CrossRef] [PubMed]
2. Zheng, Y.; Jiao, Y.; Jaroniec, M.; Qiao, S.Z. Advancing the Electrochemistry of the Hydrogen-Evolution Reaction through Combining Experiment and Theory. *Angew. Chem. Int. Ed.* **2015**, *54*, 52–65. [CrossRef] [PubMed]
3. Conway, B.E.; Tilak, B.V. Interfacial processes involving electrocatalytic evolution and oxidation of H_2 and the role of chemisorbed H. *Electrochim. Acta* **2002**, *47*, 3571–3594. [CrossRef]

4. Li, H.; Tsai, C.; Koh, A.L.; Cai, L.; Contryman, A.W.; Fragapane, A.H.; Zhao, J.; Han, H.S.; Manoharan, H.C.; Abild–Pedersen, F.; et al. Activating andoptimizing MoS_2 basal planes for hydrogen evolution through the formation ofstrained sulphur vacancies. *Nat. Mater.* **2016**, *15*, 48–53. [CrossRef]
5. Voiry, D.; Yamaguchi, H.; Li, J.; Silva, R.; Alves, D.C.B.; Fujita, T.; Chen, M.; Asefa, T.; Shenoy, V.B.; Eda, G.; et al. Enhanced catalytic activity in strained chemically exfoliated WS_2 nanosheets for hydrogen evolution. *Nat. Mater.* **2013**, *12*, 850–855. [CrossRef]
6. Jaramillo, T.F.; Jørgensen, K.P.; Bonde, J.; Nielsen, J.H.; Horch, S.; Chorkendorff, I. Identification of Active Edge Sites for Electrochemical H_2 Evolution from MoS2 Nanocatalysts. *Science* **2007**, *317*, 100–102. [CrossRef]
7. Conway, B.E.; Bockris, J.O.M. Electrolytic Hydrogen Evolution Kinetics and Its Relation to the Electronic and Adsorptive Properties of the Metal. *J. Chem. Phys.* **1957**, *26*, 532–541. [CrossRef]
8. Parsons, R. The rate of electrolytic hydrogen evolution and the heat of adsorption of hydrogen. *Trans. Faraday Soc.* **1958**, *54*, 1053–1063. [CrossRef]
9. Greeley, J.; Jaramillo, T.F.; Bonde, J.; Chorkendorff, I.; Nørskov, J.K. Computational high-throughput screening of electrocatalytic materials for hydrogen evolution. *Nat. Mater.* **2006**, *5*, 909–913. [CrossRef]
10. Nørskov, J.K.; Bligaard, T.; Logadottir, A.; Kitchin, J.R.; Chen, J.G.; Pandelov, S.; Stimming, U.J. Trends in the exchange current for hydrogen evolution. *Electrochem. Soc.* **2005**, *152*, J23–J26. [CrossRef]
11. Lv, H.; Xi, Z.; Chen, Z.; Guo, S.; Yu, Y.; Zhu, W.; Li, Q.; Zhang, X.; Pan, M.; Lu, G.; et al. A New Core/Shell NiAu/Au Nanoparticle Catalyst with Pt-like Activity for Hydrogen Evolution Reaction. *J. Am. Chem. Soc.* **2015**, *137*, 5859–5862. [CrossRef] [PubMed]
12. Gao, D.; Guo, J.; Cui, X.; Yang, L.; Yang, Y.; He, H.; Xiao, P.; Zhang, Y. Three-Dimensional Dendritic Structures of NiCoMo as Efficient Electrocatalysts for the Hydrogen Evolution Reaction. *ACS Appl. Mater. Interfaces* **2017**, *9*, 22420–22431. [CrossRef] [PubMed]
13. Li, G.; Zhang, D.; Qiao, Q.; Yu, Y.; Peterson, D.; Zafar, A.; Kumar, R.; Curtarolo, S.; Hunte, F.; Shannon, S.; et al. All the Catalytic Active Sites of MoS_2 for Hydrogen Evolution. *J. Am. Chem. Soc.* **2016**, *138*, 16632–16638. [CrossRef]
14. Vidruk, R.; Landau, M.V.; Herskowitz, M.; Talianker, M.; Frage, N.; Ezersky, V.; Froumin, N.J. Grain boundary control in nanocrystalline MgO as a novel means for significantly enhancingsurface basicity and catalytic activity. *J. Catal.* **2009**, *263*, 196–204. [CrossRef]
15. Gavrilov, A.N.; Savinova, E.R.; Simonov, P.A.; Zaikovskii, V.I.; Cherepanova, S.V.; Tsirlina, G.A.; Parmon, V.N. On the influence of the metal loading on the structure of carbon–supported PtRu catalysts and their electrocatalytic activities in CO and methanol electrooxidation. *Phys. Chem. Chem. Phys.* **2007**, *9*, 5476–5489. [CrossRef] [PubMed]
16. Li, C.W.; Ciston, J.; Kanan, M.W. Electroreduction of carbon monoxide to liquid fuel on oxide—derived nanocrystalline copper. *Nature* **2014**, *508*, 504–507. [CrossRef] [PubMed]
17. Kim, K.-S.; Kim, W.J.; Lim, H.-K.; Lee, E.K.; Kim, H. Tuned Chemical Bonding Ability of Au at Grain Boundaries for Enhanced Electrochemical CO_2 Reduction. *ACS Catal.* **2016**, *6*, 4443–4448. [CrossRef]
18. Dong, C.; Fu, J.; Liu, H.; Ling, T.; Yang, J.; Qiao, S.Z.; Du, X.-W. Tuning the selectivity and activity of Au catalysts for carbon dioxide electroreduction via grain boundary engineering: A DFT study. *Mater. Chem. A* **2017**, *5*, 7184–7190. [CrossRef]
19. Feng, X.; Jiang, K.; Fan, S.; Kanan, M.W. A Direct Grain-Boundary-Activity Correlation for CO Electroreduction on Cu Nanoparticles. *ACS Cent. Sci.* **2016**, *2*, 169–174. [CrossRef]
20. Hammer, B.; Nørskov, J.K. Why gold is the noblest of all the metals. *Nature* **1995**, *376*, 238–240. [CrossRef]
21. Kresse, G.; Joubert, D. From Ultrasoft Pseudopotentials to the Projector Augmented—Wave Method. *Phys. Rev. B Condens. Matter Mater. Phys.* **1999**, *59*, 1758. [CrossRef]
22. Blöchl, P.E. Projector augmented-wave method. *Phys. Rev. B* **1994**, *50*, 17953–17979. [CrossRef] [PubMed]
23. de Gennero Chialvo, M.R.; Chialvo, A.C. Existence of Two Sets of Kinetic Parameters in the Correlation of the Hydrogen Electrode Reaction. *J. Electrochem. Soc.* **2000**, *47*, 1619–1622. [CrossRef]
24. de Gennero Chialvo, M.R.; Chialvo, A.C. Hydrogen diffusion effects on the kinetics of the hydrogen electrode reaction. Part, I. Theoretical aspects. *Phys. Chem. Chem. Phys.* **2004**, *6*, 4009–4017. [CrossRef]
25. Quaino, P.M.; de Gennero Chialvo, M.R.; Chialvo, A.C. Hydrogen diffusion effects on the kinetics of the hydrogen electrode reaction. Part, II. Evaluation of kinetic parameters. *Phys. Chem. Chem. Phys.* **2004**, *6*, 4450–4455. [CrossRef]
26. Parsons, R. *Catalysis in Electrochemistry*; Santos, E., Schmickler, W., Eds.; Wiley: Hoboken, NJ, USA, 2011; Chapter 1; pp. 1–15.
27. Zheng, Y.; Jiao, Y.; Zhu, Y.; Li, L.H.; Han, Y.; Chen, Y.; Du, A.; Jaroniec, M.; Qiao, S.Z. Hydrogen evolution by a metal-free electrocatalyst. *Nat. Commun.* **2014**, *5*, 3783. [CrossRef]
28. Hamelin, A. Underpotential deposition of lead on single crystal faces of gold: Part, I. The influence of crystallographic orientation of the substrate. *J. Electroanal. Chem.* **1984**, *165*, 167. [CrossRef]
29. Hamelin, A.; Lipkowski, J. Underpotential deposition of lead on gold single crystal faces: Part II. General discussion. *J. Electroanal. Chem.* **1984**, *171*, 317. [CrossRef]
30. Hernaández, J.; Solla–Gulloán, J.; Herrero, E. Gold nanoparticles synthesized in a water-in-oil microemulsion: Electrochemical characterization and effect of the surface structure on the oxygen reduction reaction. *J. Electroanal. Chem.* **2004**, *574*, 185. [CrossRef]
31. Shinagawa, T.; Garcia–Esparza, A.T.; Takanabe, K. Insight on Tafel slopes from a microkinetic analysis of aqueous electrocatalysis for energy conversion. *Sci. Rep.* **2015**, *5*, 13801. [CrossRef]
32. Hammer, B.; Nørskov, J.K. Theoretical surface science and catalysis—Calculations and concepts. *Adv. Catal.* **2000**, *45*, 71–129.

33. Wu, D.; Xi, C.; Dong, C.; Liu, H.; Du, X.-W. Bond-energy-integrated coordination number: An accurate descriptor for transition—metal catalysts. *J. Phys. Chem. C* **2019**, *123*, 28248–28254. [CrossRef]
34. Xi, C.; Zou, C.; Wang, M.; Wu, D.; Liu, H.; Dong, C.; Du, X.-W. A bond-energy-integrated-based descriptor for high-throughput screening of transition metal catalysts. *J. Phys. Chem. C* **2020**, *124*, 5241–5247. [CrossRef]
35. Shustorovich, E.; Sellers, H. The UBI-QEP method: A practical theoretical approach to understanding chemistry on transition metal surfaces. *Surf. Sci. Rep.* **1998**, *31*, 1–119. [CrossRef]
36. Kleis, J.; Greeley, J.; Romero, N.A.; Morozov, V.A.; Falsig, H.; Larsen, A.H.; Lu, J.; Mortensen, J.J.; Dułak, M.; Thygesen, K.S.; et al. Finite Size Effects in Chemical Bonding: From Small Clusters to Solids. *Catal. Lett.* **2011**, *141*, 1067–1071. [CrossRef]

Review

Recent Progress in ZnO-Based Nanostructures for Photocatalytic Antimicrobial in Water Treatment: A Review

Ziming Xin [1], Qianqian He [1], Shuangao Wang [1], Xiaoyu Han [1], Zhongtian Fu [1], Xinxin Xu [2] and Xin Zhao [1,*]

[1] Department of Environmental Engineering, School of Resources and Civil Engineering, Northeastern University, Shenyang 110819, China
[2] Department of Chemistry, School of Science, Northeastern University, Shenyang 110819, China
* Correspondence: zhaoxin@mail.neu.edu.cn; Tel.: +86-2483679128

Abstract: Advances in nanotechnology have led to the development of antimicrobial technology of nanomaterials. In recent years, photocatalytic antibacterial disinfection methods with ZnO-based nanomaterials have attracted extensive attention in the scientific community. In addition, recently widely and speedily spread viral microorganisms, such as COVID-19 and monkeypox virus, have aroused global concerns. Traditional methods of water purification and disinfection are inhibited due to the increased resistance of bacteria and viruses. Exploring new and effective antimicrobial materials and methods has important practical application value. This review is a comprehensive overview of recent progress in the following: (i) preparation methods of ZnO-based nanomaterials and comparison between methods; (ii) types of nanomaterials for photocatalytic antibacterials in water treatment; (iii) methods for studying the antimicrobial activities and (iv) mechanisms of ZnO-based antibacterials. Subsequently, the use of different doping strategies to enhance the photocatalytic antibacterial properties of ZnO-based nanomaterials is also emphatically discussed. Finally, future research and practical applications of ZnO-based nanomaterials for antibacterial activity are proposed.

Keywords: nanomaterial; photocatalyst; antibacterial; zinc oxide; water treatment; antimicrobial

Citation: Xin, Z.; He, Q.; Wang, S.; Han, X.; Fu, Z.; Xu, X.; Zhao, X. Recent Progress in ZnO-Based Nanostructures for Photocatalytic Antimicrobial in Water Treatment: A Review. *Appl. Sci.* **2022**, *12*, 7910. https://doi.org/10.3390/app12157910

Academic Editor: Anna Annibaldi

Received: 15 July 2022
Accepted: 5 August 2022
Published: 7 August 2022

1. Introduction

With rapid global population growth, urbanization increasing, illicit misuse of freshwater resources, and continued destruction of the global climate, the increasing demand for clean water is becoming a global concern [1–3]. Seven billion people, more than 15% of the world's people, are facing a shortage of fresh water resources, which even causes them to not have enough fresh water to sustain normal life and productive work [4,5]. Water scarcity is exacerbated by the increasing water pollution from releases of waterborne pathogens, inorganic pollutants, organic pollutants, agricultural chemicals, derivatives of human and animal drugs, and endocrine disruptors [6–8].

Infectious diseases caused by biological contamination such as typhoid fever, dysentery, cholera, and diarrhea are a major cause of death worldwide and continue to replicate at an alarming rate [9]. The extensive use of antibiotics and antibacterial drugs has led to strong drug resistance in viruses and bacteria, which further exacerbates the spread of biological infectious diseases [10]. In addition, the recent epidemics of global security issues such as the COVID-19 virus and monkeypox virus are caused by the spread of viruses that threaten all human beings and have a great impact on human production, life, and health [11–14]. Similar to bacteria and pathogenic microorganisms, epidemic viruses are always difficult to eradicate due to the abuse of antibiotics and various disadvantages of disinfectants [15,16]. In view of the above situation, it is crucial to explore more effective solutions and approaches.

Recent advancements in semiconductor materials and new nanomaterials have blazed new trails for their applications in the fields of photocatalysis and bacterial inactivation [9].

Nanotechnology provides a variety of promising nanomaterials for the field of photocatalytic antimicrobial. Metal oxides have many advantages such as non-toxic, stable, and efficient biological properties, which make them stand out among many nanomaterials and become a research hotspot in this field. Numerous nanomaterials doped with metal oxides such as ZnO [17], Fe_2O_3 [18], TiO_2 [19], Ag_2O [20], CaO [20], MgO [21], and CuO [22] have been applied as efficient antibacterial agents for both Gram-positive (G+) and Gram-negative (G-) bacteria, such as *Escherichia coli, Salmonella enteritidis, Streptococcus pyogenes, Aeromonas hydrophila, Pseudomonas aeruginosa, Salmonella typhimurium, Fecal intestinal cocci,* etc. [23]. Among the antibacterial agents, ZnO-based nanomaterials are widely recognized as promising antibacterial agents with strong photocatalytic antibacterial activity [24,25]. Nevertheless, the wide bandgap of ZnO is approximately 3.2–3.3 eV, which affects its light absorption ability, resulting in a response only in the ultraviolet band [26]. Previous studies indicated that defects of nanomaterials in photocatalytic antibacterial processes could be effectively improved after being modified [27]. Strategies, such as loading antibacterial agents, loading oxidized nanomaterials, and adjusting the particle size, material microshape, and concentration of ZnO were employed to enhance the antibacterial properties. While exploring the antibacterial ability improvement, the antibacterial mechanism should also be in-depth investigated.

In previous studies, some mechanisms for photocatalytic antimicrobials, such as metal ion release, reactive oxygen species (ROS) generation [28], destruction of cell membranes, internalization of nanoparticles [29], interruption or blockade of transmembrane transport, etc., have been proposed [30,31]. Overall, the purpose of all antibacterial mechanisms is to disrupt the bacterial cell structure and break it down into harmless substances. However, our understanding of the specific process of substance transformation during the photocatalyst-induced antimicrobial process is still very limited, which requires further exploration.

This review is an exhaustive summary of recent research advances on the antimicrobial of ZnO-based nanomaterials. By summarizing previous studies, ZnO-based nanomaterials with excellent antibacterial effects were obtained. From the perspective of material preparation, different preparation methods are reviewed, and both the advantages and disadvantages are compared. Subsequently, the antimicrobial mechanism of ZnO-based nanomaterials is discussed in-depth from both the physical and chemical aspects. In detail, various strategies to enhance the antimicrobial ability of ZnO-based nanomaterials in recent studies are proposed. Finally, temporary deficiencies in the improvement strategy are summed up, and prospects for the future development direction and application potential are presented.

2. ZnO-Based Nanostructures Preparation

Among the numerous methods for preparing ZnO-based nanomaterials, the wet-chemical/solution technique has many advantages such as simplicity, rapid operation, and cost savings, which make it a promising method for the preparation of ZnO-based nanomaterials. The advantages and disadvantages of commonly used wet-chemical/solution techniques, such as the sol–gel method, co-precipitation method, microwave-assisted method, and hydrothermal method, are presented in detail in Table 1. The preparatory stage of ZnO nanomaterial growth is fully wetted by wet-chemical/solution techniques, which greatly improves the stability of the materials.

Table 1. Advantages and disadvantages of different antibacterial synthesis methods.

Method	Preparation Shape	Advantages	Disadvantages	References
Sol–gel	Nanorods; Nanotubes; Nanobelts; Nano springs; Nano spirals; Nano rings.	(1) Uniform doping; (2) High stability; (3) Low synthesis temperature.	(1) Expensive raw material prices; (2) Longer reaction time; (3) Organics escape.	[32–34]
Co-precipitation	Homogeneous and spherical; Nanobelts; Nano springs.	(1) Simple preparation process; (2) Low cost; (3) Short synthesis cycle.	(1) Additional precipitant; (2) High temperature calcination; (3) Uneven dispersion.	[35,36]
Microwaves-assisted	Nanorods; Nanotubes; Nanobelts; Nano springs.	(1) High synthesis efficiency; (2) Energy saving; (3) Improved material properties.	(1) Indeterminate form; (2) Large investment; (3) High requirements for equipment.	[37–39]
Hydrothermal	Nanobelts; Nano springs; Nano spirals.	(1) Less thermal stress; (2) High particle purity; (3) Controllable crystal shape; (4) Low cost.	(1) Inconvenient to observe; (2) Not intuitive; (3) High equipment requirements; (4) Technical difficulty; (5) Poor safety performance.	[40–42]

2.1. *Sol–Gel Method*

The sol–gel method is one of the most effective chemical methods for nanocomposites preparation with desired properties and advantages, such as low cost, mild reaction, environmental friendliness, reliability, and simplicity [43]. In the preparation of photocatalytic materials, the materials synthesized by this method have better photocatalytic activity [44]. ZnO nanoparticles were successfully synthesized using the gel-sol method by Hasnidawani [45], and the surface morphology was verified by Fe-SEM images (Figure 1), which was confirmed to have a rod-like structure with a dense particle structure. Varieties of ZnO nanostructures have been discovered, which are in the form of nanorods, nanotubes, nanobelts, nano springs, nano spirals, nano rings, and many more [46]. Among these structures, the rod-like structure is the best nanostructure compared to others due to their one-dimensional nanostructures (such as nanorods, nanowires, and nanotubes) that can facilitate more efficient carrier transport for the decreased grain boundaries, surface defects, disorders, and discontinuous interfaces [47,48]. In the process of preparing ZnO-based nanomaterials by the sol–gel method, the influence of factors such as solution drop acceleration rate, reaction temperature, pH, etc., will have a significant impact on the antibacterial properties of the materials [49]. Effects of various preparation influencing factors on the antibacterial properties of ZnO were tested, and pH was proven to be the most important influencing factor [50]. The reason is suggested to be that the neutral and acidic solution environment is more suitable for Zn^{2+} to function and achieve an antibacterial effect. As long as the optimal pH and preparation temperature are found, the sol–gel method will be one of the effective preparation methods with high efficiency and low cost. Therefore, in recent studies, the sol–gel method is used more in the synthetization of ZnO-based nanoparticles [51–53].

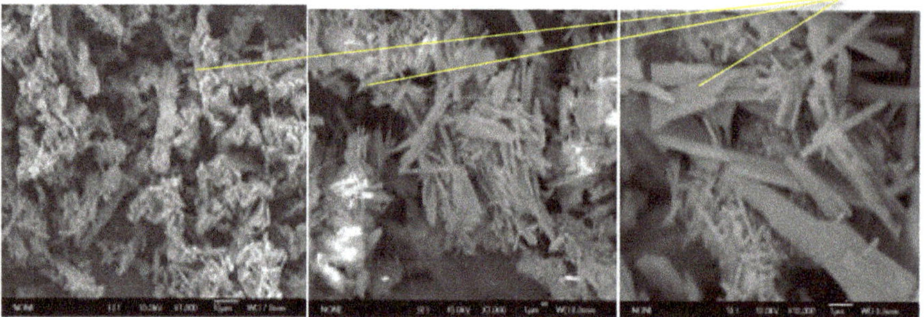

Figure 1. FE-SEM micrographs of synthesized ZnO at different magnifications. Reprinted/adapted with permission from Ref. [45]. Copyright © 2022, Elsevier B.V.

2.2. Co-Precipitation Method

The co-precipitation method does not require expensive raw materials and complicated equipment, which provides a suitable method for low-cost and large-scale production [54]. Furthermore, in addition to simpler devices, suitable metals, metal oxides, and surfactants are added to change the morphology of the materials [55]. The co-precipitation method was chosen to prepare ZnO-based nanoparticles in an aqueous solution at two different reaction temperatures (50 °C and 70 °C) by Kotresh et al. [56]. The surface morphology of ZnO nanoparticles prepared by the co-precipitation method was observed by scanning electron microscope (SEM) images as shown in Figure 2. It can be seen from the SEM images that the particles are uniformly spherical with a dense and dense structure. The spherical ZnO nanoparticles prepared by the co-precipitation method are favorable for uniform dispersion in the photocatalytic reaction and efficiency improvement of photocatalytic reactions. However, it was found that the droplet acceleration rate had the greatest impact on the antibacterial properties of ZnO-based nanomaterials synthesized by the co-precipitation method [57]. The size of the nanoparticles synthesized by the co-precipitation method is affected by the drop rate of the solution, which will affect the contact area of the nanoparticles during the antibacterial reaction, thereby greatly affecting the antibacterial effect [58]. Therefore, the droplet acceleration rate and particle size need to be carefully considered during synthesis, which is an important part of the success of the co-precipitation method. Due to the advantages of the co-precipitation method with a simple preparation process, the use of this method has gradually increased in nanomaterial synthesis research in recent years [59–61].

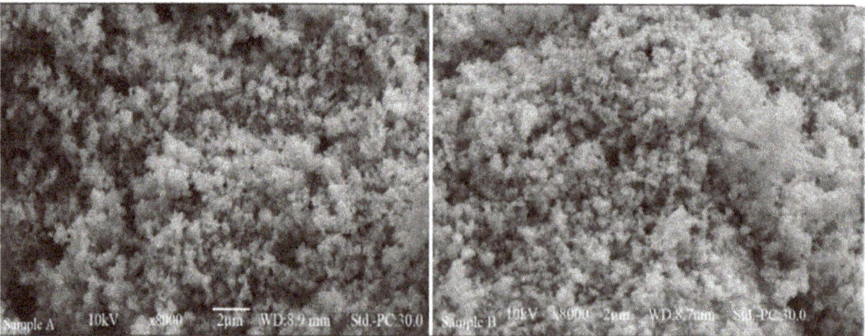

Figure 2. SEM images of sample A and sample B. Reprinted/adapted with permission from Ref. [56]. Copyright © 2022, Elsevier GmbH.

2.3. Microwave-Assisted Method

The microwave-assisted method is not only an energy-saving, environmentally friendly, and heat-free method, but also with many advantages such as fast synthesis speed and the ability to tune the particle shape [44,62]. ZnO nanoparticles with different morphologies can be synthesized by the microwave-assisted hydrothermal method by adjusting the time and power of microwave irradiation [63,64]. Through microwave-assisted chemistry techniques, ZnO nanostructures with different morphologies were synthesized in different pH reaction mixtures (acidic, basic, or neutral). Furthermore, nanomaterials are synthesized without any heating and addition of surfactants. Hence, obtaining ZnO particles with oxygen vacancies and defects is expected to improve their pollutant degradation behavior due to the fast reaction process and non-stoichiometric synthesis [65]. As shown in Figure 3, the microscopic morphology of the microwave-synthesized ZnO nanostructures was observed by SEM. It can be seen intuitively that the appearance of ZnO nanoparticles changes dramatically with the pH change. In addition, from the XRD analysis in Figure 4, it was demonstrated that the change in intensity and peak width of the two sets of samples (prepared with NaOH and KOH as pH control agents) can be observed as the solution pH changes. The above results showed that the shape of synthesized ZnO nanoparticles is affected by pH changes, which has important implications for the directional synthesis of nanoparticles with diverse morphologies. In addition, several studies have shown that the power of the microwave is the most important factor affecting the synthesis of ZnO-based nanomaterials by microwave-assisted method [66,67]. Moderate-power microwave have been shown to be suitable for the synthesis of ZnO-based nanomaterials with stronger antibacterial capabilities [68,69]. However, the exact power influence mechanism needs to be further explored.

Figure 3. Typical SEM images of synthesized ZnO nanostructures at different reaction pH values. Reprinted/adapted with permission from Ref. [65]. Copyright © 2022, Elsevier Ltd and Techna Group S.r.l.

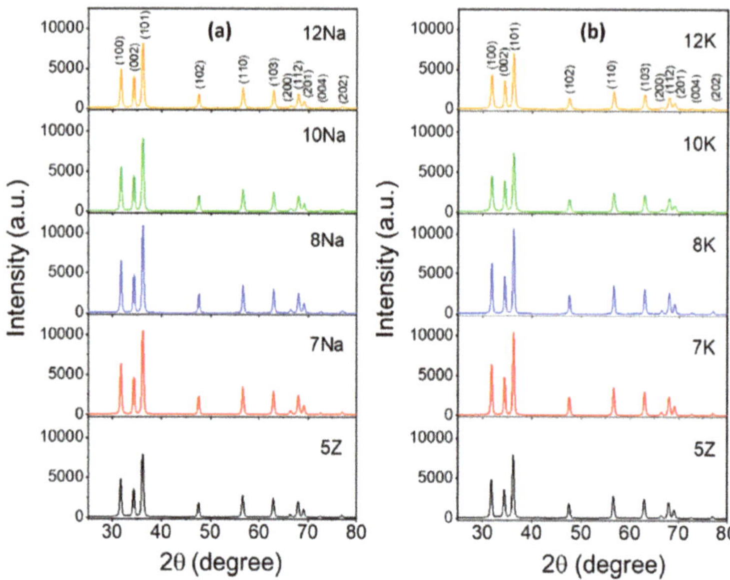

Figure 4. Typical XRD patterns of ZnO nanostructures synthesized with different reaction pH ((**a**) NaOH and (**b**) KOH as pH controlling agents, 7, 8, 10, 12 correspond to their respective concentrations). Reprinted/adapted with permission from Ref. [65]. Copyright © 2022, Elsevier Ltd. and Techna Group S.r.l.

2.4. Hydrothermal Method

The hydrothermal method is a convenient and highly efficient method, which requires a lower reaction temperature and saves costs [70]. In addition, by adjusting the duration, density, and reaction temperature of the contained substances, the morphology and size of the particles can be controlled [71]. As shown in Figure 5a–c, TEM images of synthesized ZnO particles at precursor concentrations of 5, 10, and 20 mM are revealed. The morphological features of the poorly dispersed nano-ZnO crystals are clearly demonstrated by the TEM images in all cases. At precursor concentrations of 5, 10, and 20 mM, ZnO nanoparticles were observed to have diameters of approximately 4.5, 6, and 8–9 nm, respectively. Furthermore, it was also observed from the images that the size distribution of the nanoparticles was fairly uniform [72]. As shown in the XRD pattern (Figure 6), the crystalline structure of the synthesized ZnO particles after hydrothermal treatment was confirmed. Simultaneously, no impurity peaks were detected from the XRD pattern, indicating that the target substance with a higher purity was successfully synthesized [72]. Furthermore, the preferred orientation of the ZnO particle samples was not seen from XRD patterns, suggesting that ZnO crystals may have the most shapes other than rods or sheets. Based on the above conclusions, the hydrothermal method is a suitable method to prepare ZnO-based nanoparticles with different shapes. In addition, the antibacterial effect of ZnO-based nanomaterials prepared by hydrothermal method is affected by several factors, such as pH, reaction temperature, and dosage ratio [73,74]. The reaction temperature directly affects the structure of the material and changes the antibacterial ability, while the pH changes the surface properties and shape of the material to affect the antibacterial ability [75].

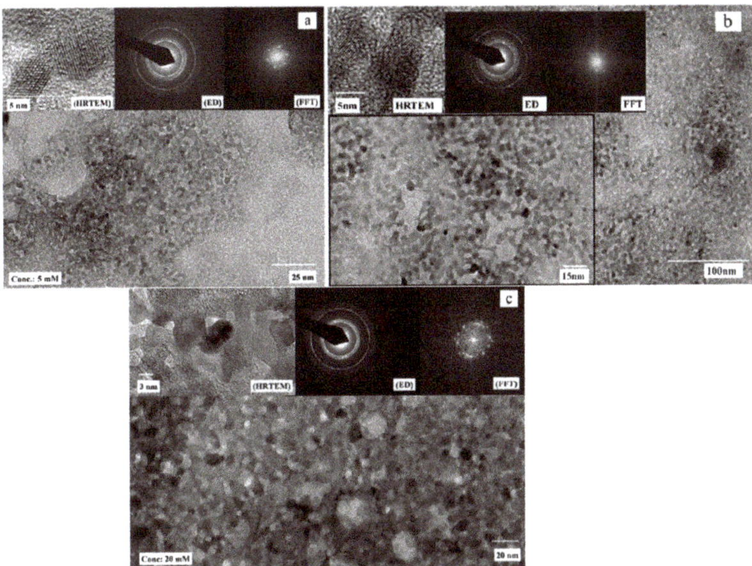

Figure 5. Electron microscope image of ZnO nanoparticles: the concentration of precursor is (**a**) 5 mM, (**b**) 10 mM, and (**c**) 20 mM. Reprinted/adapted with permission from Ref. [72]. Copyright © 2022, Elsevier Ltd.

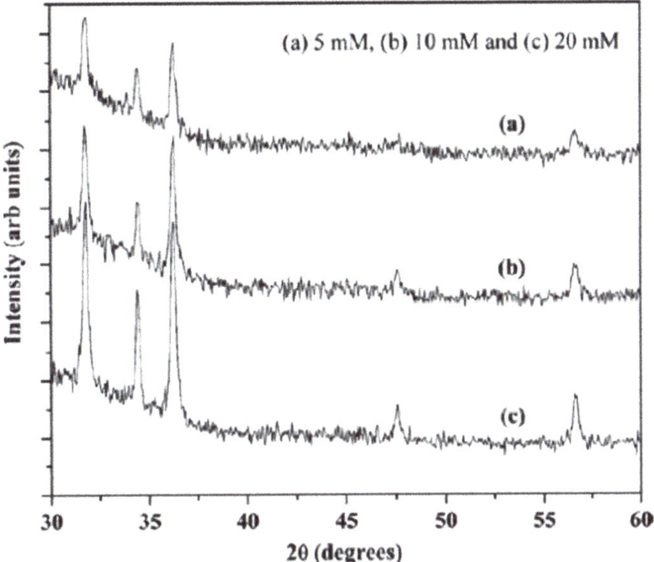

Figure 6. XRD patterns of ZnO samples synthesized with different precursor concentrations. Reprinted/adapted with permission from Ref. [72]. Copyright © 2022, Elsevier Ltd.

3. Types of Nanomaterials for Photocatalytic Antimicrobials in Water Treatment

A variety of nanomaterials as efficient adsorption materials and catalytic degradation and purification of wastewater will be discussed in this section. As shown in Figure 7, various microbial contaminants and their sources, as well as different nanomaterials for removal applications are revealed [76]. Limitations of single nanomaterials can be overcome by various strategies such as polymer/metal oxide nanocomposite [77], metal oxide/metal-based nanomaterials [78,79], polymer/metal oxide nanocomposite [80,81], and polymer-structure-based materials [82,83].

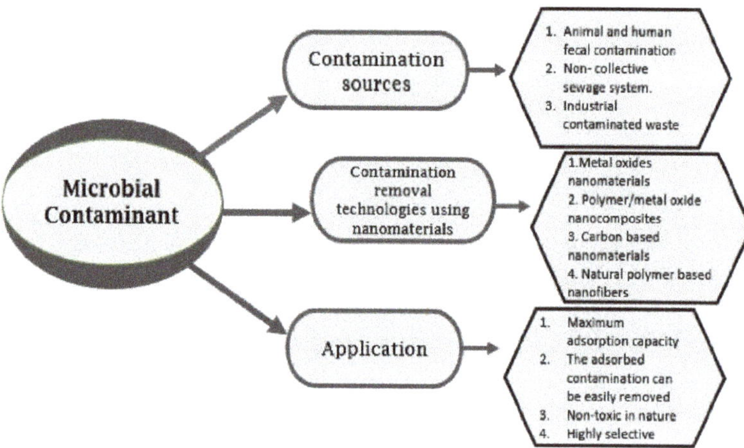

Figure 7. Schematic representation of microbial contamination sources and their various nanomaterials for water treatment. Reprinted/adapted with permission from Ref. [76]. Copyright © 2022, Elsevier Ltd.

The nanocomposite is a safe, non-toxic, green and environmentally friendly nanomaterial, usually prepared with ZnO and TiO_2 as substrates. The addition of TiO_2 and ZnO NPs in some polymers such as polypropylene matrix would increase the dielectric constant of the nanocomposite, thereby enhancing the photocatalytic ability of the material [84]. In addition, the polymer boundary layer transition zone forms a crystalline structure, which increases conductivity and acts as a tuning surfactant [85], which greatly enhances the photoreactivity of the catalytic material. A variety of composite nanomaterials have been reported that can be applied to organic pollutants and microbial contamination removal from water [86]. Moreover, the addition of Ag, zinc, zeolite, and titanium is obtained having better efficiencies in pathogenic pathogens and microorganisms removal from water [87,88]. For example, silver NPs with polyurethane will flexibly remove almost 100% of *B. subtilis* and *E. coli* from water [89,90].

Among the composite nanomaterial antibacterial agents, ZnO-based nanomaterials have the advantages of strong compatibility, green friendliness, lower costs, and simple preparation, so they occupy a large proportion of the related studies in this field in recent years. Studies showed that ZnO has stronger direct interactions with bacterial and pathogen cell surfaces than other semiconductor materials [91]. In addition, ZnO nanomaterials leak Zn^{2+} in solution, which can exacerbate toxicity to bacteria, pathogens, and viruses. Compared with Cu^{+}, Fe^{2+}, and Al^{3+}, Zn^{2+} showed a better ability to fight microbial contaminants [92]. Therefore, ZnO-based nanomaterials have a better potential to combat microbial contamination, which is of practical significance for the study of antimicrobial contamination. Among types of microbial pollution in the water environment, bacterial pollution is still the most important problem to be solved. Thus, in the next section, the antimicrobial properties and mechanisms of ZnO-based nanomaterials, as well as methods for studying antimicrobial properties are introduced, and the methods for improving their catalytic properties are summarized and discussed.

4. ZnO-Based Nanomaterials for Antimicrobial Application in Water Treatment

The antibacterial activity of ZnO-based nanomaterials is greatly affected by the morphology and particle diameter. Considerable methods for the preparation of ZnO-based nanomaterials with different morphologies have been reported in the literature, such as nanorods [93–96], nano/micro flowers [97–101], microspheres [102], nano powders [103–105], nanotubes [4], quantum dots [106,107], films [107], nanoparticles [108], and capped nanoparticles [109], to understand their application prospects for antibacterial agents. To conduct an in-depth exploration of the antibacterial properties of ZnO-based nanomaterials, the antibacterial research methods in this chapter are firstly introduced, and then the antibacterial mechanism and material improvement strategies are expounded to discuss the latest studies in this field.

4.1. Research Methods for Antimicrobial Activities of ZnO Nanostructures

To better study the effect of antibacterial agents, many techniques have been adopted to test antibacterial properties in recent years. As shown in Table 2, various kinds of adsorbents for microbes removal in different water resources are listed, which has great reference significance for antibacterial application studies. In the process of antibacterial ability testing, the accuracy of the results is affected by many factors, such as types of bacterial species and the type of data to be tested, etc. In addition, external factors such as the experimental environment also affect antibacterial detection. Commonly, the temperature of 37 °C and incubation time of 24 h are selected for the tests. Moreover, culture media of Tryptic Soy Broth, Luria–Bertani (LB) broth, Nutrient Agar, and Tryptic Soy Agar (TSA) are commonly used. Meanwhile, depending on the technique used, parameters such as minimum inhibitory concentration (MIC), zone of inhibition (ZOI), colony count, or optical density of bacterial cultures are selected as the basis for the evaluation. The specific application methods and practical application cases of various antibacterial detection technologies are introduced below, which are beneficial for better understanding the antibacterial properties of ZnO-based nanomaterials.

Table 2. Various kinds of adsorbents for microbial removal in different water resources.

Water Type	Sample Type	Microbial Pollution Target	Types of Adsorbents	References
sDrinking water	Domestic drinking water	Bacteria; virus; protozoan;	Activated carbon;	[110]
			Carbon-graphene;	[111]
			Nano adsorbent;	[112]
		Microbial contaminants.	Carbon-graphene.	[113]
	Drinking water source	Biological contaminants;	Nano adsorbent;	[114]
		Microbial contaminants;	Iron Oxide Nanoparticles;	[115]
		E. coli.	Mixed matrix.	[116]
Wastewater	Lake wastewater	Pathogenic microorganisms.	Zinc oxide; Iron oxide; Silver oxide nanoparticles.	[117,118]
	Domestic wastewater	Antibiotics;	Biochar;	[119]
		Pathogenic microorganism;	Activated carbon;	[120]
		V. fischeri, B. subtilis, *E. coli*;	Carbon nanotube;	[121]
		Pathogenic microorganism.	Copper oxide; Zinc oxide; Silver oxide; Titanium oxide.	[122,123]
	Industrial wastewater	Pathogenic microorganisms;	Graphene oxide nanosheets; Silver nanoparticles;	[124]
		Virus.	Carbon nanotubes; TiO_2.	[125]
Laboratory water	Laboratory simulated water	*Enterobacter, Citrobacter, Hafnia*;	Biochar stabilized; Iron oxide; Copper oxide nanoparticles;	[126]
		Klebsiella, Escherichia;	Fe_3O_4; SnO_2; NiO; PAN/boehmite nanofibers.	[127]
		S. aureus, E. coli;	Electro spun; Nanofibers;	[128]
		Bacteria; virus.	Silver @ Eggshell; Nanocomposite.	[83,129]
	Laboratory pure water	*S. aureus, E. coli, C. albicans*;	Ag nano-embedded pebbles;	[85]
			Nano cellulose;	
		E. coli.	Nanofibers; Granular activated carbon;	[130,131]
			Graphite flake.	[132]

4.1.1. Disk-Diffusion Method

The disk-diffusion method is a simple and efficient antimicrobial test, also known as the Kirby–Bauer antibiotic test (KB test) [133]. Mueller–Hinton agar and Brucella blood agar are commonly used as media for this method [133]. During the disk-diffusion method, the pH of the medium is usually controlled at around 7.2. The bacterial suspension containing a specific concentration was spread on the above agar medium and the experimental environment was required to be absolutely dry [22,24,93,94,97,134–136]. Under sterile conditions, a certain number of ZnO-based nanomaterials were soaked on the filter paper disk using the selected solvent according to the specific requirements of the experiment. Let the disk dry and carefully mount it on the medium in a Petri dish. Solvent-soaked disks were used as controls to ensure the accuracy of the test. Next, the dishes were incubated at 37 °C for 24 h, providing the right conditions for bacterial growth. Subsequently, due to the bactericidal activity of the ZnO-based nanomaterials, no bacterial growth was observed around the disks at specific distances. The minimum concentration at which ZnO-based exhibited antibacterial activity was called the minimum inhibitory concentration (MIC) and the area around the disk with no bacterial growth observed was called the zone of inhibition. The lower the MIC value, the larger diameter of the inhibition zone and the higher antibacterial activity. Based on this, the antibacterial properties of ZnO-based nanomaterials were judged [137,138].

4.1.2. Well-Diffusion Method

The medium used in the well-diffusion method is similar to the method described in Section 4.1.1, and the two methods can be used together analogously. In contrast to the disk-diffusion method, the filter paper disk is installed by drilling holes in the media plate. The wells were filled with various concentrations of ZnO nanoparticle suspensions for testing. In addition, the calculation method of MIC and ZOI can also refer to the disk-diffusion method [108,139–142]. Consistent with the disk-diffusion method, sterile conditions are also one of the most necessary environmental conditions for antimicrobial performance testing. Based on MIC and ZOI, the pros and cons of antibacterial properties of ZnO-based nanomaterials can be studied and the efficiency can be evaluated.

4.1.3. Antimicrobial Measurements in Liquid Culture Media

During the incubation period, the turbidity of the growth solution increases with bacterial growth. The liquid turbidity and cell proliferation can be measured by periodically measuring the optical density. The technique does not require reagents and special handling [143–147]. The basic principle of this method is to judge the quality of antibacterial performance by observing the absorbance at a specific wavelength in a spectrophotometer and regularly monitoring the corresponding bacterial growth. Umar et al. used ZnO nanomaterials as antibacterial agents to conduct growth inhibition experiments on *E. coli* and achieved excellent experimental results as shown in Figure 8 [101].

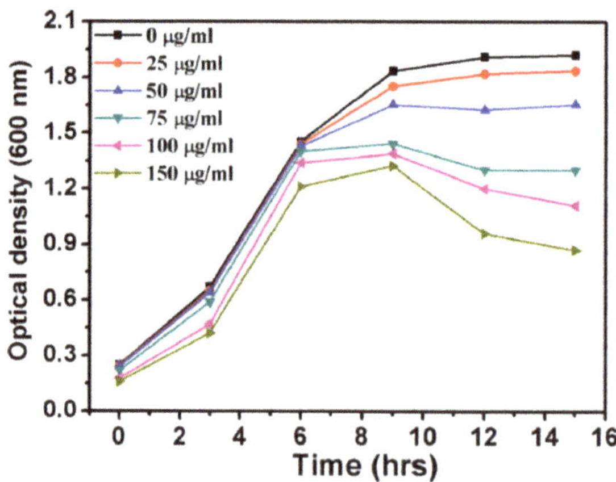

Figure 8. Bacterial growth curve of *E. coli* under different ZnO-NFs concentrations.

4.1.4. Colony Unit Measurements

Colony unit measurement, also known as the diffusion plate technique, is often used to count the amount of living bacteria cells. ZnO-based nanomaterials are introduced into agar plates or dispersed as suspensions in specific liquid media. The strains were mounted on agar plates and incubated at 37 °C for a specific time. Colony forming units are counted using an appropriate counting method. In addition, the colony forming unit (CFU) value can be used to judge the antibacterial ability of ZnO-based nanomaterials [148–151]. Stankovic et al. [152] calculated the percentage of bacterial cell reduction (R%) using Equation (1).

$$\mathrm{R\%} = \frac{CFU_{control} - CFU_{sample}}{CFU_{sample}} \quad (1)$$

where $CFU_{control}$ = numbers of *CFUs* per milliliter for the negative control, and CFU_{sample} = *CFUs* per milliliter in the presence of ZnO dispersion.

Within a certain range, the above formula can be used to calculate the bacterial concentration, and the antibacterial performance can be investigated based on the calculation results.

4.1.5. Microtiter Plate Method

The microtiter plate method, also known as the microplate method, is a method of performing antimicrobial testing by observing changes in a variable number of small test tubes or plates of microwells. Resazurin [153], 2,3,5-triphenyltetrazolium chloride (TTC), crystal violet [4] and 3-(4,5-dimethylthiazol-2-yl)-2,5-diphenyltetra-zolium bromide (MTT) [154], etc., are put into the wells as indicator solutions. Next, known or different concentrations of the test strain was dispersed into the wells of a microtiter plate. A known concentration of ZnO nanomaterials was then dispersed into the wells after dilution in a sterile broth medium. The palate was then incubated at 37 °C for timed intervals. The bacterial cell activity can be judged by the change of absorbance to detect the antibacterial ability.

4.2. Mechanisms of ZnO-Based Antimicrobial Nanomaterials

4.2.1. Mechanisms of ZnO-Based Photocatalytic

In photocatalysis, an electron–hole couple is created under light force by reduction or oxidation reactions on the catalyst surface. The photocatalytic degradation mechanism of ZnO-based photocatalyst on pollutants is shown in Figure 9. Photocatalysis occurs when a ZnO-based photocatalyst is illuminated by light with energy greater than its band gap energy [27]. Charge separation is triggered by a light energy absorption process, which excites electrons from VB to CB, leaving holes in VB [155]. Subsequently, the photogenerated e−/h+ supports a move to the ZnO photocatalyst surface. Simultaneously, e− and h+ recombine, which reduces the quantum yield. The level of this recombination rate is affected by many factors, such as the structure of the photocatalyst and the surface modification process of the photocatalysts [156,157]. The ZnO surface is aggregated with reactive e− and h+, which promote oxidation and reduction reactions that generate excess ROS, including superoxide anion ($\cdot O_2^-$) hydroxyl radicals ($\cdot OH$). Furthermore, the redox potential of the CB bottom of ZnO is more negative than that of O_2/O_2^-. Therefore, these excited electrons can generate $O_2^-\cdot$. Simultaneously, the top of the VB of ZnO is more positive than the redox potential of $\cdot OH/H_2O$. Consequently, H_2O molecules can be oxidized by these holes to form hydroxyl radicals. These highly reactive radicals ($\cdot OH$, $O_2^-\cdot$) directly oxidize organic pollutant molecules in solutions.

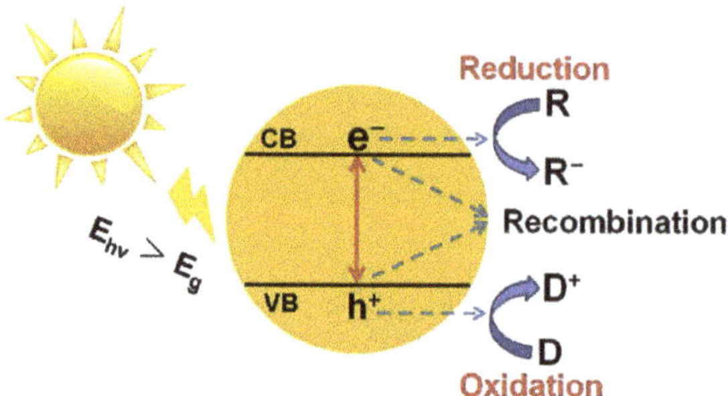

Figure 9. Basic mechanism of ZnO photocatalysis. Reprinted/adapted with permission from Ref. [27]. Copyright© 2022, Elsevier B.V.

4.2.2. Chemical Effect of ZnO-Based Nanomaterials on Antibacterial

In the process of exploring the antibacterial mechanism of ZnO-based nanomaterials, three main chemical antibacterial mechanisms were obtained: generation of reactive oxygenated species (ROS) [158], release of Zn^{2+} ions [31], and photoinduced production of H_2O_2 [159]. The ROS mechanism is basically the same as the photocatalytic mechanism mentioned in Section 4.2.1.

Similar exhaustive studies carried out by Li et al. [31] and Song et al. [160] demonstrated that the toxicity of Zn^{2+} ions to cells is one of the mechanisms for the sterilization of ZnO-based nanomaterials. The concentration of ZnO NPs in ultrapure water in the toxicity test was 5 mg/L. To compare the cytotoxicity, a Zn^{2+} ion solution also prepared in ultrapure water was used (concentration below 0.1 mg/L). TEM images of the treated G− strains of *E. coli* are shown in Figure 10. It is evident that the morphology of *E. coli* was deformed after modification with ZnO-based nanomaterials or Zn^{2+} ion solution. Figure 10b,c shows intracellular fluid leakage due to Zn^{2+} ions and osmotic stress. The experimental results showed that the cytotoxic effects of ZnO NPs and Zn^{2+} ion-treated solutions on *E. coli* were comparable. This fully proves that the release of Zn^{2+} has a positive effect on antibacterial, and it also shows that it is one of the antibacterial mechanisms of ZnO-based nanomaterials (Figure 11).

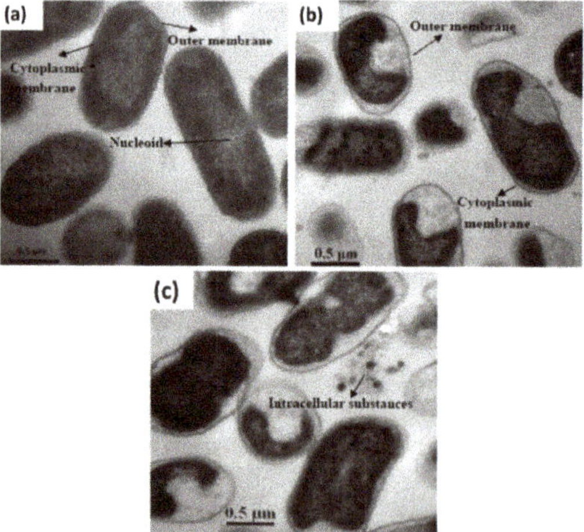

Figure 10. TEM images of (**a**) untreated *E. coli* cells, (**b**) treated with ZnO nanoparticles, and (**c**) treated with Zn^{2+} ions solution in ultrapure water. Reprinted/adapted with permission from Ref. [31]. Copyright © 2022, American Chemical Society.

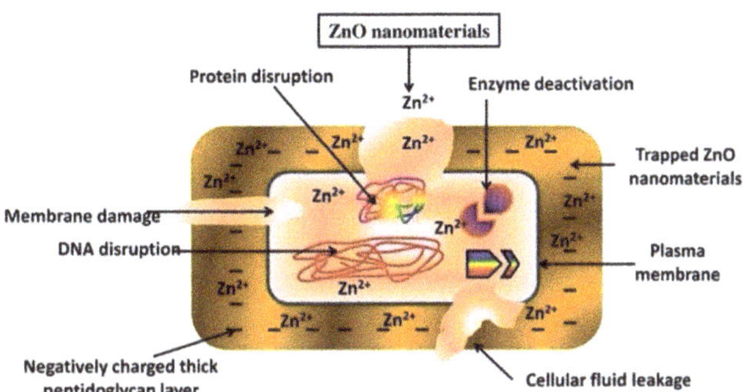

Figure 11. Schematic diagram of cell damage to G+ bacteria by Zn^{2+} ions. Reprinted/adapted with permission from Ref. [27]. Copyright© 2022 Elsevier B.V.

Besides the light-induced reactive oxygen species produced by ZnO NPs, many pieces of literature consider H_2O_2 as the main substance for antibacterial activity against dermo bacteria [28,161]. Negatively charged ROS cannot penetrate bacterial cell walls, but H_2O_2 can also easily penetrate bacterial cell walls. Sawai et al. [162] believed that the H_2O_2 produced by ZnO slurry was the main reason for the biocidal mechanism. It can also be hypothesized that after H_2O_2 or HO disrupts the membrane, ROS can penetrate the cell wall and enter the intracellular space, thereby enhancing the biocidal effect (Figure 12).

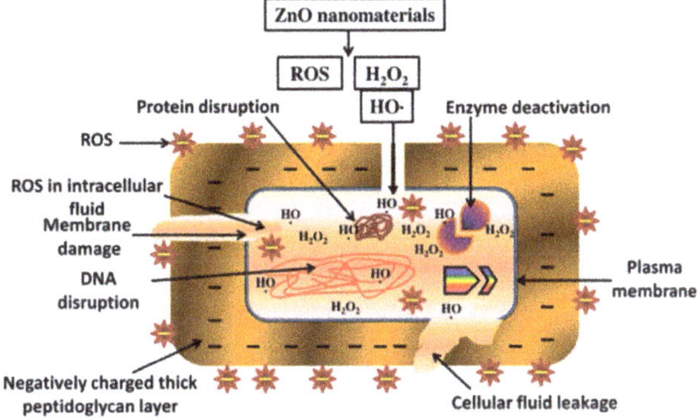

Figure 12. Schematic diagram of the damage to bacterial cells caused by ZnO nanomaterials producing H_2O_2. Reprinted/adapted with permission from Ref. [27]. Copyright© 2022, Elsevier B.V.

4.2.3. Influence of Physical Effects of ZnO-Based Nanomaterials on Antibacterial Performance

In the process of exploring the antibacterial mechanism of ZnO-based nanomaterials, there are three main chemical antibacterial mechanisms: plasma membrane disruption through ZnO interactions [163], cellular internalization of ZnO-based nanoparticles [164], and mechanical damage of the cell envelope [165].

At suitable pH, the bacterial surface is negatively charged due to the dissociation of carboxyl and other functional groups. Meanwhile, ZnO-based nanomaterials are positively charged with a zeta potential of +24 mV [161]. As shown in Figure 13, the opposite charges carried by bacterial cells and ZnO NPs are the reasons for the strong electrostatic attraction between them. Strong electrostatic interactions force particles larger than 10 nm in size to accumulate on the outer surface of the plasma membrane and neutralize the surface potential of the bacterial membrane, resulting in increased surface tension and membrane depolarization. In addition, strong electrostatic interactions can induce bacterial cell changes such as changes in cell membrane and membrane vesicle structure, rupture, morphological changes, and components, leading to bacterial cell death [166,167]. Since interactions play an important role in the bactericidal effect, surface modifiers and templates of ZnO-based nanomaterials would enhance the interaction with bacterial cell walls.

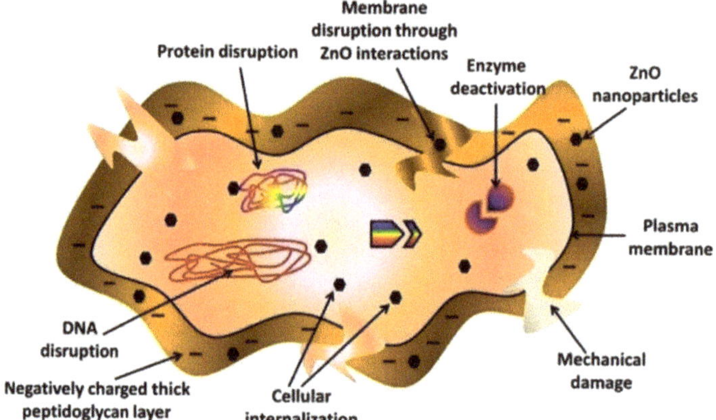

Figure 13. Schematic diagram of the physical action of ZnO nanomaterials for sterilization. Reprinted/adapted with permission from Ref. [27]. Copyright© 2022, Elsevier B.V.

Another important mechanism is cellular internalization. Cellular internalization is simply summarized as those nanostructures with a size of less than 10 nm pass through the plasma membrane, accumulate in bacterial cells, and destroy intracellular components such as nucleic acids [164,168–170]. In addition, it has also been suggested that the cellular interaction of ZnO with bacteria can enhance cell permeability (Figure 14) [171]. In conclusion, cellular internalization is one of the physical methods that plays an important role in the antibacterial process of ZnO-based NPs.

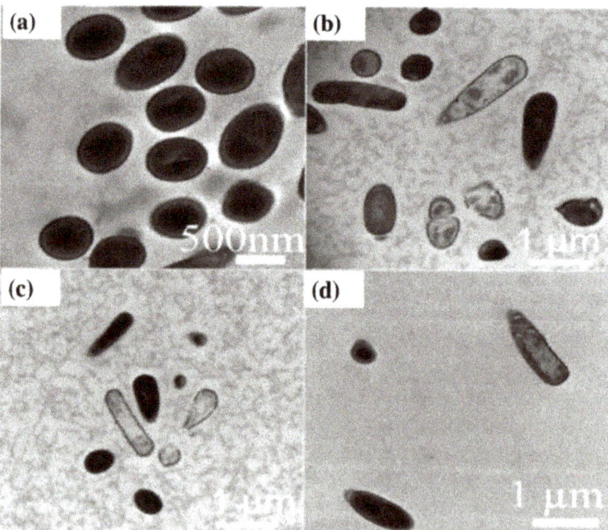

Figure 14. TEM images of *B. atrophaeus* (**a**) control, (**b**) ZnO powders, (**c**) ZnO nanorods, and (**d**) ZnO nanoparticles. Reprinted/adapted with permission from Ref. [172]. Copyright © 2022, Elsevier B.V.

The last physical mechanism is to use ZnO NPs to create cell membrane damage to destroy bacterial cells and achieve antibacterial effects. Compared with bulk ZnO materials, the presence of surface defects, uneven surface texture, and rough edges and corners on the surface of ZnO-based nanomaterials can lead to effective abrasiveness, resulting in excessive mechanical damage to bacterial cell membranes [163].

4.3. Effects of Radiation Types on the Antibacterial Activity of ZnO-Based Nanomaterials

In the process of photocatalytic antibacterial studies, ultraviolet (UV) light, sunlight, and other visible light are the most common types of radiation. In previous studies, ZnO nanomaterials have always been used for antibacterial testing under UV light due to their high band constraints [173–175]. From the work of Joe et al. [174], an important conclusion was found that the oxygen vacancy of ZnO crystals enhanced the photogeneration of ROS, and ZnO nanoparticles (NPs) with polar facets exhibited the most significant effect of antibacterial activity under UV light stimulation. Furthermore, Ma et al. creatively combined N-halamine-based materials with ZnO to improve the stability of ZnO's antibacterial performance under UV light. As a simple and effective method for nanoparticle modification, this technique can be further extended to the application of ZnO nanoparticles in other polar substrates for antibacterial functionalization [176].

Despite the promising antibacterial performance of ZnO-based nanomaterials under UV-driven radiation, photocatalysis using UV-active semiconductors is difficult due to the limited use of the solar spectrum. Simultaneously, significant progress has been made in photocatalysis using visible-light-active heteronanostructured semiconductors due to their simplicity of use, practicality, reproducibility, reliability, and commercialization [177,178]. Recently, several studies have found that photocatalytic efficiency can be improved by promoting the surface charge transfer reaction of ZnO. In addition, it may also affect the absorption spectrum of many metal oxide nanoparticles including ZnO, which enables the composites to undergo photocatalytic reactions under visible light [179–183]. The antibacterial activity of ZnO-based nanomaterials driven by visible light is of great significance for the development of photocatalytic antibacterials. Compared with ultraviolet light, visible light is more accessible, which makes visible light photocatalysis one of the hottest research fields. In addition, studies have shown that the ability of ZnO-based nanomaterials to excite ROS under visible light will be greatly improved, which also leads to the improvement of antibacterial ability under visible light [184,185]. Therefore, the design and preparation of visible-light-driven ZnO-based nanomaterials will become a meaningful research direction in future explorations.

4.4. Strategies for Enhancing ZnO-Based Nanomaterials Antibacterial Activity

4.4.1. Alkaline Earth Metal Doping into ZnO

Common alkaline earth metals including Ca, Mg, Al, and Sr, are frequently used to introduce significantly altered NPs, such as lattice defects and ionic radius differences between metal ions and Zn^{2+} ions, which can improve optical properties and photodegradation activity of catalysts. Taking Sr as an example, it played an important role in enhancing the catalytic degradation ability of commonly used metal oxides such as ZnO, TiO_2, which can be used for photocatalytic degradation of organic pollutants in wastewater and photocatalytic antibacterial [186]. Due to the lack of local d orbitals in alkaline earth metals and the presence of local d orbitals in transition metals, the doping of alkaline earth metals is more effective in reducing the optical threshold energy of semiconductors than that of transition metals [187].

Antibacterial activity of Mg-doped ZnO nanostructures was investigated by Okeke et al. [188] towards *E. coli*, *P. aeruginosa* and *Staphylococcus aureus*. The zones of inhibition diameter of the Zn1−xMgxO sample against the selected bacteria pathogen are displayed in Figure 15. The results showed that all samples were susceptible to bacteria. The presence of more reactive sites allows surface defects to create space for ZnO to interact with microorganisms [189]. Therefore, the increase of surface defects in ZnO nanostructures increases the reaction sites and the rate of interaction with microorganisms. Bacteria are 250 times larger than nanoparticles [105], while the bacterias have a larger relative surface area, which makes it easier for nanoparticles of much smaller size to enter the interior of bacterial cells and cause damage.

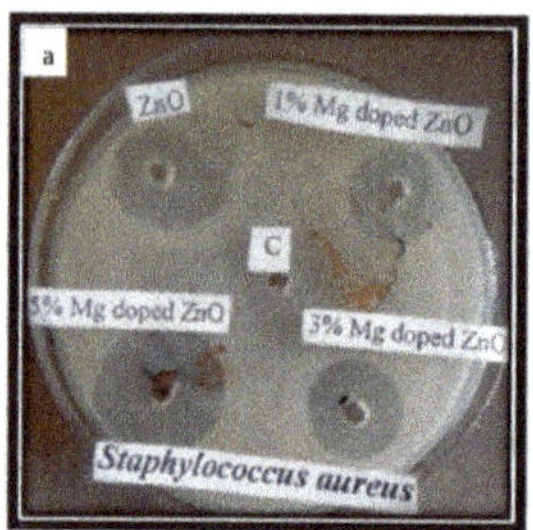

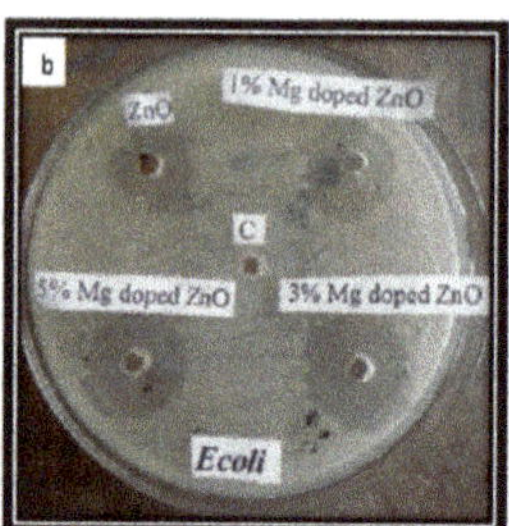

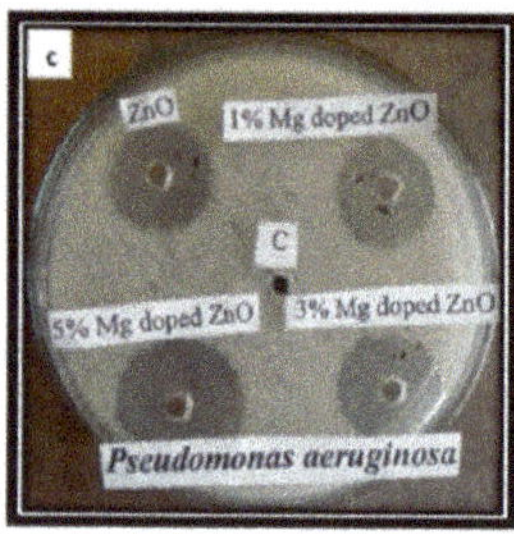

Figure 15. (**a**–**c**) Diameter of inhibition zones of Zn1−xMgxO against bacteria pathogen. Reprinted/adapted with permission from Ref. [188]. Copyright© 2022, Elsevier Ltd.

4.4.2. Transition Metals Doping into ZnO

The addition of transition metal ions can generate electronic states in the intermediate bandgap region to change charge separation and recombination kinetics, which is beneficial to enhancing the ability of photocatalytic antibacterial [190]. Numerous common transition metals, such as Co, Cu, Ni, Fe, and Mn, are often doped into ZnO to enhance its photocatalytic antibacterial ability [191].

Fe is the most used metal for enhancing the photocatalytic antibacterial ability of ZnO. Iron has two oxidation states with ionic radii of 0.61 Å and 0.55 Å, which are used as dopants for zinc lattice sites, respectively, since their ionic radii are smaller than those of zinc +2 (0.74 Å) radius. Hence, the doping can be alternative or interstitial to increase the conductivity of the product. Chandramouli et al. [192] doped Fe into ZnO and investigated the antibacterial properties of the composites. As shown in Figure 16, TEM images of Figure 16a pure, Figure 16b Fe doped, and Figure 16c capped ZnO NPs are observed. For undoped and doped ZnO NPs, they are more spherical with dimensions of 17–19 nm. Simultaneously, the agglomeration phenomenon occurs when ZnO is doped with Fe, which is due to the greater surface area and energy. Furthermore, the antibacterial activity against *E. coli* showed that the capped ZnO NPs were less toxic to the organism than ZnO NPs. As shown in Figure 17, the XRD patterns associated with various different ZnO-based nanomaterials are displayed. Iron doping of ZnO reduces the grain size, resulting in a further increase in the grain size of glucose-terminated ZnO nanoparticles, which is consistent with previous reports [193]. Therefore, Fe-ZnO nanoparticles have good antibacterial activity.

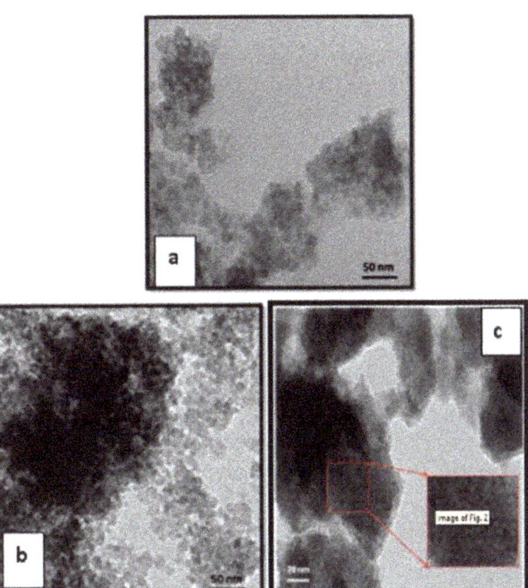

Figure 16. TEM images of (**a**) pure, (**b**) Fe doped, and (**c**) capped ZnO nanoparticles. Reprinted/adapted with permission from Ref. [192]. Copyright© 2022, Elsevier B.V.

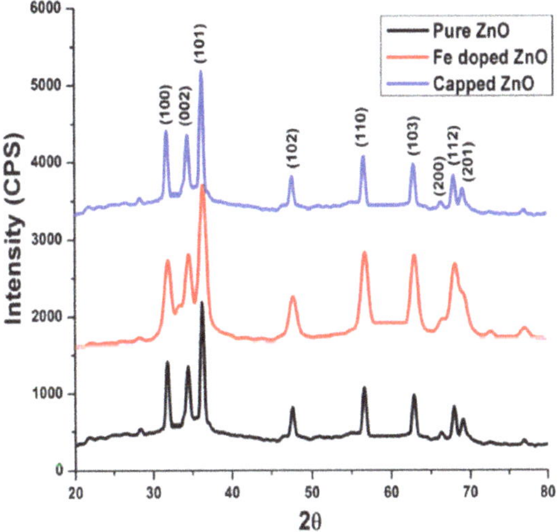

Figure 17. XRD patterns of various ZnO-based nanomaterials. Reprinted/adapted with permission from Ref. [192]. Copyright© 2022, Elsevier B.V.

4.4.3. Noble Metals Doping into ZnO

Due to the formation of Schottky barriers at the metal–semiconductor interface, noble metal ions (such as Au, Ag, Pd, etc.) doped on the ZnO surface are considered to be excellent photogenerated electron traps. In addition, noble metals delay electron–hole recombination by preventing photoexcited electrons from returning to the ZnO surface, which greatly enhances the photocatalytic antibacterial ability of the composites [194]. Of all the precious metals, silver is the most stable and suitable dopant with good thermal conductivity and electrical conductivity, which will better play the photocatalytic effect of the composite material. Therefore, it has potential as a catalyst. The surface plasmon resonance (SPR) properties of silver also contribute to visible light absorption and subsequent electron–hole pair generation for the degradation of contaminants in water and for antibacterial [195].

Ye et al. [184] reported the synthesis of a series of $ZnO/Ag_2MoO_4/Ag$(ZAA) samples with theoretical molar ratios of ZnO and Ag_2MoO_4 of 20:1, 30:1, and 60:1 by ultrasonic-assisted hydrothermal synthesis, and named the corresponding products as ZAA-20, ZAA-30, and ZAA-60 to investigate the optimal Ag_2MoO_4/Ag loadings. As shown in Figure 18, the characteristic diffraction peaks of ZnO and Ag2Mo4 can be clearly found in the XRD patterns, which indicates that the ZnO/Ag_2Mo_4 composite was successfully synthesized. The sharp diffraction peaks reveal the ultra-high crystallinity of the ZnO-based nanocomposites. In addition, with the increase of the molar ratio of ZnO to Ag_2MoO_4, the diffraction peak intensity of ZnO on the (002) and (110) crystal planes gradually weakened, while the diffraction peak intensity of Ag_2MoO_4 gradually increased. The antibacterial properties of ZnO nanosheets and ZAA nanocomposites against different contents of G− *E. coli* and G+ *S. aureus* were evaluated by the visible light electroplating counting method. In Figure 19, the bacterial cell numbers of all nanocomposites were shown to decrease with increasing contact time, and the photocatalytic antibacterial activities of the four ternary ZAA nanocomposites were much better than that of pure ZnO sheets. The experimental results demonstrate that the addition of noble metal Ag will significantly improve the antibacterial properties of ZnO nanomaterials.

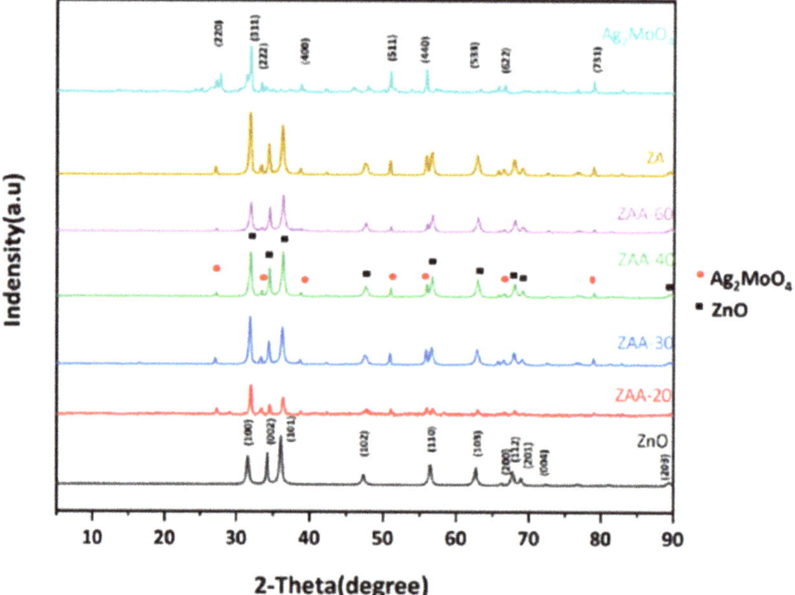

Figure 18. XRD patterns of ZnO, Ag_2MoO_4/Ag, ZA, and ZAA nanocomposite with different proportions. Reprinted/adapted with permission from Ref. [184]. Copyright© 2022, Elsevier B.V.

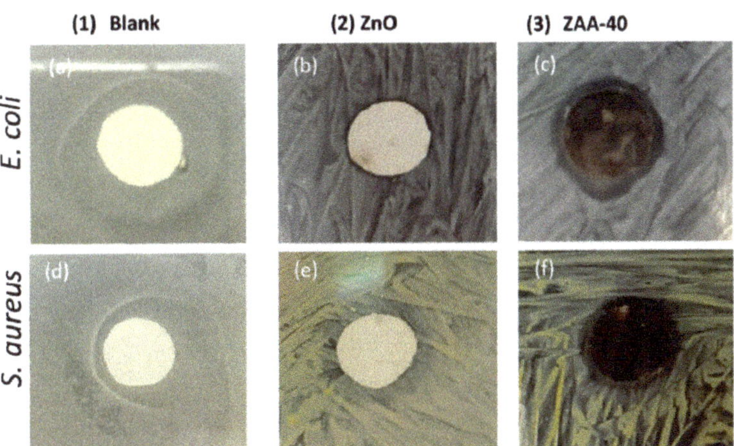

Figure 19. Photographs of antibacterial test results of ZnO, Ag2MoO4/Ag, and ZAA samples against Escherichia coli (**a**–**c**) and Staphylococcus aureus (**d**–**f**). Reprinted/adapted with permission from Ref. [184]. Copyright© 2022, Elsevier B.V.

4.4.4. Rare Earth Metal Doping into ZnO

Doping rare earth metals into ZnO can improve the ability of the composite in trapping photogenerated carriers and reducing electron–hole recombination, which can enhance the photocatalytic antibacterial ability. In the rare earth doping process, f-orbital doping is the most common and efficient way, which can improve the photocatalytic activity by enhancing the adsorption of pollutants on the catalyst surface, while reducing the band gap energy to the visible light range [196]. Lanthanide ion doping is considered a versatile strategy to tune the optical response and improve the photocatalytic performance of ZnO. Lanthanides are composed of 17 elements in the periodic table, including Sc, Y, La, Ce, Pr, Nd, Pm, Sm, Eu, Gd, Tb, Dy, Ho, Er, Tm, Yb, and Lu. Lanthanides have attracted much attention due to their multifunctional properties resulting from their unique f-orbital structures, and due to the f-f or f-d intra-electron transitions, lanthanides are considered candidates for luminescent centers in doped materials, which is beneficial to prolonging the effective response time [197–199].

Doping ZnO with Ln^{3+} and Ce^{4+} ions can convert the magnetism from diamagnetism to ferromagnetism, improve the n-type conductivity, enhance the photo response, increase the concentration of free electrons in the CB, and increase the electron mobility [200–202]. A novel Z-type ZnO–CeO_2-Yb_2O_3 heterojunction photocatalyst was prepared for the first time by Tauseef et al. [203], and its physical, photocatalytic, and antibacterial properties were investigated. Growth samples were tested for antimicrobial properties against *E. coli* and *S. aureus*. The effects of operating parameters such as catalyst dosage, dye concentration, and solution pH on the photocatalytic performance of the nanocomposites were investigated. The ZOIs of *S. aureus* and *E. coli* along with the standard antibiotic ciprofloxacin are shown in Figure 20a,b. The synthesized nanocomposites exhibited good activity against both bacteria with a ZOI > 6 mm, but higher activity against *E. coli* with a ZOI of 14 mm shown in Figure 21a,b. Positively charged heavy metal ions such as Zn^{2+}, Ce^{4+}, and Yb^{3+} can be released from the surface of the nanocomposite to interact with negatively charged microbial cell membranes. The entry of these metal ions into the cell membrane reduces the capacity and permeability of proteins, which in turn leads to the death of microorganisms such as bacteria and viruses. The above antibacterial action mechanism can be visualized in Figure 22C. In conclusion, the nanocomposites doped with rare earth ions are effective materials for preventing diseases caused by *S. aureus* and *E. coli*.

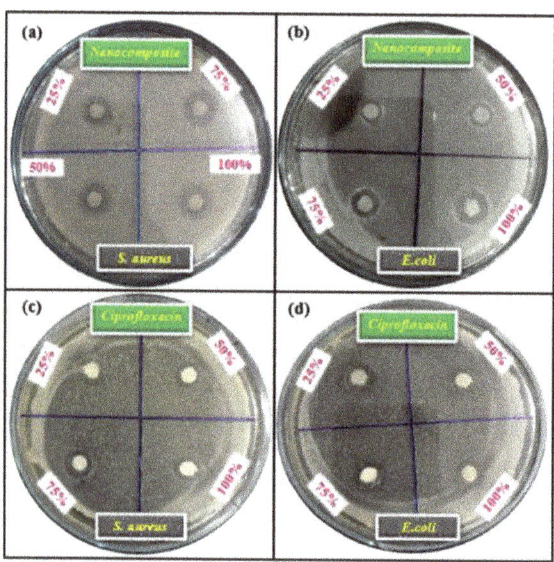

Figure 20. Antibacterial effect of $ZnO\text{-}CeO_2\text{-}Yb_2O_3$ nanocomposites: (**a**) gram-positive *Staphylococcus aureus* bacteria, (**b**) gram-negative *Escherichia coli* bacteria at different concentrations, and (**c**,**d**) Standard antibiotics Ciprofloxacin against *S. aureus* and *E. coli*. Reprinted/adapted with permission from Ref. [203]. Copyright© 2022, Elsevier Masson SAS.

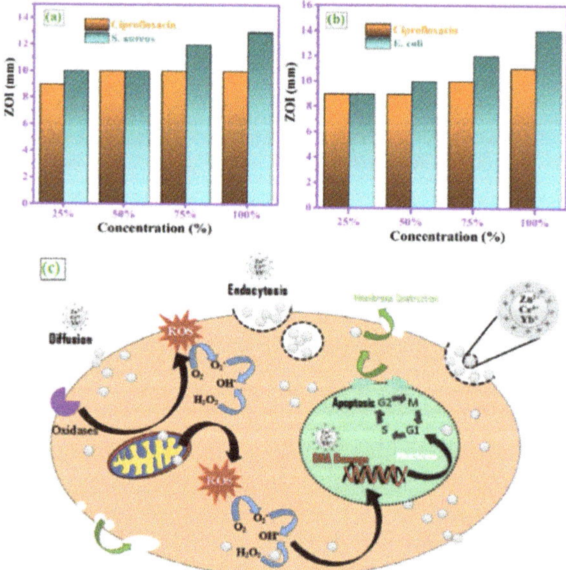

Figure 21. Comparison of the zone of inhibition (ZOI) against different species of bacteria (**a**,**b**), mechanism of antibacterial activity of $ZnO\text{–}CeO_2\text{-}Yb_2O_3$ nanocomposite (**c**). Reprinted/adapted with permission from Ref. [203]. Copyright© 2022, Elsevier Masson SAS.

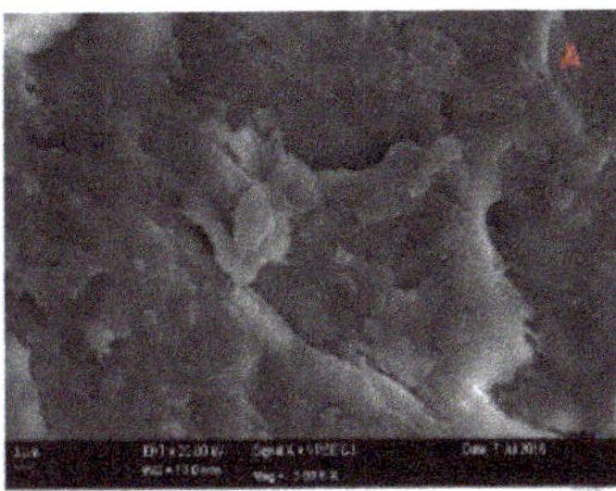

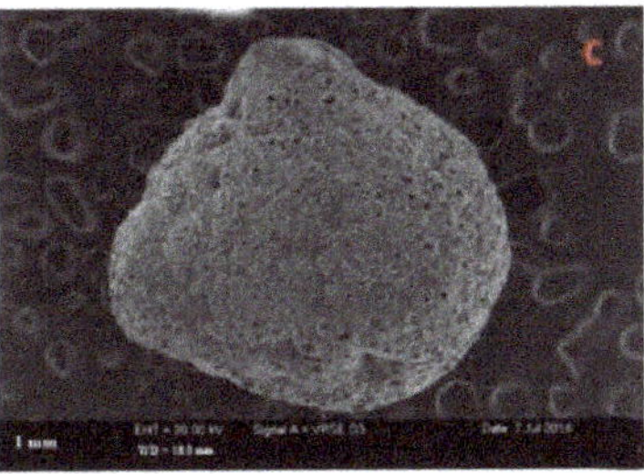

Figure 22. SEM images of various prepared CS/PVA/ZnO-related materials and precursors: (**A**) pure chitosan, (**B**) ZnO, and (**C**) CS/PVA/ZnO. Reprinted/adapted with permission from Ref. [204]. Copyright© 2022, Elsevier B.V.

4.4.5. Organic Antimicrobial Agents Doping into ZnO

Studies have shown that the composites obtained by co-doping and fusion of organic antibacterial agents and ZnO nanoparticles exhibited stronger antibacterial activity than ZnO nanoparticles alone [205,206]. The organic antimicrobial agents are usually immobilized or embedded on the ZnO surface. Taking chitosan (CS) as an example, it is an abundant natural biopolymer derived from the deacetylation of chitin in crustacean shells and can be made into films, fibers, beads, and powders. Cationic polymers are generally antimicrobial [207,208]. In general, antibacterial activity depends on molecular weight (Mw), degree of deacetylation, temperature, and solution pH [209,210].

Cutha et al. [204] used CS and ZnO as raw materials to prepare a new composite material chitosan/poly(vinyl alcohol)/zinc oxide (CS/PVA/ZnO), which was used as a novel antibacterial agent with wound healing properties. CS/PVA/ZnO was proved to be an effective antibacterial nanomaterial after being analyzed by various characterization methods. SEM images of various prepared CS/PVA/ZnO-related materials and precursors are shown in Figure 22A–C. The surface of sole chitosan was obtained to be smooth. The sole ZnO nanoparticles showed nanosheet-like morphology. The surface of sole CS/PVA/ZnO microbeads presents a certain pore structure, and the surface of the microbeads is rough, which is conducive to exerting the ability of photocatalytic antibacterial. The antibacterial activities of CS, CS/PVA, and CS/PVA/ZnO are shown in Figure 23. The diameter of the inhibition zone against *E. coli* cultures (G-) was 10 mm in the CS group, 14 mm in the CS/PVA group, and 19 mm in the CS/PVA/ZnO group (Figure 23A). Likewise, *S. aureus* cultures (G+) had a diameter of 12 mm in the CS group, 15 mm in the CS/PVA group, and 20 mm in the CS/PVA/ZnO group (Figure 23A).

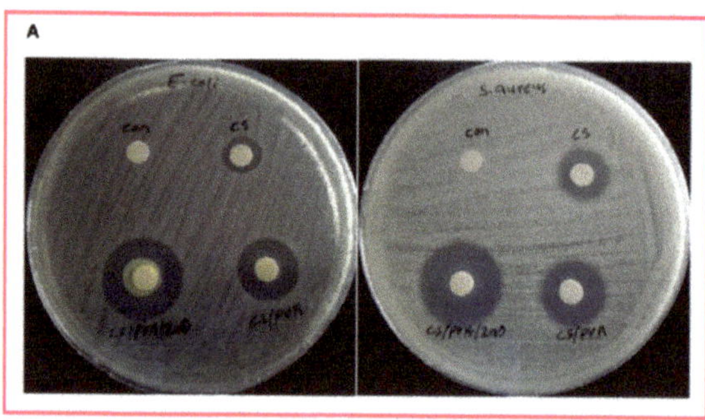

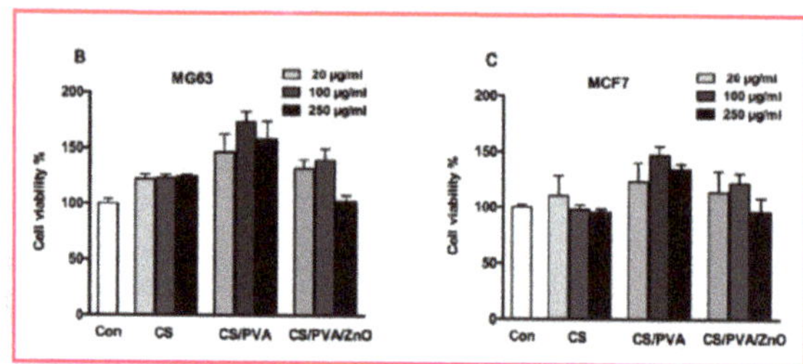

Figure 23. The antibacterial activities of CS, CS/PVA, and CS/PVA/ZnO: (**A**) Diameter of zone of inhibition on *E. coli* culture (gram negative bacteria); (**B**) Effect of mammalian cell viability of freshly prepared CS, CS/PVA, and CS/PVA/ZnO on MG63 (human osteosarcoma cell line) and (**C**) MCF7 (Human breast cancer cell line) cells. Reprinted/adapted with permission from Ref. [204]. Copyright© 2022, Elsevier B.V.

5. Future Scope and Conclusions

5.1. Future Scope

Although ZnO-based nanomaterials have been applied in organic pollutants removal from water and conduct antibacterial reactions in water, there is still plenty of space for improvement. The following points are the aspects that can be improved and strengthened in the application process of ZnO-based nanomaterials in the future:

(1) Exploring strategies for changing the weak toxicity of ZnO-based nanomaterials so that they can be better used in drinking water treatment, clinical medicine, virus killing, and other fields closely related to human beings.
(2) Enhancing the ability of ZnO-based nanomaterials to respond to visible light enables them to have a wider range of applications. Visible light is one of the most abundant light sources, and a better response under visible light can maximize the use of existing energy and reduce investment.
(3) Exploring the use of ZnO-based nanomaterials for photocatalytic removal of resistant bacteria, cancer cells, and other difficult-to-remove microorganisms and pathogen cells to improve the availability of the material.
(4) Exploring stronger ZnO-based nanomaterial structures and carriers to improve recyclability and improve existing problems such as high solubility and difficulty in recycling.

5.2. Conclusions

This manuscript is based on recent developments in antibacterial water treatment with ZnO-based nanomaterials. Due to the increasing global requirements for water environment quality and drinking water quality, especially for the prevention and control of various epidemics, new ideas and directions are provided for our study. Therefore, the existence of bacteria and harmful microorganisms in water is introduced in detail, and various commonly used antibacterial agents and antibacterial methods are summarized. In conclusion, different morphologies of ZnO-based nanomaterials can be effectively used against various Gram-positive and Gram-negative strains by physicochemical interactions with bacterial cells. Cell membrane damage and biocidal activity are thought to be triggered by the collective action of chemical and physical interactions. Chemical interactions leading to the production of ROS and H_2O_2 and the release of Zn^{2+} ions from ZnO solubility have been proved to be the main cause of the above activities. Subsequently, based on this theory, an in-depth study of the antibacterial mechanism was carried out. Finally, the review also summarizes the following synthetic strategies to improve the antibacterial properties of ZnO: (1) doping of alkaline earth metals to ZnO; (2) doping of transition metals to ZnO; (3) doping of noble metals to ZnO; (4) doping of rare earth metal to ZnO; and (5) loading organic antimicrobial agents.

It can be expected that the antibacterial potential of ZnO-based nanomaterials in water treatment is very promising. Studies on ZnO-based nanomaterials continuously increased in recent years, although they still have many aspects that can be improved. In the future, we can expect more perfect ZnO-based nanomaterials to be prepared to solve more antimicrobial-related problems in water treatment.

Author Contributions: Z.X.: Investigation, Writing—Original Draft. Q.H.: Investigation. S.W.: Investigation. X.H.: Supervision, Writing—Reviewing and Editing. Z.F.: Supervision, Writing—Reviewing and Editing. X.X.: Supervision, Writing—Reviewing and Editing. X.Z.: Supervision, Writing—Reviewing and Editing. All authors have read and agreed to the published version of the manuscript.

Funding: This research was supported by the National Key Research and Development Plan, China (2019YFC1907204). We are grateful for the test services from the Analytical and Testing Center of Northeastern University.

Institutional Review Board Statement: Not applicable.

Informed Consent Statement: Not applicable.

Data Availability Statement: Not applicable.

Conflicts of Interest: The authors declare no conflict of interest.

References

1. Fuzil, N.S.; Othman, N.H.; Alias, N.H.; Marpani, F.; Othman, M.H.D.; Ismail, A.F.; Lau, W.J.; Li, K.; Kusworo, T.D.; Ichinose, I.; et al. A review on photothermal material and its usage in the development of photothermal membrane for sustainable clean water production. *Desalination* **2021**, *517*, 115259. [CrossRef]
2. Tang, W.; Pei, Y.; Zheng, H.; Zhao, Y.; Shu, L.; Zhang, H. Twenty years of China's water pollution control: Experiences and challenges. *Chemosphere* **2022**, *295*, 133875. [CrossRef] [PubMed]
3. Yan, C.; Qu, Z.; Wang, J.; Cao, L.; Han, Q. Microalgal bioremediation of heavy metal pollution in water: Recent advances, challenges, and prospects. *Chemosphere* **2022**, *286*, 131870. [CrossRef] [PubMed]
4. Aal, N.A.; Al-Hazmi, F.; Al-Ghamdi, A.A.; Alghamdi, A.A.; El-Tantawy, F.; Yakuphanoglu, F. Novel rapid synthesis of zinc oxide nanotubes via hydrothermal technique and antibacterial properties. *Spectrochim. Acta Part A Mol. Biomol. Spectrosc.* **2015**, *135*, 871–877. [CrossRef]
5. Coelho, E.F.; Santos, D.L.; Lima, L.W.F.d.; Castricini, A.; Barros, D.L.; Filgueiras, R.; da Cunha, F.F. Water regimes on soil covered with plastic film mulch and relationships with soil water availability, yield, and water use efficiency of papaya trees. *Agric. Water Manag.* **2022**, *269*, 107709. [CrossRef]
6. Sheikh, M.; Pazirofteh, M.; Dehghani, M.; Asghari, M.; Rezakazemi, M.; Valderrama, C.; Cortina, J.-L. Application of ZnO nanostructures in ceramic and polymeric membranes for water and wastewater technologies: A review. *Chem. Eng. J.* **2019**, *391*, 123475. [CrossRef]
7. Sondermann, M.N.; Proença de Oliveira, R. Using the WEI+ index to evaluate water scarcity at highly regulated river basins with conjunctive uses of surface and groundwater resources. *Sci. Total Environ.* **2022**, *836*, 155754. [CrossRef]
8. Ouassanouan, Y.; Fakir, Y.; Simonneaux, V.; Kharrou, M.H.; Bouimouass, H.; Najar, I.; Benrhanem, M.; Sguir, F.; Chehbouni, A. Multi-decadal analysis of water resources and agricultural change in a Mediterranean semiarid irrigated piedmont under water scarcity and human interaction. *Sci. Total Environ.* **2022**, *834*, 155328. [CrossRef]
9. Kumar, R.; Umar, A.; Kumar, G.; Nalwa, H.S. Antimicrobial properties of ZnO nanomaterials: A review. *Ceram. Int.* **2017**, *43*, 3940–3961. [CrossRef]

10. Witte, W. International dissemination of antibiotic resistant strains of bacterial pathogens. *Infect. Genet. Evol.* **2004**, *4*, 187–191. [CrossRef]
11. Shahin, O.R.; Alshammari, H.H.; Taloba, A.I.; El-Aziz, R.M.A. Machine Learning Approach for Autonomous Detection and Classification of COVID-19 Virus. *Comput. Electr. Eng.* **2022**, *101*, 108055. [CrossRef] [PubMed]
12. Wang, H.; Paulson, K.R.; Pease, S.A.; Watson, S.; Comfort, H.; Zheng, P.; Aravkin, A.Y.; Bisignano, C.; Barber, R.M.; Alam, T.; et al. Estimating excess mortality due to the COVID-19 pandemic: A systematic analysis of COVID-19-related mortality, 2020–2021. *Lancet* **2022**, *399*, 1513–1536. [CrossRef]
13. Flor, L.S.; Friedman, J.; Spencer, C.N.; Cagney, J.; Arrieta, A.; Herbert, M.E.; Stein, C.; Mullany, E.C.; Hon, J.; Patwardhan, V.; et al. Quantifying the effects of the COVID-19 pandemic on gender equality on health, social, and economic indicators: A comprehensive review of data from March, 2020, to September, 2021. *Lancet* **2022**, *399*, 2381–2397. [CrossRef]
14. Pan, D.; Sze, S.; Nazareth, J.; Martin, C.A.; Al-Oraibi, A.; Baggaley, R.F.; Nellums, L.B.; Hollingsworth, T.D.; Tang, J.W.; Pareek, M. Monkeypox in the UK: Arguments for a broader case definition. *Lancet* **2022**, *399*, 2345–2346. [CrossRef]
15. Liu, Y.; Shao, Y.; Wang, L.; Lu, W.; Li, S.; Xu, D.; Fu, Y.V. Inactivation of porcine epidemic diarrhea virus with electron beam irradiation under cold chain conditions. *Environ. Technol. Innov.* **2022**, *27*, 102715. [CrossRef]
16. Yang, S.; Dong, Q.; Li, S.; Cheng, Z.; Kang, X.; Ren, D.; Xu, C.; Zhou, X.; Liang, P.; Sun, L.; et al. Persistence of SARS-CoV-2 RNA in wastewater after the end of the COVID-19 epidemics. *J. Hazard. Mater.* **2022**, *429*, 128358. [CrossRef] [PubMed]
17. Guo, Y.; Lin, S.; Li, X.; Liu, Y. Amino acids assisted hydrothermal synthesis of hierarchically structured ZnO with enhanced photocatalytic activities. *Appl. Surf. Sci.* **2016**, *384*, 83–91. [CrossRef]
18. Chang, Y.-N.; Zhang, M.; Xia, L.; Zhang, J.; Xing, G.J.M. The toxic effects and mechanisms of CuO and ZnO nanoparticles. *Creat. Commons Attrib. Licens.* **2012**, *5*, 2850–2871. [CrossRef]
19. Gupta, K.; Singh, R.P.; Pandey, A.; Pandey, A. Photocatalytic antibacterial performance of TiO_2 and Ag-doped TiO_2 against *S.* aureus. P. aeruginosa and E. coli. *J. Nanotechnol.* **2013**, *4*, 345–351. [CrossRef]
20. Allahverdiyev, A.M.; Abamor, E.S.; Bagirova, M.; Rafailovich, M. Antimicrobial effects of TiO_2 and Ag_2O nanoparticles against drug-resistant bacteria and leishmania parasites. *Future Med.* **2011**, *6*, 933–940. [CrossRef]
21. Jin, T.; He, Y. Antibacterial activities of magnesium oxide (MgO) nanoparticles against foodborne pathogens. *J. Nanopart. Res.* **2011**, *13*, 6877–6885. [CrossRef]
22. Azam, A.; Ahmed, A.S.; Oves, M.; Khan, M.S.; Habib, S.S.; Memic, A. Antimicrobial activity of metal oxide nanoparticles against Gram-positive and Gram-negative bacteria: A comparative study. *Int. J. Nanomed.* **2012**, *7*, 6003. [CrossRef]
23. Luo, Z.; Wu, Q.; Xue, J.; Ding, Y. Selectively enhanced antibacterial effects and ultraviolet activation of antibiotics with ZnO nanorods against Escherichia coli. *J. Biomed. Nanotechnol.* **2013**, *9*, 69–76. [CrossRef] [PubMed]
24. Şahin, B.; Aydin, R.; Soylu, S.; Türkmen, M.; Kara, M.; Akkaya, A.; Çetin, H.; Ayyıldız, E. The effect of thymus syriacus plant extract on the main physical and antibacterial activities of ZnO nanoparticles synthesized by SILAR method. *Inorg. Chem. Commun.* **2022**, *135*, 109088. [CrossRef]
25. Negi, A.; Gangwar, R.; Kumar Vishwakarma, R.; Singh Negi, D. Antibacterial, antioxidant and photodegradation potential of ZnO nanoparticles mediated via roots of Taraxacum officinale radix. *Mater. Today Proc.* **2022**, *57*, 2435–2443. [CrossRef]
26. Etacheri, V.; Di Valentin, C.; Schneider, J.; Bahnemann, D.; Pillai, S.C. Visible-light activation of TiO_2 photocatalysts: Advances in theory and experiments. *J. Photochem. Photobiol. C Photochem. Rev.* **2015**, *25*, 1–29. [CrossRef]
27. Qi, K.; Cheng, B.; Yu, J.; Ho, W. Review on the improvement of the photocatalytic and antibacterial activities of ZnO. *J. Alloys Compd.* **2017**, *727*, 792–820. [CrossRef]
28. Zhang, L.; Li, Y.; Liu, X.; Zhao, L.; Ding, Y.; Povey, M.; Cang, D. The properties of ZnO nanofluids and the role of H_2O_2 in the disinfection activity against *Escherichia coli*. *Water Res.* **2013**, *47*, 4013–4021. [CrossRef]
29. Ravindranadh, M.R.K.; Mary, T.R. Development of ZnO nanoparticles for clinical applications. *J. Chem. Biol. Phys. Sci.* **2013**, *4*, 469.
30. Zhang, L.; Jiang, Y.; Ding, Y.; Daskalakis, N.; Jeuken, L.; Povey, M.; O'Neill, A.J.; York, D.W. Mechanistic investigation into antibacterial behaviour of suspensions of ZnO nanoparticles against *E. coli*. *J. Nanopart. Res.* **2009**, *12*, 1625–1636. [CrossRef]
31. Li, M.; Zhu, L.; Lin, D. Toxicity of ZnO Nanoparticles to *Escherichia coli*: Mechanism and the Influence of Medium Components. *Environ. Sci. Technol.* **2011**, *45*, 1977–1983. [CrossRef] [PubMed]
32. Choubari, M.S.; Mazloom, J.; Ghodsi, F.E. Supercapacitive properties, optical band gap, and photoluminescence of CeO_2–ZnO nanocomposites prepared by eco-friendly green and citrate sol-gel methods: A comparative study. *Ceram. Int.* **2022**, *48*, 21385–21395. [CrossRef]
33. Li, Y.; Li, Y.; Fei, Y.; Xie, A.; Li, Y.; Sun, D. Investigation of properties of ZnO and $Mg_xZn_{1-x}O$ films prepared by sol-gel method. *J. Mol. Struct.* **2022**, *1261*, 132959. [CrossRef]
34. Castro-Lopes, S.; Guerra, Y.; Silva-Sousa, A.; Oliveira, D.M.; Gonçalves, L.A.P.; Franco, A.; Padrón-Hernández, E.; Peña-Garcia, R. Influence of pH on the structural and magnetic properties of Fe-doped ZnO nanoparticles synthesized by sol gel method. *Solid State Sci.* **2020**, *109*, 106438. [CrossRef]
35. Selvaraj, S.; Vangari, G.A.; Mohan, M.K.; Ponnusamy, S.; Muthamizchelvan, C. Facile synthesis of Sm doped ZnO nanoflowers by Co-precipitation method for enhanced photocatalytic degradation of MB dye under sunlight irradiation. *Ceram. Int.* **2022**. [CrossRef]

36. Shivaraj, B.; Prabhakara, M.C.; Bhojya Naik, H.S.; Indrajith Naik, E.; Viswanath, R.; Shashank, M.; Kumara Swamy, B.E. Optical, bio-sensing, and antibacterial studies on Ni-doped ZnO nanorods, fabricated by chemical co-precipitation method. *Inorg. Chem. Commun.* **2021**, *134*, 109049. [CrossRef]
37. Soltani, S.; Roodbar Shojaei, T.; Khanian, N.; Shean Yaw Choong, T.; Asim, N.; Zhao, Y. Artificial neural network method modeling of microwave-assisted esterification of PFAD over mesoporous TiO_2–ZnO catalyst. *Renew. Energy* **2022**, *187*, 760–773. [CrossRef]
38. Saravanan, P.; SenthilKannan, K.; Vimalan, M.; Tamilselvan, S.; Sankar, D. Biofriendly and competent domestic microwave assisted method for the synthesis of ZnO nanoparticles from the extract of Azadirachta indica leaves. *Mater. Today Proc.* **2020**, *33*, 3160–3163. [CrossRef]
39. Ashok, C.H.; Venkateswara Rao, K. ZnO/TiO_2 nanocomposite rods synthesized by microwave-assisted method for humidity sensor application. *Superlattices Microstruct.* **2014**, *76*, 46–54. [CrossRef]
40. Li, M.; Xu, Z.P.; Sultanbawa, Y.; Chen, W.; Liu, J.; Qian, G. Potent and durable antibacterial activity of ZnO-dotted nanohybrids hydrothermally derived from ZnAl-layered double hydroxides. *Colloids Surf. B Biointerfaces* **2019**, *181*, 585–592. [CrossRef]
41. Krajian, H.; Abdallah, B.; Kakhia, M.; AlKafri, N. Hydrothermal growth method for the deposition of ZnO films: Structural, chemical and optical studies. *Microelectron. Reliab.* **2021**, *125*, 114352. [CrossRef]
42. Ma, J.; Liu, J.; Bao, Y.; Zhu, Z.; Wang, X.; Zhang, J. Synthesis of large-scale uniform mulberry-like ZnO particles with microwave hydrothermal method and its antibacterial property. *Ceram. Int.* **2013**, *39*, 2803–2810. [CrossRef]
43. Shastri, L.; Qureshi, M.S.; Malik, M.M. Photoluminescence study of ZnO–SiO_2 nanostructures grown in silica matrix obtained via sol–gel method. *J. Phys. Chem. Solids* **2013**, *74*, 595–598. [CrossRef]
44. Hassena, H. Photocatalytic degradation of methylene blue by using Al_2O_3/Fe_2O_3 nano composite under visible light. *Mod. Chem. Appl.* **2016**, *2016*. [CrossRef]
45. Hasnidawani, J.N.; Azlina, H.N.; Norita, H.; Bonnia, N.N.; Ratim, S.; Ali, E.S. Synthesis of ZnO Nanostructures Using Sol-Gel Method. *Procedia Chem.* **2016**, *19*, 211–216. [CrossRef]
46. Bari, A.R.; Shinde, M.; Deo, V.; Patil, L. Effect of solvents on the particle morphology of nanostructured ZnO. *Indian J. Pure Appl. Phys.* **2009**, *47*, 24.
47. Morkoç, H.; Özgür, Ü. *Zinc Oxide: Fundamentals, Materials and Device Technology*; John Wiley & Sons: Hoboken, NJ, USA, 2008.
48. Moezzi, A.; McDonagh, A.M.; Cortie, M.B. Zinc oxide particles: Synthesis, properties and applications. *Chem. Eng. J.* **2012**, *185*, 1–22. [CrossRef]
49. Naik, E.I.; Naik, H.S.B.; Viswanath, R.; Kirthan, B.R.; Prabhakara, M.C. Effect of zirconium doping on the structural, optical, electrochemical and antibacterial properties of ZnO nanoparticles prepared by sol-gel method. *Chem. Data Collect.* **2020**, *29*, 100505. [CrossRef]
50. Chen, Z.; Zhao, X.; Wei, S. Comparative study on sol-gel combined with a hydrothermal synthesis of $ZnAl_2O_4$ and ZnO/$ZnAl_2O_4$ nanocomposites and its photoluminescence properties and antibacterial activity. *Optik* **2021**, *242*, 167151. [CrossRef]
51. El-Katori, E.E.; Kasim, E.A.; Ali, D.A.J.C.; Physicochemical, S.A.; Aspects, E. Sol–gel synthesis of mesoporous NiO/ZnO heterostructure nanocomposite for photocatalytic and anticorrosive applications in aqueous media. *Colloids Surf. A Physicochem. Eng. Asp.* **2022**, *636*, 128153. [CrossRef]
52. Ahmad, S.; Aadil, M.; Ejaz, S.R.; Akhtar, M.U.; Noor, H.; Haider, S.; Alsafari, I.A.; Yasmin, G.J.C.I. Sol-gel synthesis of nanostructured ZnO/$SrZnO_2$ with boosted antibacterial and photocatalytic activity. *Ceram. Int.* **2022**, *48*, 2394–2405. [CrossRef]
53. Jing, J.; Lin, L.; Yang, K.; Hu, H.; Guo, T.; Li, F.J.O.E. Highly efficient inverted quantum dot light-emitting diodes employing sol-gel derived Li-doped ZnO as electron transport layer. *Org. Electron.* **2022**, *103*, 106466. [CrossRef]
54. Devi, P.G.; Velu, A.S. Synthesis, structural and optical properties of pure ZnO and Co doped ZnO nanoparticles prepared by the co-precipitation method. *J. Theor. Appl. Phys.* **2016**, *10*, 233–240. [CrossRef]
55. Suwanboon, S.; Amornpitoksuk, P.; Sukolrat, A.; Muensit, N.J.C.I. Optical and photocatalytic properties of La-doped ZnO nanoparticles prepared via precipitation and mechanical milling method. *Ceram. Int.* **2013**, *39*, 2811–2819. [CrossRef]
56. Kotresh, M.; Patil, M.; Inamdar, S.J.O. Reaction temperature based synthesis of ZnO nanoparticles using co-precipitation method: Detailed structural and optical characterization. *Optik* **2021**, *243*, 167506. [CrossRef]
57. Anitha, S.; Muthukumaran, S. Structural, optical and antibacterial investigation of La, Cu dual doped ZnO nanoparticles prepared by co-precipitation method. *Mater. Sci. Eng. C* **2020**, *108*, 110387. [CrossRef]
58. Pavithra, M.; Raj, M.J. Influence of ultrasonication time on solar light irradiated photocatalytic dye degradability and antibacterial activity of Pb doped ZnO nanocomposites. *Ceram. Int.* **2021**, *47*, 32324–32331. [CrossRef]
59. Aziz, F.; Abo-Dief, H.M.; Warsi, A.-Z.; Warsi, M.F.; Shahid, M.; Ahmad, T.; Mersal, G.A.; Ibrahim, M.M. Facile synthesis of NiO/ZnO nano-composite by Co-precipitation, characterization and photocatalytic study of colored and colorless organic pollutants by solar irradiation. *Phys. B Condens. Matter* **2022**, *640*, 413858. [CrossRef]
60. Bhogaita, M.; Devaprakasam, D. Hybrid photoanode of TiO_2-ZnO synthesized by co-precipitation route for dye-sensitized solar cell using phyllanthus reticulatas pigment sensitizer. *Sol. Energy* **2021**, *214*, 517–530. [CrossRef]
61. Patel, N.; Patel, I.; Bharat, V. Structural, thermal and chemical studies of Mn doped ZnO nanoparticles synthesized by Co-precipitation method. *Mater. Proc.* **2021**, *46*, 2277–2280. [CrossRef]
62. Saleh, S.M. ZnO nanospheres based simple hydrothermal route for photocatalytic degradation of azo dye. *Spectrochim. Acta Part A Mol. Biomol. Spectrosc.* **2019**, *211*, 141–147. [CrossRef] [PubMed]

63. Ahammed, K.R.; Ashaduzzaman, M.; Paul, S.C.; Nath, M.R.; Bhowmik, S.; Saha, O.; Rahaman, M.; Bhowmik, S.; Aka, T. Microwave assisted synthesis of zinc oxide (ZnO) nanoparticles in a noble approach: Utilization for antibacterial and photocatalytic activity. *SN Appl. Sci.* **2020**, *2*, 955. [CrossRef]
64. Sanakousar, F.; Vidyasagar, C.; Jiménez-Pérez, V.; Prakash, K. Recent progress on visible-light-driven metal and non-metal doped ZnO nanostructures for photocatalytic degradation of organic pollutants. *Mater. Sci. Semicond. Processing* **2022**, *140*, 106390. [CrossRef]
65. Arellano-Cortaza, M.; Ramirez-Morales, E.; Pal, U.; Pérez-Hernández, G.; Rojas-Blanco, L. pH dependent morphology and texture evolution of ZnO nanoparticles fabricated by microwave-assisted chemical synthesis and their photocatalytic dye degradation activities. *Ceram. Int.* **2021**, *47*, 27469–27478. [CrossRef]
66. Jin, C.; Lu, Y.; Tong, G.; Che, R.; Xu, H. Excellent microwave absorbing properties of ZnO/$ZnFe_2O_4$/Fe core-shell microrods prepared by a rapid microwave-assisted hydrothermal-chemical vapor decomposition method. *Appl. Surf. Sci.* **2020**, *531*, 147353. [CrossRef]
67. Skoda, D.; Urbanek, P.; Sevcik, J.; Munster, L.; Antos, J.; Kuritka, I. Microwave-assisted synthesis of colloidal ZnO nanocrystals and their utilization in improving polymer light emitting diodes efficiency. *Mater. Sci. Eng. B* **2018**, *232–235*, 22–32. [CrossRef]
68. Chong, W.J.; Shen, S.; Li, Y.; Trinchi, A.; Pejak Simunec, D.; Kyratzis, I.; Sola, A.; Wen, C. Biodegradable PLA-ZnO nanocomposite biomaterials with antibacterial properties, tissue engineering viability, and enhanced biocompatibility. *Smart Mater. Manuf.* **2022**, 100004. [CrossRef]
69. Okeke, I.S.; Agwu, K.K.; Ubachukwu, A.A.; Ezema, F.I. Influence of transition metal doping on physiochemical and antibacterial properties of ZnONanoparticles: A review. *Appl. Surf. Sci. Adv.* **2022**, *8*, 100227. [CrossRef]
70. Kamarulzaman, N.; Kasim, M.F.; Rusdi, R. Band gap narrowing and widening of ZnO nanostructures and doped materials. *Nanoscale Res. Lett.* **2015**, *10*, 346. [CrossRef]
71. Shahmoradi, B.; Ibrahim, I.; Namratha, K.; Sakamoto, N.; Ananda, S.; Somashekar, R.; Byrappa, K. Surface modification of indium doped ZnO hybrid nanoparticles with n-butylamine. *Int. J. Chem. Eng. Res.* **2010**, *2*, 107–117.
72. Sahoo, T.; Kim, M.; Baek, J.H.; Jeon, S.-R.; Kim, J.S.; Yu, Y.-T.; Lee, C.-R.; Lee, I.-H. Synthesis and characterization of porous ZnO nanoparticles by hydrothermal treatment of as pure aqueous precursor. *Mater. Res. Bull.* **2011**, *46*, 525–530. [CrossRef]
73. Yuvarani, A.; Vaideeswaran, R.; Sophia Rani, I.; Reeta Mary, I. Physical characterization and antibacterial activity of zinc oxide nanostructures synthesized via facile hydrothermal method. *Mater. Today Proc.* **2022**. [CrossRef]
74. Siwińska-Stefańska, K.; Kubiak, A.; Piasecki, A.; Dobrowolska, A.; Czaczyk, K.; Motylenko, M.; Rafaja, D.; Ehrlich, H.; Jesionowski, T. Hydrothermal synthesis of multifunctional TiO_2-ZnO oxide systems with desired antibacterial and photocatalytic properties. *Appl. Surf. Sci.* **2019**, *463*, 791–801. [CrossRef]
75. Ashar, A.; Bhatti, I.A.; Siddique, T.; Ibrahim, S.M.; Mirza, S.; Bhutta, Z.A.; Shoaib, M.; Ali, M.; Taj, M.B.; Iqbal, M.; et al. Integrated hydrothermal assisted green synthesis of ZnO nano discs and their water purification efficiency together with antimicrobial activity. *J. Mater. Res. Technol.* **2021**, *15*, 6901–6917. [CrossRef]
76. Chithra, A.; Sekar, R.; Kumar, P.S.; Padmalaya, G.J.C. A review on removal strategies of microorganisms from water environment using nanomaterials and their behavioural characteristics. *Chemosphere* **2022**, *295*, 133915. [CrossRef] [PubMed]
77. Hassouna, M.; ElBably, M.; Mohammed, A.N.; Nasser, M.A.G. Assessment of carbon nanotubes and silver nanoparticles loaded clays as adsorbents for removal of bacterial contaminants from water sources. *J. Water Health* **2017**, *15*, 133–144. [CrossRef] [PubMed]
78. Kokkinos, P.; Mantzavinos, D.; Venieri, D.J.M. Current trends in the application of nanomaterials for the removal of emerging micropollutants and pathogens from water. *Molecules* **2020**, *25*, 2016. [CrossRef]
79. Raghunath, A.; Perumal, E. Metal oxide nanoparticles as antimicrobial agents: A promise for the future. *Int. J. Antimicrob. Agents* **2017**, *49*, 137–152. [CrossRef]
80. Soytaş, S.H.; Oğuz, O.; Menceloğlu, Y.Z. Polymer nanocomposites with decorated metal oxides. In *Polymer Composites with Functionalized Nanoparticles*; Elsevier: Amsterdam, The Netherlands, 2019; pp. 287–323.
81. Chen, G.-Q.; Wu, Y.-H.; Wang, Y.-H.; Chen, Z.; Tong, X.; Bai, Y.; Luo, L.-W.; Xu, C.; Hu, H.-Y. Effects of microbial inactivation approaches on quantity and properties of extracellular polymeric substances in the process of wastewater treatment and reclamation: A review. *J. Hazard. Mater.* **2021**, *413*, 125283. [CrossRef]
82. Jayaramudu, T.; Ko, H.-U.; Zhai, L.; Li, Y.; Kim, J. Preparation and characterization of hydrogels from polyvinyl alcohol and cellulose and their electroactive behavior. *Soft Mater.* **2017**, *15*, 64–72. [CrossRef]
83. Dayanidhi, K.; Vadivel, P.; Jothi, S.; Eusuff, N.S. Facile synthesis of Silver@ Eggshell nanocomposite: A heterogeneous catalyst for the removal of heavy metal ions, toxic dyes and microbial contaminants from water. *J. Environ. Manag.* **2020**, *271*, 110962. [CrossRef] [PubMed]
84. Li, J.-F.; Xu, Z.-L.; Yang, H.; Yu, L.-Y.; Liu, M. Effect of TiO_2 nanoparticles on the surface morphology and performance of microporous PES membrane. *Appl. Surf. Sci.* **2009**, *255*, 4725–4732. [CrossRef]
85. Suman; Kardam, A.; Gera, M.; Jain, V. A novel reusable nanocomposite for complete removal of dyes, heavy metals and microbial load from water based on nanocellulose and silver nano-embedded pebbles. *Environ. Technol.* **2015**, *36*, 706–714. [CrossRef]
86. Gao, F.; Hou, X.; Wang, A.; Chu, G.; Wu, W.; Chen, J.; Zou, H. Preparation of polypyrrole/TiO_2 nanocomposites with enhanced photocatalytic performance. *Particuology* **2016**, *26*, 73–78. [CrossRef]
87. Kalló, D. Applications of natural zeolites in water and wastewater treatment. *Rev. Miner. Geochem.* **2001**, *45*, 519–550. [CrossRef]

88. Inoue, Y.; Hoshino, M.; Takahashi, H.; Noguchi, T.; Murata, T.; Kanzaki, Y.; Hamashima, H.; Sasatsu, M. Bactericidal activity of Ag–zeolite mediated by reactive oxygen species under aerated conditions. *J. Inorg. Biochem.* **2002**, *92*, 37–42. [CrossRef]
89. Lin, S.; Huang, R.; Cheng, Y.; Liu, J.; Lau, B.L.; Wiesner, M.R. Silver nanoparticle-alginate composite beads for point-of-use drinking water disinfection. *Water Res.* **2013**, *47*, 3959–3965. [CrossRef]
90. Song, J.; Kong, H.; Jang, J. Adsorption of heavy metal ions from aqueous solution by polyrhodanine-encapsulated magnetic nanoparticles. *J. Colloid Interface Sci.* **2011**, *359*, 505–511. [CrossRef]
91. Bellanger, X.; Schneider, R.; Dezanet, C.; Arroua, B.; Balan, L.; Billard, P.; Merlin, C. Zn^{2+} leakage and photo-induced reactive oxidative species do not explain the full toxicity of ZnO core Quantum Dots. *J. Hazard. Mater.* **2020**, *396*, 122616. [CrossRef]
92. Elaaraj, I.; Raouan, S.E.R.; Nakkabi, A.; Es-sounni, B.; Koraichi, I.; El moualij, N.; Fahim, M. Synthesis, characterization and antioxidant, antibacterial activity Zn^{2+}, Cu^{2+}, Ni^{2+} and Co^{2+}, complexes of ligand [2-(thiophen-2-yl)-1-(thiophen-2-ylmethyl)-1H-benzo[d]imidazole]. *J. Indian Chem. Soc.* **2022**, *99*, 100404. [CrossRef]
93. Ramani, M.; Ponnusamy, S.; Muthamizhchelvan, C.; Cullen, J.; Krishnamurthy, S.; Marsili, E.J.C.; Biointerfaces, S.B. Morphology-directed synthesis of ZnO nanostructures and their antibacterial activity. *Colloids Surf. B Biointerfaces* **2013**, *105*, 24–30. [CrossRef] [PubMed]
94. Jain, A.; Bhargava, R.; Poddar, P. Probing interaction of Gram-positive and Gram-negative bacterial cells with ZnO nanorods. *Mater. Sci. Eng. C Mater. Biol. Appl.* **2013**, *33*, 1247–1253. [CrossRef] [PubMed]
95. Talebian, N.; Amininezhad, S.M.; Doudi, M. Controllable synthesis of ZnO nanoparticles and their morphology-dependent antibacterial and optical properties. *J. Photochem. Photobiol. B Biol.* **2013**, *120*, 66–73. [CrossRef] [PubMed]
96. Jansson, T.; Clare-Salzler, Z.J.; Zaveri, T.D.; Mehta, S.; Dolgova, N.V.; Chu, B.-H.; Ren, F.; Keselowsky, B.G. Antibacterial effects of zinc oxide nanorod surfaces. *J. Nanosci. Nanotechnol.* **2012**, *12*, 7132–7138. [CrossRef] [PubMed]
97. Elumalai, K.; Velmurugan, S.; Ravi, S.; Kathiravan, V.; Ashokkumar, S. RETRACTED: Facile, eco-friendly and template free photosynthesis of cauliflower like ZnO nanoparticles using leaf extract of *Tamarindus indica* (L.) and its biological evolution of antibacterial and antifungal activities. *Spectrochim. Part A Mol. Biomol. Spectrosc.* **2015**, *136*, 1052–1057. [CrossRef]
98. Wahab, R.; Kim, Y.-S.; Mishra, A.; Yun, S.-I.; Shin, H.-S. Formation of ZnO micro-flowers prepared via solution process and their antibacterial activity. *Nanoscale Res. Lett.* **2010**, *5*, 1675–1681. [CrossRef]
99. Kumar, K.M.; Mandal, B.K.; Naidu, E.A.; Sinha, M.; Kumar, K.S.; Reddy, P.S. Synthesis and characterisation of flower shaped zinc oxide nanostructures and its antimicrobial activity. *Spectrochim. Acta Part A-Mol. Biomol. Spectrosc.* **2013**, *104*, 171–174. [CrossRef]
100. Khan, M.F.; Hameedullah, M.; Ansari, A.H.; Ahmad, E.; Lohani, M.; Khan, R.H.; Alam, M.M.; Khan, W.; Husain, F.M.; Ahmad, I. Flower-shaped ZnO nanoparticles synthesized by a novel approach at near-room temperatures with antibacterial and antifungal properties. *Int. J. Nanomed.* **2014**, *9*, 853. [CrossRef]
101. Umar, A.; Chauhan, M.; Chauhan, S.; Kumar, R.; Sharma, P.; Tomar, K.J.; Wahab, R.; Al-Hajry, A.; Singh, D. Applications of ZnO nanoflowers as antimicrobial agents for Escherichia coli and enzyme-free glucose sensor. *J. Biomed. Nanotechnol.* **2013**, *9*, 1794–1802. [CrossRef]
102. Shinde, V.V.; Dalavi, D.S.; Mali, S.S.; Hong, C.K.; Kim, J.H.; Patil, P.S. Surfactant free microwave assisted synthesis of ZnO microspheres: Study of their antibacterial activity. *Appl. Surf. Sci.* **2014**, *307*, 495–502. [CrossRef]
103. Liu, Y.; He, L.; Mustapha, A.; Li, H.; Hu, Z.; Lin, M. Antibacterial activities of zinc oxide nanoparticles against *Escherichia coli* O157:H7. *J. Appl. Microbiol.* **2009**, *107*, 1193–1201. [CrossRef] [PubMed]
104. Abboud-Abi Saab, M.; Hassoun, A.E.R. Effects of organic pollution on environmental conditions and the phytoplankton community in the central Lebanese coastal waters with special attention to toxic algae. *Reg. Stud. Mar. Sci.* **2017**, *10*, 38–51. [CrossRef]
105. Yousef, J.M.; Danial, E.N. In Vitro Antibacterial Activity and Minimum Inhibitory Concentration of Zinc Oxide and Nano-particle Zinc oxide Against Pathogenic Strains. *Int. J. Health Sci.* **2012**, *2*, 38–42. [CrossRef]
106. Wahab, R.; Khan, F.; Lutfullah; Singh, R.; Khan, A. Enhance antimicrobial activity of ZnO nanomaterial's (QDs and NPs) and their analytical applications. *Phys. E: Low-Dimens. Syst. Nanostruct.* **2014**, *62*, 111–117. [CrossRef]
107. Carvalho, P.; Sampaio, P.; Azevedo, S.; Vaz, C.; Espinos, J.P.; Teixeira, V.; Carneiro, J. Influence of thickness and coatings morphology in the antimicrobial performance of zinc oxide coatings. *Appl. Surf. Sci.* **2014**, *307*, 548–557. [CrossRef]
108. Palanikumar, L.; Ramasamy, S.N.; Balachandran, C. Size-dependent antimicrobial response of zinc oxide nanoparticles. *IET Nanobiotechnol.* **2014**, *8*, 111–117. [CrossRef]
109. Dutta, R.K.; Nenavathu, B.P.; Gangishetty, M.K. Correlation between defects in capped ZnO nanoparticles and their antibacterial activity. *J. Photochem. Photobiol. B Biol.* **2013**, *126*, 105–111. [CrossRef]
110. Sharma, S.; Bhattacharya, A. Drinking water contamination and treatment techniques. *Appl. Water Sci.* **2017**, *7*, 1043–1067. [CrossRef]
111. Dinesha, B.; Sharanagouda, H.; Udaykumar, N.; Ramachandr, C.; Dandekar, A.B. Removal of Pollutants from Water/Waste Water Using Nano-Adsorbents: A Potential Pollution Mitigation. *Int. J. Curr. Microbiol. Appl. Sci.* **2017**, *6*, 4868–4872. [CrossRef]
112. Abu Hasan, H.; Muhammad, M.H.; Ismail, N. A review of biological drinking water treatment technologies for contaminants removal from polluted water resources. *J. Water Process Eng.* **2019**, *33*, 101035. [CrossRef]
113. Mukherjee, M.; De, S. Reduction of microbial contamination from drinking water using an iron oxide nanoparticle-impregnated ultrafiltration mixed matrix membrane: Preparation, characterization and antimicrobial properties. *Environ. Sci. Water Res. Technol.* **2015**, *1*, 204–217. [CrossRef]

114. Valbonesi, P.; Profita, M.; Vasumini, I.; Fabbri, E. Contaminants of emerging concern in drinking water: Quality assessment by combining chemical and biological analysis. *Sci. Total Environ.* **2020**, *758*, 143624. [CrossRef] [PubMed]
115. Hartmann, J.; van Driezum, I.; Ohana, D.; Lynch, G.; Berendsen, B.; Wuijts, S.; van der Hoek, J.P.; Husman, A.M.D.R. The effective design of sampling campaigns for emerging chemical and microbial contaminants in drinking water and its resources based on literature mining. *Sci. Total Environ.* **2020**, *742*, 140546. [CrossRef] [PubMed]
116. Sanaeepur, H.; Ebadi Amooghin, A.; Shirazi, M.M.A.; Pishnamazi, M.; Shirazian, S. Water desalination and ion removal using mixed matrix electrospun nanofibrous membranes: A critical review. *Desalination* **2022**, *521*, 115350. [CrossRef]
117. Ali, A.; Gul, A.; Mannan, A.; Zia, M. Efficient metal adsorption and microbial reduction from Rawal Lake wastewater using metal nanoparticle coated cotton. *Sci. Total Environ.* **2018**, *639*, 26–39. [CrossRef]
118. Ahmed, M.B.; Zhou, J.L.; Ngo, H.H.; Guo, W. Adsorptive removal of antibiotics from water and wastewater: Progress and challenges. *Sci. Total Environ.* **2015**, *532*, 112–126. [CrossRef]
119. Marcińczyk, M.; Ok, Y.S.; Oleszczuk, P. From waste to fertilizer: Nutrient recovery from wastewater by pristine and engineered biochars. *Chemosphere* **2022**, *306*, 135310. [CrossRef]
120. Zahmatkesh, S.; Far, S.S.; Sillanpää, M. RSM-D-optimal modeling approach for COD removal from low strength wastewater by microalgae, sludge, and activated carbon- case study Mashhad. *J. Hazard. Mater. Adv.* **2022**, *7*, 100110. [CrossRef]
121. Azam, K.; Shezad, N.; Shafiq, I.; Akhter, P.; Akhtar, F.; Jamil, F.; Shafique, S.; Park, Y.-K.; Hussain, M. A review on activated carbon modifications for the treatment of wastewater containing anionic dyes. *Chemosphere* **2022**, 135566. [CrossRef]
122. Srivastava, N.; Chattopadhyay, J. Effective utilization of biofiltration techniques for removal of pathogenic microorganisms from wastewater treatment plants. In *An Innovative Role of Biofiltration in Wastewater Treatment Plants (WWTPs)*; Shah, M., Rodriguez-Couto, S., Biswas, J., Eds.; Elsevier: Amsterdam, The Netherlands, 2022; pp. 207–216. [CrossRef]
123. Rani, M.; Paul, B.; Bhattacharjee, A.; Das, K.; Singh, P.; Basu, S.; Pandey, S.; Tripathi, D.; Kumar, A. Detection and removal of pathogenic bacteria from wastewater using various nanoparticles. In *Development in Wastewater Treatment Research and Processes*; Shah, M., Rodriguez-Couto, S., Biswas, J., Eds.; Elsevier: Amsterdam, The Netherlands, 2022; pp. 311–322. [CrossRef]
124. Rikta, S.Y. Application of nanoparticles for disinfection and microbial control of water and wastewater. *Nanotechnol. Water Wastewater Treat. Theory Appl.* **2019**, 159–176. [CrossRef]
125. Amin, M.; Alazba, A.; Manzoor, U. A review of removal of pollutants from water/wastewater using different types of nanomaterials. *Adv. Mater. Sci. Eng.* **2014**, *2014*, 825910. [CrossRef]
126. Priyadarshni, N.; Nath, P.; Chanda, N. Sustainable removal of arsenate, arsenite and bacterial contamination from water using biochar stabilized iron and copper oxide nanoparticles and associated mechanism of the remediation process. *J. Water Process Eng.* **2020**, *37*, 101495. [CrossRef]
127. Gebre, S.H.; Sendeku, M.G. New frontiers in the biosynthesis of metal oxide nanoparticles and their environmental applications: An overview. *Appl. Sci.* **2019**, *1*, 928. [CrossRef]
128. Borhani, S.; Asadi, A.; Dabbagh, H.A. Preparation and characterization of PAN nanofibers containing boehmite nanoparticles for the removal of microbial contaminants and cadmium ions from water. *J. Water Health* **2020**, *18*, 106–117. [CrossRef] [PubMed]
129. Fahimirad, S.; Fahimirad, Z.; Sillanpää, M. Efficient removal of water bacteria and viruses using electrospun nanofibers. *Sci. Total Environ.* **2021**, *751*, 141673. [CrossRef]
130. Rivera-Utrilla, J.; Bautista-Toledo, I.; Ferro-García, M.A.; Moreno-Castilla, C. Activated carbon surface modifications by adsorption of bacteria and their effect on aqueous lead adsorption. *J. Chem. Technol. Biotechnol.* **2001**, *76*, 1209–1215. [CrossRef]
131. Hassan, M.; Abou-Zeid, R.; Hassan, E.; Berglund, L.; Aitomäki, Y.; Oksman, K.J.P. Membranes based on cellulose nanofibers and activated carbon for removal of *Escherichia coli* bacteria from water. *Polymers* **2017**, *9*, 335. [CrossRef]
132. Hussain, S.; de Las Heras, N.; Asghar, H.; Brown, N.; Roberts, E. Disinfection of water by adsorption combined with electrochemical treatment. *Water Res.* **2014**, *54*, 170–178. [CrossRef]
133. Dubreuil, L.; Jehl, F.; Cattoen, C.; Bonnet, R.; Bru, J.P.; Caron, F.; Cattoir, V.; Courvalin, P.; Jarlier, V.; Lina, G.; et al. Improvement of a disk diffusion method for antibiotic susceptibility testing of anaerobic bacteria. French recommendations revisited for 2020. *Anaerobe* **2020**, *64*, 102213. [CrossRef]
134. Jan, T.; Azmat, S.; Rahman, A.U.; Ilyas, S.; Mehmood, A. Experimental and DFT study of Al doped ZnO nanoparticles with enhanced antibacterial activity. *Ceram. Int.* **2022**, *48*, 20838–20847. [CrossRef]
135. Chennimalai, M.; Vijayalakshmi, V.; Senthil, T.; Sivakumar, N. One-step green synthesis of ZnO nanoparticles using Opuntia humifusa fruit extract and their antibacterial activities. *Mater. Today Proc.* **2021**, *47*, 1842–1846. [CrossRef]
136. Pauzi, N.; Zain, N.M.; Kutty, R.V.; Ramli, H. Antibacterial and antibiofilm properties of ZnO nanoparticles synthesis using gum arabic as a potential new generation antibacterial agent. *Mater. Today Proc.* **2020**, *41*, 1–8. [CrossRef]
137. Ramesh, M.; Anbuvannan, M.; Viruthagiri, G. Green synthesis of ZnO nanoparticles using Solanum nigrum leaf extract and their antibacterial activity. *Spectrochim. Acta Part A Mol. Biomol. Spectrosc.* **2015**, *136*, 864–870. [CrossRef] [PubMed]
138. Huang, Z.; Zheng, X.; Yan, D.; Yin, G.; Liao, X.; Kang, Y.; Yao, Y.; Huang, D.; Hao, B. Toxicological Effect of ZnO Nanoparticles Based on Bacteria. *Langmuir* **2008**, *24*, 4140–4144. [CrossRef]
139. Suresh, D.; Udayabhanu; Nethravathi, P.; Lingaraju, K.; Rajanaika, H.; Sharma, S.; Nagabhushana, H. EGCG assisted green synthesis of ZnO nanopowders: Photodegradative, antimicrobial and antioxidant activities. *Spectrochim. Acta Part A Mol. Biomol. Spectrosc.* **2015**, *136*, 1467–1474. [CrossRef]

140. Roy, S.; Barua, N.; Buragohain, A.K.; Ahmed, G.A. Study of ZnO nanoparticles: Antibacterial property and light depolarization property using light scattering tool. *J. Quant. Spectrosc. Radiat. Transf.* **2013**, *118*, 8–13. [CrossRef]
141. Ambika, S.; Sundrarajan, M. Antibacterial behaviour of Vitex negundo extract assisted ZnO nanoparticles against pathogenic bacteria. *J. Photochem. Photobiol. B Biol.* **2015**, *146*, 52–57. [CrossRef]
142. Gudkov, S.V.; Burmistrov, D.E.; Serov, D.A.; Rebezov, M.B.; Semenova, A.A.; Lisitsyn, A.B. A Mini Review of Antibacterial Properties of ZnO Nanoparticles. *Front. Phys.* **2021**, *9*, 641481. [CrossRef]
143. Kairyte, K.; Kadys, A.; Luksiene, Z. Antibacterial and antifungal activity of photoactivated ZnO nanoparticles in suspension. *J. Photochem. Photobiol. B Biol.* **2013**, *128*, 78–84. [CrossRef]
144. Raghupathi, K.R.; Koodali, R.T.; Manna, A.C. Size-Dependent Bacterial Growth Inhibition and Mechanism of Antibacterial Activity of Zinc Oxide Nanoparticles. *Langmuir* **2011**, *27*, 4020–4028. [CrossRef]
145. Jaber, G.S.; Khashan, K.S.; Abbas, M.J. Study the antibacterial activity of zinc oxide nanoparticles synthesis by laser ablation in liquid. *Mater. Today Proc.* **2021**, *42*, 2668–2673. [CrossRef]
146. Rao, S.M.; Kotteeswaran, S.; Visagamani, A.M. Green synthesis of zinc oxide nanoparticles from camellia sinensis: Organic dye degradation and antibacterial activity. *Inorg. Chem. Commun.* **2021**, *134*, 108956. [CrossRef]
147. Nallal, V.U.; Prabha, K.; Muthupandi, S.; Razia, M. Synergistic antibacterial potential of plant-based zinc oxide nanoparticles in combination with antibiotics against Pseudomonas aeruginosa. *Mater. Proc.* **2022**, *49*, 2632–2635. [CrossRef]
148. Ng, Y.; Leung, Y.; Liu, F.; Ng, A.; Gao, M.; Chan, C.; Djurišić, A.; Leung, F.; Chan, W. Antibacterial activity of ZnO nanoparticles under ambient illumination—The effect of nanoparticle properties. *Thin Solid Films* **2013**, *542*, 368–372. [CrossRef]
149. Weldegebrieal, G.K. Synthesis method, antibacterial and photocatalytic activity of ZnO nanoparticles for azo dyes in wastewater treatment: A review. *Inorg. Chem. Commun.* **2020**, *120*, 108140. [CrossRef]
150. Dyshlyuk, L.; Babich, O.; Ivanova, S.; Vasilchenco, N.; Prosekov, A.; Sukhikh, S. Suspensions of metal nanoparticles as a basis for protection of internal surfaces of building structures from biodegradation. *Case Stud. Constr. Mater.* **2019**, *12*, e00319. [CrossRef]
151. Zudyte, B.; Luksiene, Z. Visible light-activated ZnO nanoparticles for microbial control of wheat crop. *J. Photochem. Photobiol. B Biol.* **2021**, *219*, 112206. [CrossRef]
152. Stanković, A.; Dimitrijević, S.; Uskoković, D. Influence of size scale and morphology on antibacterial properties of ZnO powders hydrothemally synthesized using different surface stabilizing agents. *Colloids Surfaces B Biointerfaces* **2013**, *102*, 21–28. [CrossRef]
153. Premanathan, M.; Karthikeyan, K.; Jeyasubramanian, K.; Manivannan, G. Selective toxicity of ZnO nanoparticles toward Gram-positive bacteria and cancer cells by apoptosis through lipid peroxidation. *Nanomed.-Nanotechnol. Biol. Med.* **2011**, *7*, 184–192. [CrossRef]
154. Papavlassopoulos, H.; Mishra, Y.K.; Kaps, S.; Paulowicz, I.; Abdelaziz, R.; Elbahri, M.; Maser, E.; Adelung, R.; Röhl, C. Toxicity of Functional Nano-Micro Zinc Oxide Tetrapods: Impact of Cell Culture Conditions, Cellular Age and Material Properties. *PLoS ONE* **2014**, *9*, e84983. [CrossRef]
155. Gupta, S.M.; Tripathi, M. A review of TiO_2 nanoparticles. *Chin. Sci. Bull.* **2011**, *56*, 1639–1657. [CrossRef]
156. Yu, M.; Ma, Y.; Liu, J.; Li, X.; Li, S.; Liu, S. Sub-coherent growth of ZnO nanorod arrays on three-dimensional graphene framework as one-bulk high-performance photocatalyst. *Appl. Surf. Sci.* **2016**, *390*, 266–272. [CrossRef]
157. Bera, S.; Pal, M.; Naskar, A.; Jana, S. J Hierarchically structured ZnO-graphene hollow microspheres towards effective reusable adsorbent for organic pollutant via photodegradation process. *J. Alloys Compd.* **2016**, *669*, 177–186. [CrossRef]
158. Lakshmi Prasanna, V.; Vijayaraghavan, R. Insight into the mechanism of antibacterial activity of ZnO: Surface defects mediated reactive oxygen species even in the dark. *Langmuir* **2015**, *31*, 9155–9162. [CrossRef] [PubMed]
159. Pasquet, J.; Chevalier, Y.; Pelletier, J.; Couval, E.; Bouvier, D.; Bolzinger, M.-A.J.C.; Physicochemical, S.A.; Aspects, E. The contribution of zinc ions to the antimicrobial activity of zinc oxide. *Colloids Surf. A Physicochem. Eng. Asp.* **2014**, *457*, 263–274. [CrossRef]
160. Song, W.; Zhang, J.; Guo, J.; Zhang, J.; Ding, F.; Li, L.; Sun, Z. Role of the dissolved zinc ion and reactive oxygen species in cytotoxicity of ZnO nanoparticles. *Toxicol. Lett.* **2010**, *199*, 389–397. [CrossRef] [PubMed]
161. Zhang, L.; Jiang, Y.; Ding, Y.; Povey, M.; York, D. Investigation into the antibacterial behaviour of suspensions of ZnO nanoparticles (ZnO nanofluids). *J. Nanopart. Res.* **2007**, *9*, 479–489. [CrossRef]
162. Sawai, J.; Shoji, S.; Igarashi, H.; Hashimoto, A.; Kokugan, T.; Shimizu, M.; Kojima, H. Hydrogen peroxide as an antibacterial factor in zinc oxide powder slurry. *J. Ferment. Bioeng.* **1998**, *86*, 521–522. [CrossRef]
163. Stoimenov, P.K.; Klinger, R.L.; Marchin, G.L.; Klabunde, K. Metal oxide nanoparticles as bactericidal agents. *Langmuir* **2002**, *18*, 6679–6686. [CrossRef]
164. Choi, O.; Deng, K.K.; Kim, N.-J.; Ross Jr, L.; Surampalli, R.Y.; Hu, Z. The inhibitory effects of silver nanoparticles, silver ions, and silver chloride colloids on microbial growth. *Water Res.* **2008**, *42*, 3066–3074. [CrossRef]
165. Padmavathy, N.; Vijayaraghavan, R. Enhanced bioactivity of ZnO nanoparticles—An antimicrobial study. *Sci. Technol. Adv. Mater.* **2008**, *9*, 035004. [CrossRef] [PubMed]
166. Sharma, V.K.; Yngard, R.A.; Lin, Y. Silver nanoparticles: Green synthesis and their antimicrobial activities. *Adv. Colloid Interface Sci.* **2009**, *145*, 83–96. [CrossRef] [PubMed]
167. Lok, C.-N.; Ho, C.-M.; Chen, R.; He, Q.-Y.; Yu, W.-Y.; Sun, H.; Tam, P.K.-H.; Chiu, J.-F.; Che, C.-M. Proteomic analysis of the mode of antibacterial action of silver nanoparticles. *J. Proteome Res.* **2006**, *5*, 916–924. [CrossRef] [PubMed]
168. Rai, M.; Yadav, A.; Gade, A. Silver nanoparticles as a new generation of antimicrobials. *Biotechnol. Adv.* **2009**, *27*, 76–83. [CrossRef]

169. Alabresm, A.; Decho, A.W.; Lead, J. A novel method to estimate cellular internalization of nanoparticles into gram-negative bacteria: Non-lytic removal of outer membrane and cell wall. *Nanoimpact* **2021**, *21*, 100283. [CrossRef]
170. Liu, L.; Xu, K.; Zhang, B.; Ye, Y.; Zhang, Q.; Jiang, W. Cellular internalization and release of polystyrene microplastics and nanoplastics. *Sci. Total Environ.* **2021**, *779*, 146523. [CrossRef]
171. Zhao, Z.; Yang, S.; Yang, P.; Lin, J.; Fan, J.; Zhang, B. Study of oxygen-deficient W18O49-based drug delivery system readily absorbed through cellular internalization pathways in tumor-targeted chemo-/photothermal therapy. *Biomater. Adv.* **2022**, *136*, 212772. [CrossRef]
172. Tam, K.H.; Djurišić, A.B.; Chan, C.M.N.; Xi, Y.Y.; Tse, C.W.; Leung, Y.H.; Chan, W.K.; Leung, F.C.C.; Au, D.W.T. Antibacterial activity of ZnO nanorods prepared by a hydrothermal method. *Thin Solid Film.* **2008**, *516*, 6167–6174. [CrossRef]
173. Sun, J.e.; Cai, S.; Li, Q.; Li, Z.; Xu, G. UV-irradiation induced biological activity and antibacterial activity of ZnO coated magnesium alloy. *Mater. Sci. Eng. C* **2020**, *114*, 110997. [CrossRef]
174. Joe, A.; Park, S.-H.; Kim, D.-J.; Lee, Y.-J.; Jhee, K.-H.; Sohn, Y.; Jang, E.-S. Antimicrobial activity of ZnO nanoplates and its Ag nanocomposites: Insight into an ROS-mediated antibacterial mechanism under UV light. *J. Solid State Chem.* **2018**, *267*, 124–133. [CrossRef]
175. Sroila, W.; Kantarak, E.; Kumpika, T.; Bovonsombut, S.; Thongmanee, P.; Singjai, P.; Thongsuwan, W. Antibacterial activity absence UV irradiation of Ag, TiO_2 and ZnO NPs prepared by sparking method. *Mater. Today Proc.* **2019**, *17*, 1569–1574. [CrossRef]
176. Ma, W.; Li, L.; Liu, Y.; Sun, Y.; Kim, I.S.; Ren, X. Tailored assembly of vinylbenzyl N-halamine with end-activated ZnO to form hybrid nanoparticles for quick antibacterial response and enhanced UV stability. *J. Alloy. Compd.* **2019**, *797*, 692–701. [CrossRef]
177. Nenavathu, B.P.; Kandula, S.; Verma, S. Visible-light-driven photocatalytic degradation of safranin-T dye using functionalized graphene oxide nanosheet (FGS)/ZnO nanocomposites. *RSC Adv.* **2018**, *8*, 19659–19667. [CrossRef] [PubMed]
178. Sudrajat, H.; Babel, S. A novel visible light active N-doped ZnO for photocatalytic degradation of dyes. *J. Water Process Eng.* **2017**, *16*, 309–318. [CrossRef]
179. Jin, Y.; Long, J.; Ma, X.; Zhou, T.; Zhang, Z.; Lin, H.; Long, J.; Wang, X. Synthesis of caged iodine-modified ZnO nanomaterials and study on their visible light photocatalytic antibacterial properties. *Appl. Catal. B Environ.* **2019**, *256*, 117873. [CrossRef]
180. Liang, Y.; Li, W.; Wang, X.; Zhou, R.; Ding, H. TiO_2–ZnO/Au ternary heterojunction nanocomposite: Excellent antibacterial property and visible-light photocatalytic hydrogen production efficiency. *Ceram. Int.* **2022**, *48*, 2826–2832. [CrossRef]
181. Chen, L.; Chuang, Y.; Chen, C.-W.; Dong, C.-D. Facile synthesis of MoS2/ZnO quantum dots for enhanced visible-light photocatalytic performance and antibacterial applications. *Nano-Struct. Nano-Objects* **2022**, *30*, 100873. [CrossRef]
182. Yang, Y.; Wu, Z.; Yang, R.; Li, Y.; Liu, X.; Zhang, L.; Yu, B. Insights into the mechanism of enhanced photocatalytic dye degradation and antibacterial activity over ternary ZnO/ZnSe/MoSe2 photocatalysts under visible light irradiation. *Appl. Surf. Sci.* **2021**, *539*, 148220. [CrossRef]
183. Xin, Z.; Wang, S.; He, Q.; Han, X.; Fu, Z.; Xu, X.; Zhao, X. Preparation of a novel photocatalytic catalyst PW9@ZnO/Ag and the photocatalytic degradation of butyl xanthate under visible light. *Environ. Res.* **2022**, *214*, 113776. [CrossRef]
184. Ye, W.; Jiang, Y.; Liu, Q.; Xu, D.; Zhang, E.; Cheng, X.; Wan, Z.; Liu, C. The preparation of visible light-driven ZnO/Ag_2MoO_4/Ag nanocomposites with effective photocatalytic and antibacterial activity. *J. Alloys Compd.* **2022**, *891*, 161898. [CrossRef]
185. Samuel, J.; Suresh, S.; Shabna, S.; Sherlin Vinita, V.; Joslin Ananth, N.; Shajin Shinu, P.M.; Mariappan, A.; simon, T.; Samson, Y.; Biju, C.S. Characterization and antibacterial activity of Ti doped ZnO nanorods prepared by hydrazine assisted wet chemical route. *Phys. E Low-Dimens. Syst. Nanostruct.* **2022**, *143*, 115374. [CrossRef]
186. Shkir, M.; Al-Shehri, B.M.; Pachamuthu, M.; Khan, A.; Chandekar, K.V.; AlFaify, S.; Hamdy, M.S. A remarkable improvement in photocatalytic activity of ZnO nanoparticles through Sr doping synthesized by one pot flash combustion technique for water treatments. *Colloids Surf. A-Physicochem. Eng. Asp.* **2020**, *587*, 124340. [CrossRef]
187. Kumaran, N.N.; Muraleedharan, K. Photocatalytic activity of ZnO and Sr^{2+} doped ZnO nanoparticles. *J. Water Process Eng.* **2017**, *17*, 264–270. [CrossRef]
188. Okeke, I.; Agwu, K.; Ubachukwu, A.; Madiba, I.; Maaza, M.; Whyte, G.; Ezema, F. Impact of particle size and surface defects on antibacterial and photocatalytic activities of undoped and Mg-doped ZnO nanoparticles, biosynthesized using one-step simple process. *Vacuum* **2021**, *187*, 110110. [CrossRef]
189. Grace, A.N.; Pandian, K. Antibacterial efficacy of aminoglycosidic antibiotics protected gold nanoparticles—A brief study. *Colloids Surf. A Physicochem. Eng. Asp.* **2007**, *297*, 63–70. [CrossRef]
190. Ma, Q.; Yang, X.; Lv, X.; Jia, H.; Wang, Y. Cu doped ZnO hierarchical nanostructures: Morphological evolution and photocatalytic property. *J. Mater. Sci.-Mater. Electron.* **2019**, *30*, 2309–2315. [CrossRef]
191. Kuriakose, S.; Satpati, B.; Mohapatra, S. Highly efficient photocatalytic degradation of organic dyes by Cu doped ZnO nanostructures. *Phys. Chem. Chem. Phys.* **2015**, *17*, 25172–25181. [CrossRef]
192. Chandramouli, K.; Suryanarayana, B.; Babu, T.A.; Raghavendra, V.; Parajuli, D.; Murali, N.; Malapati, V.; Mammo, T.W.; Shanmukhi, P.; Gudla, U.R.J.S.; et al. Synthesis, structural and antibacterial activity of pure, Fe doped, and glucose capped ZnO nanoparticles. *Surf. Interfaces* **2021**, *26*, 101327. [CrossRef]
193. Moussa, D.; Bakeer, D.E.-S.; Awad, R.; Abdel-Gaber, A. Physical properties of ZnO nanoparticles doped with Mn and Fe. In *Journal of Physics: Conference Series*; IOP Publishing: Bristol, UK, 2017; p. 012021.
194. Ahmad, M.; Ahmed, E.; Zafar, F.; Khalid, N.; Niaz, N.; Hafeez, A.; Ikram, M.; Khan, M.A.; Zhanglian, H. Enhanced photocatalytic activity of Ce-doped ZnO nanopowders synthesized by combustion method. *J. Rare Earths* **2015**, *33*, 255–262. [CrossRef]

195. Mohammadzadeh Kakhki, R.; Tayebee, R.; Ahsani, F. New and highly efficient Ag doped ZnO visible nano photocatalyst for removing of methylene blue. *J. Mater. Sci.-Mater. Electron.* **2017**, *28*, 5941–5952. [CrossRef]
196. Raza, W.; Faisal, S.M.; Owais, M.; Bahnemann, D.; Muneer, M. Facile fabrication of highly efficient modified ZnO photocatalyst with enhanced photocatalytic, antibacterial and anticancer activity. *RSC Adv.* **2016**, *6*, 78335–78350. [CrossRef]
197. Weber, A.S.; Grady, A.M.; Koodali, R.T. Lanthanide modified semiconductor photocatalysts. *Catal. Sci. Technol.* **2012**, *2*, 683–693. [CrossRef]
198. Daksh, D.; Agrawal, Y.K. Rare earth-doped zinc oxide nanostructures: A review. *Rev. Nanosci. Nanotechnol.* **2016**, *5*, 1–27. [CrossRef]
199. Mazierski, P.; Mikolajczyk, A.; Bajorowicz, B.; Malankowska, A.; Zaleska-Medynska, A.; Nadolna, J.J.A.C.B.E. The role of lanthanides in TiO_2-based photocatalysis: A review. *Appl. Catal. B-Environ.* **2018**, *233*, 301–317. [CrossRef]
200. Verma, K.; Goyal, N.; Kotnala, R. Lattice defect-formulated ferromagnetism and UV photo-response in pure and Nd, Sm substituted ZnO thin films. *Phys. Chem. Chem. Phys.* **2019**, *21*, 12540–12554. [CrossRef]
201. Vakili, B.; Shahmoradi, B.; Maleki, A.; Safari, M.; Yang, J.; Pawar, R.R.; Lee, S.-M. Synthesis of immobilized cerium doped ZnO nanoparticles through the mild hydrothermal approach and their application in the photodegradation of synthetic wastewater. *J. Mol. Liq.* **2019**, *280*, 230–237. [CrossRef]
202. Yuan, C.; Cui, W.; Sun, Y.; Wang, J.; Chen, R.; Zhang, J.; Zhang, Y.; Dong, F. Inhibition of the toxic byproduct during photocatalytic NO oxidation via La doping in ZnO. *Chin. Chem. Lett.* **2020**, *31*, 751–754. [CrossRef]
203. Munawar, T.; Mukhtar, F.; Nadeem, M.S.; Riaz, M.; ur Rahman, M.N.; Mahmood, K.; Hasan, M.; Arshad, M.; Hussain, F.; Hussain, A. Novel photocatalyst and antibacterial agent; direct dual Z-scheme ZnO–CeO_2-Yb_2O_3 heterostructured nanocomposite. *Solid State Sci.* **2020**, *109*, 106446. [CrossRef]
204. Gutha, Y.; Pathak, J.L.; Zhang, W.; Zhang, Y.; Jiao, X. Antibacterial and wound healing properties of chitosan/poly(vinyl alcohol)/zinc oxide beads (CS/PVA/ZnO). *Int. J. Biol. Macromol.* **2017**, *103*, 234–241. [CrossRef]
205. Bai, H.; Liu, Z.; Sun, D.D. Hierarchical ZnO/Cu "corn-like" materials with high photodegradation and antibacterial capability under visible light. *Phys. Chem. Chem. Phys.* **2011**, *13*, 6205–6210. [CrossRef]
206. Sudheesh Kumar, P.; Lakshmanan, V.-K.; Anilkumar, T.; Ramya, C.; Reshmi, P.; Unnikrishnan, A.; Nair, S.V.; Jayakumar, R. Flexible and microporous chitosan hydrogel/nano ZnO composite bandages for wound dressing: In vitro and in vivo evaluation. *Appl. Mater. Interfaces* **2012**, *4*, 2618–2629. [CrossRef] [PubMed]
207. Sudarshan, N.; Hoover, D.; Knorr, D. Antibacterial action of chitosan. *Food Biotechnol.* **1992**, *6*, 257–272. [CrossRef]
208. Tsai, G.-J.; Su, W.-H. Antibacterial activity of shrimp chitosan against *Escherichia coli*. *J. Food Prot.* **1999**, *62*, 239–243. [CrossRef] [PubMed]
209. Shin, Y.; Yoo, D.; Jang, J. Molecular weight effect on antimicrobial activity of chitosan treated cotton fabrics. *J. Appl. Polym. Sci.* **2001**, *80*, 2495–2501. [CrossRef]
210. Fei Liu, X.; Lin Guan, Y.; Zhi Yang, D.; Li, Z.; De Yao, K. Antibacterial action of chitosan and carboxymethylated chitosan. *J. Appl. Polym. Sci.* **2001**, *79*, 1324–1335. [CrossRef]

Review

A Review of Treatment Techniques for Short-Chain Perfluoroalkyl Substances

Yang Liu [1,*], Tingyu Li [1], Jia Bao [1,*], Xiaomin Hu [2], Xin Zhao [2], Lixin Shao [1], Chenglong Li [1] and Mengyuan Lu [1]

[1] School of Environmental and Chemical Engineering, Shenyang University of Technology, Shenyang 110870, China; LITingyu990109@163.com (T.L.); shaolixin0503@163.com (L.S.); lcl960315@163.com (C.L.); lumengyuan2022@163.com (M.L.)

[2] School of Resources and Civil Engineering, Northeastern University, Shenyang 110819, China; hxmin_jj@163.com (X.H.); zhaoxin@mail.neu.edu.cn (X.Z.)

* Correspondence: liuyang@sut.edu.cn (Y.L.); baojia@sut.edu.cn (J.B.)

Abstract: In recent years, an increasing amount of short-chain perfluoroalkyl substance (PFAS) alternatives has been used in industrial and commercial products. However, short-chain PFASs remain persistent, potentially toxic, and extremely mobile, posing potential threats to human health because of their widespread pollution and accumulation in the water cycle. This study systematically summarized the removal effect, operation conditions, treating time, and removal mechanism of various low carbon treatment techniques for short-chain PFASs, involving adsorption, advanced oxidation, and other practices. By the comparison of applicability, pros, and cons, as well as bottlenecks and development trends, the most widely used and effective method was adsorption, which could eliminate short-chain PFASs with a broad range of concentrations and meet the low-carbon policy, although the adsorbent regeneration was undesirable. In addition, advanced oxidation techniques could degrade short-chain PFASs with low energy consumption but unsatisfied mineralization rates. Therefore, combined with the actual situation, it is urgent to enhance and upgrade the water treatment techniques to improve the treatment efficiency of short-chain PFASs, for providing a scientific basis for the effective treatment of PFASs pollution in water bodies globally.

Keywords: low carbon; short-chain PFASs; water treatment; adsorption; advanced oxidation

Citation: Liu, Y.; Li, T.; Bao, J.; Hu, X.; Zhao, X.; Shao, L.; Li, C.; Lu, M. A Review of Treatment Techniques for Short-Chain Perfluoroalkyl Substances. *Appl. Sci.* **2022**, *12*, 1941. https://doi.org/10.3390/app12041941

Academic Editor: Dino Musmarra

Received: 18 January 2022
Accepted: 9 February 2022
Published: 12 February 2022

1. Introduction

Since the 1950s, perfluoroalkyl substances (PFASs) have been widely used in industrial production and commercial products, involving chrome plating, foam extinguishing agents, aviation hydraulic oil, and food packaging paper [1,2]. They are a class of man-made chemicals with all the hydrogen atoms on the carbon skeleton replaced by fluorine atoms, together with a terminal functional group [3]. Abbreviations for different PFASs are shown in Table 1. Due to the strong energy of the C-F bond (536 kJ/mol), PFASs possess exclusive physio-chemical characteristics, including environmental persistence, extraordinary resistance to both environmental and biological degradation, high thermal and chemical stability against oxidation, photolysis, and hydrolysis reactions, hydrophobicity and oleophobicity, as well as multiple toxicities [4]. Moreover, many PFAS (precursors) can easily degrade into persistent PFAS (acids). Therefore, long-chain PFASs (C8-C14) and their sodium, as well as ammonium, salts were added into the candidate list of regulatory substances in the EU, and PFOA and PFOS were added in the Stockholm Convention on Persistent Organic Pollutants (POPs) list [5]. With the ban of long-chain PFASs, short-chain PFASs (PFCAs < C8, PFSAs < C7) have been produced and used as substitutes in large quantities.

With the improvement of modern analytical techniques such as high-resolution mass spectrometry in non-target and suspect screening approaches in recent years, the researchers found that long-chain PFASs tended to be adsorbed by solid matter (soil, sediment, etc.), which made them less mobile. However, short-chain PFASs showed a high polarity and solubility, which rendered them difficult to be removed through environmental adsorption and water treatment processes, thus contributing to long-term mobility in the water cycle through migration of surface water, groundwater, and natural and urban water systems [6–9]. However, short-chain PFASs show similar properties with long-chain congeners, including being persistent, bioaccumulative, and toxic to a certain extent [10]. Therefore, short-chain PFASs could be classified as persistent and mobile organic compounds (PMOCs) [6].

Table 1. Abbreviation for different PFASs.

Abbreviation			
TFA	Trifluoroacetic acid	GenX	2,3,3,3-Tetrafluoro-2-(1,1,2,2,3,3,3-heptafluoropropoxy) propanoic acid
PFSAs	Perfluoroalkane sulfonates	PFCAs	Perfluoroalkyl carboxylic acids
PFBS	Perfluorobutane sulfonic acid	PFBA	Perfluorobutanoic acid
PFPrA	Pentafluoropropionic acid	PFPeA	Perfluoropentanoic acid
PFHxS	Perfluorohexane sulfonic acid	PFHxA	Perfluorohexanoic acid
PFOS	Perfluorooctane sulfonic acid	PFOA	Perfluorooctanoic acid
PFHpA	Perfluoroheptanoic acid	PFNA	Perfluorononanoic acid
PFDA	Perfluorodecanoic acid	PFDoA	Perfluorododecanoic acid

Large quantity usage of short-chain PFASs could lead to accumulation in the water cycle and water pollution, thereafter threatening drinking water quality. For instance, the concentrations of PFBS and PFBA in Tangxun Lake of Wuhan in China reach up to 3.66 μg/L and 4.77 μg/L, respectively [11]. There was a serious PFBS contamination in groundwater around fluorochemical plants in Fuxin in China, with a concentration up to 31 μg/L [12], exceeding 10 times the health risk limits (HRLs) in drinking water (3 μg/L) issued by the Minnesota Department of Health (MDH). Thus, it is an urgent incident to effectively develop suitable water treatment techniques to regulate short-chain PFAS contaminations in waters.

At present, activated carbon (AC) adsorption and ozonation are the commonly applied techniques for organic contaminant elimination in drinking water treatment. However, AC presented low efficiency for very polar compounds [6]. Meanwhile, ozonation generally exhibited poor reactive activity to polar compounds containing acidic functional groups [13], which might be a source of smaller and more polar by-products than the parent compounds [14]. Consequently, the removal efficiency of conventional treatment for short-chain PFASs was undesirable. Moreover, various techniques presented different removal efficiencies for short-chain PFASs, even requiring high energy consumption, strict operating conditions, and releasing a large number of by-pollutants [15].

Recently, the C40 Cities Network of 91 large cities committed to low carbon infrastructure to ensure carbon emissions peak by 2020 and almost halve by 2030 [16–18]. In 2020, China announced that it would take more aggressive policy measures to achieve peak carbon dioxide emissions by 2030 and achieve carbon neutrality by 2060 [19]. Specifically, the regulation instruments include applying forced power to administer high energy consumption and emission of pollutants [20,21]. Therefore, it is significant to explore effective low carbon treatment techniques to eliminate short-chain PFAS contaminations under convenient conditions.

This review aims to select suitable treatment techniques for short-chain PFASs. To achieve this aim, the low carbon treatment techniques for PFASs involving adsorption, electrochemical oxidation, photocatalytic oxidation, membrane separation, pyrolysis, and ultrasonic chemical degradation, and their individual removal efficiency, operating conditions, and removal mechanisms were systematically summarized. Thereafter, based upon the characteristics of short-chain PFASs, the suitable treatment techniques were determined by the comparison of the applicability, as well as the pros and cons of various techniques. This will provide a scientific basis for the effective treatment and regulation of short-chain PFAS contaminations in different waters.

2. Methodology of Literature Sources

To obtain an overview of short-chain PFAS chemical usage, the present review initially focused on the risk profiles and risk management assessments. Reports that addressed fluorosurfactants and fluoropolymers were also involved. Literature related to certain use categories was retrieved for more information either on the substances used, or to understand why PFAS are, or were, necessary for a given use.

In addition, databases, patents, information from PFAS manufacturers, and scientific studies were examined via "Web of Science", "PubMed", and "CNKI". The retrieved keywords involved but were not limited to "per- and polyfluoroalkyl substances", "PFAS", "PFAAs", "short-chain PFAS", "water treatment", "adsorption", "anion-exchange", "advanced oxidation", "persistent organic pollutant", "emerging contaminant", "low carbon". The searches were not exhaustive in any of the sources described, and there are still many more reports, scientific studies, patents, safety data sheets, and databases with information on the usage of PFASs than the ones cited here [22].

3. Treatment Techniques for Short-Chain PFASs

3.1. Adsorption Technique

Low carbon technique of adsorption could remove contaminations effectively in the waters, which were widely applied in the treatment of PFASs. This technique uses a porous solid as the adsorbent to adsorb one or several contaminations in the wastewater (WW) that does not change physicochemical property, thus achieving the purification purpose [4]. The commonly used adsorption materials mainly include carbons, anion-exchange resins, flocculants, etc. The key influencing factors of adsorption efficiency involved ionic strength, pH, organic matter (OM) concentration, physicochemical properties of PFASs in solution, and adsorbent characteristics (such as particle size, porosity rate, and functional group on surface). Comparisons on the adsorption capacities of different adsorbents for short-chain PFASs are shown in Table 2.

3.1.1. Adsorption by Carbon-Based Adsorbents

AC adsorption. AC is widely used to remove contaminants in wastewater due to low cost, high efficiency, and convenient operation [4]. Ochoa-Herrera et al. adopted granular activated carbon (GAC) to adsorb PFBS, with an adsorption capacity of 98.7 mg/g [23]. Hansen et al. used GAC and powdered activated carbon (PAC) to carry out batch adsorption experiments for short-chain PFASs; PAC achieved higher removal rates within 10 min (20–40% for GAC, 60–90% for PAC) due to shorter internal diffusion distance and higher BET surface area (S_{BET}) of PAC [24]. This phenomenon demonstrated that the particle size of AC was a significant factor affecting the removal efficiency of short-chain PFASs, and smaller particle sizes presented superior removal efficiency. The effect of pore size on the removal efficiency of bamboo-derived AC (BdAC) and coal-based AC (microporous type) for PFHxA were investigated, revealing that the BdAC adsorption capacity was 13-fold lower than microporous ACs [25,26]. In addition, the GAC was explored to eliminate different carbon chain length PFASs [27], the removal efficiency of short-chain PFCAs was lower than congeners PFSAs, in which the removal rates of PFBA, PFPeA, PFHxA, and PFHpA were all below 19%, especially 10% for PFBS. Furthermore, the regeneration capacity of adsorbed AC was also inferior when eluting. Therefore, it is necessary to explore new catalysts to modify AC for promoting adsorption and regeneration capacity [1].

Biocarbon adsorption. Biocarbon is a pollution-free solid biofuel produced by pyrolysis of biomass under aerobic or anaerobic conditions, which contains abundant voids, high carbon content, and high calorific value. The nature of biocarbon is different due to various raw materials and parameters in the production process [28]. Inyang and Dickenson [29] explored the adsorption capacity of hardwood biochar (HWC) and pinewood biochar (PWC) for PFBA and PFOA, demonstrating that the PFBA adsorption capacity was 3–4 times lower than PFOA. Meanwhile, the removal efficiency of HWC vaporized at 900 °C could be improved due to high S_{BET}. However, the batch adsorption kinetics experiments showed that the removal efficiency of GAC for PFBA was superior to HWC, indicating a low adsorption capacity of biocarbon for short-chain PFASs.

Carbon nanotube (CNTs) adsorption. CNTs have the advantages of easy reaction process control, convenient operation, and low cost of raw material. Deng et al. [30] used single-walled carbon nanotubes (SWCNT) and multi-walled carbon nanotubes (MWCNT) to remove short-chain PFASs (PFBA, PFHxA, PFHpA, PFBA, and PFBS) and long-chain PFASs (PFOA, PFHxS, and PFOS) under neutral conditions. It showed that 95% of PFOS and PFOA were removed by SWCNT within 5 h, but only 7.5% of PFBA was eliminated by SWCNT within 48 h, and PFSAs were more easily adsorbed than PFCAs. Moreover, in the comparison study of MWCNT functional groups, owing to the deprotonation that occurred on the carboxyl and hydroxyl functional groups, contributing to lower hydrophobicity and more negative surface potential, the adsorption efficiency of non-functional MWCNT for short-chain PFASs was improved. Therefore, based upon the above studies, the adsorption efficiency and technical maturity of CNTs were similar to ACs.

The main adsorption mechanism of carbon materials was hydrophobic and electrostatic effects, as well as possible hydrogen bonds and covalent bonds effects [4]. The hydrophobic effect would be improved with an increased C-F chain length; thus, long-chain PFASs were better adsorbed than short-chain congeners. Meanwhile, the removal efficiency of adsorbents for PFASs was also depended on the terminal functional groups; for instance, the removal efficiency of PFSAs was better than PFCAs. In addition, the electrostatic attraction could occur between the anionic PFASs and positive charge adsorbents. Therefore, the changes of ionic strength involving cations or anions and pH in the solution would influence the adsorption efficiency. For example, the increase of ionic strength caused by monovalent or divalent cations (Na^+, K^+, Ca^{2+}, etc.) might enhance PFAS removal efficiency, while the pH increase would reduce the adsorption capacity of most adsorbents [31,32]. However, the electrostatic repulsion between anionic PFASs and a negatively charged adsorbent could be overcome by the hydrophobic effect of the C-F chain [33]. Therefore, the removal of short-chain PFASs was mainly dependent on the electrostatic effect, while long-chain PFASs mainly tended to hydrophobic effects. Adsorption mechanisms of carbon-based adsorbents for PFASs removal are shown in Figure 1.

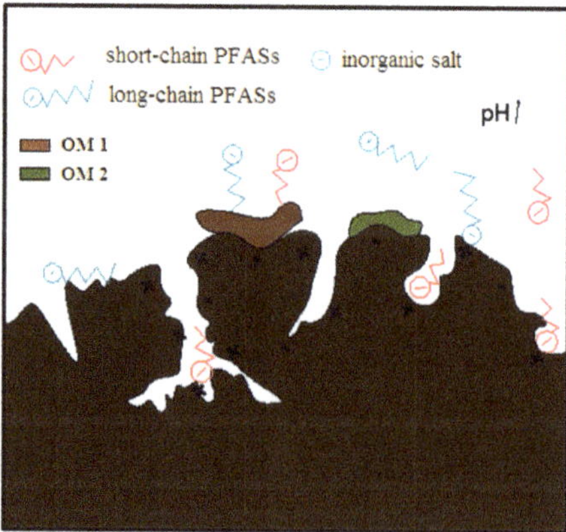

Figure 1. Adsorption mechanism of carbon-based adsorbents for PFAS removal.

3.1.2. Anion-Exchange Resin Adsorption

Resin adsorption gradually attracted researchers' attention due to its strong adsorption, regeneration ability, and convenient operation. The carbon chain length (or hydrophobicity) and terminal functional groups of PFASs could influence the adsorption of anion-exchange resin for PFASs. Maimaiti et al. explored the adsorption of large pore anion-exchange resins about IRA910 for single PFASs (PFBA, PFHxA, PFOA, PFBS, PFHxS, and PFOS), showing that the chain length had a great influence on PFCAs adsorption compared with PFSAs [34]. The adsorption efficiency of PFSAs was better than PFCAs, in which the optimum adsorption efficiency of PFBS could be up to 1023.32 mg/g. Moreover, in order to investigate the treatment effect and regeneration capacity of anion-exchange resin, Du et al. eluted the IRA67 resin that saturated adsorption with PFOA and PFHxA using the mixture solution of NaCl and methanol, the recovery could achieve 98% and 40%, respectively. This phenomenon indicated that short-chain PFASs were difficult to remove from the resin. In addition, the properties of ion-exchange resin also played an effect on the adsorption for short-chain PFASs [25]. For instance, the adsorption capacity of the ion-exchange resin was superior to the non-ion exchange resin, and most of them were better than ACs.

The mechanism of anion-exchange resin mainly included hydrophobic effects, electrostatic effects, and ion-exchange effects. Generally, the resin with stronger hydrophobicity possessed a virtuous adsorption ability; however, the regeneration capacity was deprived by contrast [4]. Furthermore, the adsorption of short-chain PFASs might be influenced by pH via changing the resin surface potential or morphology [34]. For instance, in various ranges of pH, strong base anion (SBA) resin was impregnable due to its ionization form. In contrast, weak base anion (WBA) resin was influenced significantly [31], which could take effect when the amine group was protonated under acidic conditions. Moreover, the main mechanism for short-chain PFAS removal was single-molecule anion-exchange by the analysis of transmission electron microscope; thus, the inorganic anions in solution would compete with PFAS anions for the ion-exchange sites and then decrease the removal efficiency. The ion-exchange reaction equations are shown in (1) and (2), where $[R_4N^+]$ and $[R_3N]$ indicate the ion-exchange site [15].

$$[R_4N^+]Cl^- + PFAS^- \leftrightarrow [R_4N+]PFAS^- + Cl^- \quad (1)$$

$$[R_3N] + H^+ + PFAS^- \leftrightarrow [R_3NH^+]PFAS^- \quad (2)$$

3.1.3. Coagulation and Electrocoagulation

Coagulation possessed the advantage of low price and high adsorption efficiency [4]. Deng et al. [35] discovered that the removal efficiency of PFOA exceeded 90% when the dosage of polymer aluminum chloride (PACl) was 10 mg/L. However, the removal efficiency of PFOA was reduced significantly when multiple PFASs existed simultaneously, and the removal rates followed the order of PFBA > PFHxA > PFOA > PFDoA > PFOS, which demonstrated that the removal efficiency of short-chain PFASs was superior, compared with long-chain congeners.

Electrocoagulation received widespread attention because of its higher removal efficiency for PFASs and short treatment period [4]. The electrocoagulation technique mainly produced a large number of cations and then generated flocs by sacrificing the anode; thus, the dissolved contaminants could be purified by condensation and adsorption of flocs, which was subsequently carried to the surface of the solution by the H_2 and O_2 produced by the electrodes through electrical floating. The Electrocoagulation mechanism for PFAS removal is shown in Figure 2. Liu et al. [36] adopted the periodically reversing electrocoagulation (PREC) technique to treat contaminated groundwater around fluorochemical plants, indicating that the PREC was effective for the removal of PFASs with different lengths of carbon chains. Subsequently, the above group approved that the PREC technique with Al-Zn electrodes for multiple PFSA removal was impactful, the removal rates of PFBS, PFHxS, and PFOS could reach up to 87.4%, 95.6%, and 100% within 10 min under optimal conditions (12.0 V, pH = 7, 400 r/min) [37]. In general, the removal mechanism of electrocoagulation was a hydrophobic effect and might exist with electrophoretic, polarized, and electric fields [38]. In addition, several influencing factors involving current density, inorganic ions, and OM were also crucial for the removal of short-chain PFASs.

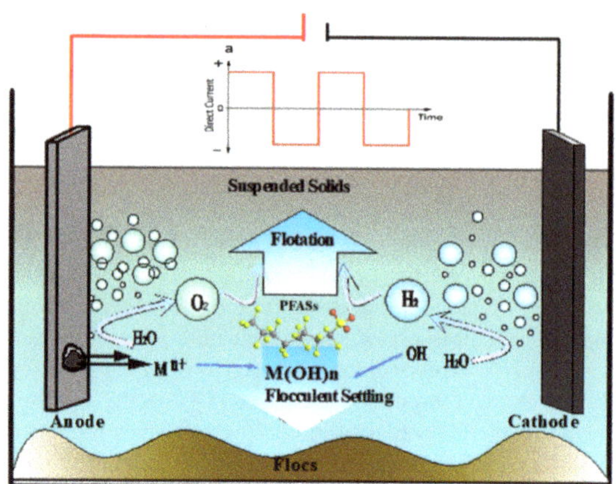

Figure 2. Electrocoagulation Mechanism for PFASs removal.

Table 2. Comparisons on adsorption capacities of various adsorbents for short-chain PFASs.

Adsorbent	Adsorbent Dose/(mg/L)	PFASs	PFAS Concentration (mg/L)	Experiment Condition	Removal Efficiency	References
GAC (F400)	1000	PFBS	15–150	DI, 30 °C, pH = 7.2	98.7 mg/g	[23]
BdAC	200	PFHxA	31.4	WW, 25 °C pH = 4, 48 h 170 r/min	18.84 mg/g	[25]
IRA67	100	PFHxA	31.4	WW, 25 °C pH = 4, 48 h 170 r/min	37.68 mg/g	[25]
AC (micropore)	250	PFBA	6.5–204	DI, room temperature pH = 6, 3 d	51.36 ± 4.28 mg/g	[26]
		PFBS	6–247		51.01 ± 3 mg/g	
		PFHxA	7.2–217		235.54 ± 72.23 mg/g	
CTF	250	PFBA	6.5–204	DI, room temperature pH = 6, 3 d	92.03 ± 4.28 mg/g	[26]
SWCNT	250	PFBA	106.4	DI, 25 °C pH = 7, 2 d 200 r/min	7.5%	[30]
IRA910	100	PFBS	50–400	DI, 25 °C, pH = 6, 240 h 160 r/min	1023.32 mg/g	[34]
		PFBA			635.69 mg/g	
Electrocoagulation	–	PFBS	0.031	GW, pH = 7.0 10 min, 400 r/min Al-Zn, 12 V	87.4%	[36]

3.1.4. Adsorption with Other Materials

In recent years, the performance and construction of new materials that could be controlled and modified easily were synthesized by researchers. Wang et al. used a covalent triazine-based framework (CTF) to eliminate PFBS, with an adsorption capacity of 92.03 mg/g [25]. Subsequently, Zaggia et al. explored the adsorption capacity of AC, CTF, and IRA910 for PFBA and found that the order followed the rules of micropore AC < CTF < IRA910, which demonstrated the adsorption mechanism was an electrostatic effect between the triazine group and the PFAS anion head [32]. Ionic strength might be another major factor of other materials that influenced the adsorption efficiency. For example, the adsorption efficiency of poly-styrene carboxylic acid (PS-COOH) for short-chain PFCAs in seawater was better than river water due to the large ionic strength in seawater [39].

Therefore, in various kinds of adsorption materials, the adsorption efficiency of anion exchange resin and electrocoagulation for short-chain PFASs were remarkable. However, the elution ability of the anion exchange resin was inferior when the short-chain PFASs were adsorbed, and the flocs of electrocoagulation were still required for further treatment. Secondly, AC owned the property of low cost and high removal efficiency, while the regeneration ability was unsatisfactory. Finally, new adsorption materials could improve electrostatic attraction effectively, but the application should be further explored.

3.2. Advanced Oxidation/Reduction Techniques

Advanced oxidation/reduction techniques have been used for the degradation of PFASs, with the advantages of high conversion efficiency and simple operation, and some techniques could achieve complete mineralization. However, these techniques generally put emphasis on long-chain PFASs, including PFOA and PFOS. Whether these techniques could remove short-chain PFASs was still lacking in studies. This section provides a systematic summary of the application of the degradation of short-chain PFASs about the techniques of electrochemical oxidation and photocatalytic degradation, etc.

3.2.1. Electrochemical Oxidation

Electrochemical oxidation is an emerging advanced oxidation technique due to the advantages of high removal efficiency, strong oxidative ability, and low energy consumption. This technique was found to degrade long-chain PFASs effectively. The most commonly used oxidation anodes included Ti/SnO_2, Ce/PbO_2, boron-doped diamond (BDD) electrodes, and their modified electrodes. Comparisons on the removal efficiency of short-chain PFASs by different electrode materials are shown in Table 3.

Table 3. Removal efficiency of short-chain PFASs by different electrode materials.

Anode	PFASs	PFASs Concentration (mg/L)	Experiment Condition	Removal Efficiency/(%)	References
Ti/SnO_2-Sb/PbO_2-Ce	PFBA PFPeA PFHxA PFHpA	100	20 mA/cm^2, 10 mmol/L $NaClO_4$	31.8 41.4 78.2 97.9	[40]
BDD	PFHxA	870	100 mA/cm^2, OM, inorganic salt	98	[41]
Si/BDD	PFBS	>150	rotating disk electrode	90	[42]

Niu et al.'s research group found that the degradation efficiency of Ti/SnO_2-Sb for 100 mg/L PFOA could reach 98.8% when the current density was 40 mA/cm^2, pH at 5, and the electrolyte with 10 mmol/L $NaClO_4$ [43]. Subsequently, the research group explored the Ti/SnO_2-Sb/PbO_2-Ce electrode to degrade short-chain PFASs of PFHpA and PFBA, with removal rates of 97.9% and 31.8%, respectively. This might be related to the high resistance of short-chain PFASs and the co-existence of multiple PFASs [40]. In addition, Soriano et al. used a BDD electrode to remove high concentrations of 870 mg/L PFHxA in a solution containing OM and inorganic salt, finding that the removal rate was 98% within 2 h when the current density was at 100 mA/cm^2, and energy consumption was 45 Wh/L [41]. Liao et al. explored a Si/BDD electrode for the elimination of high concentration PFBS (>150 mg/L) at low current density; the removal rate was 90% within 1 h [42]. Based upon the above studies, the BDD and its modified electrodes achieved remarkable results compared with other electrodes involving SnO_2 and PbO_2. However, the base materials of Si were too weak and performed unfortunate electrical conductivity. Therefore, high manufacturing costs and lack of suitable base materials limited the large-scale application of BDD electrodes. At present, it is important to discover a cheap and stable base material for the industrial application of BDD electrodes [37].

The degradation of PFASs on the electrode surface was related to electron transfer and free radical oxidation; the proposed pathways for electrochemical oxidation of PFOS in water are shown in Figure 3.

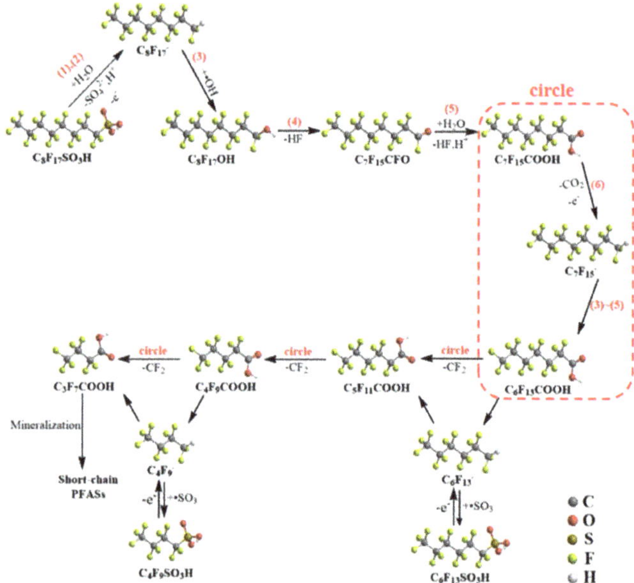

Figure 3. Proposed pathways for electrochemical oxidation of PFOS in water.

Initially, the electrons are transferred from the terminal functional group of PFASs to the anode driven by the electric field and formed PFAS radicals ($\cdot C_nF_{2n+1}COO$ or $\cdot C_nF_{2n+1}SO_3$), the extremely unstable PFAS radicals undergo decarboxylation or desulfation to form $\cdot C_nF_{2n+1}\cdot$. The generated $\cdot C_nF_{2n+1}$ could react with H_2O, OH^-, or $\cdot OH$, and finally $-CF_2$ groups were cut down gradually to generate short-chain PFASs with the specific reaction equations shown in (3)–(7) [44,45].

$$C_nF_{2n+1}COO^- \rightarrow \cdot C_nF_{2n+1}COO + e^- \quad (3)$$

$$C_nF_{2n+1}SO_3{}^- \rightarrow \cdot C_nF_{2n+1}SO_3 + e^- \quad (4)$$

$$C_nF_{2n+1} + H_2O \rightarrow \cdot CnF_{2n+1}OH + H^+ \quad (5)$$

$$C_nF_{2n+1}OH \rightarrow C_{n-1}F_{2n-1}COF + H^+ + F^- \quad (6)$$

$$C_{n-1}F_{2n-1}COF+H_2O \rightarrow C_{n-1}F_{2n-1}COOH+H^{+}+F^{-} \quad (7)$$

Plenty of studies have demonstrated that the degradation efficiency of short-chain PFASs was still lower than long-chain congeners. In order to improve the removal efficiency of short-chain PFASs and reduce the energy consumption effectively, the studies of influence factors, including the current density, pH, electrolyte, OM, and microorganisms in solution, must be carried out. Meanwhile, the electrode modification could improve the removal efficiency and the defluoridation efficiency of PFASs, as well as improve the electrode life successfully, but the metal doping on electrodes might lead to partial contamination during treatment processes [4].

3.2.2. Photocatalytic Degradation

Direct photolysis. Photocatalytic degradation is an advanced oxidation technique that can mineralize target compounds by UV light and photocatalysts. However, PFASs cannot absorb light over 220 nm directly, and the degradation rate of PFBA and PFPeA was only 16.3% and 24.3% by direct photolysis, respectively [15]. Hori et al. found that PFPeA could strongly absorb the light from the vacuum UV region to 220 nm due to the degradation of PFASs under VUV irradiation (<190 nm) by the active substances produced by homolysis and ionization of water, involving a hydrated electron (e_{aq}^{-}), hydrogen radical (H·), and hydroxyl radical (·OH) [46]. Therefore, how to use catalysts to promote the generation of active substances about free radicals for the photodegradation of PFASs was imperative. The photocatalytic efficiency of PFASs with different catalysts is shown in Table 4.

The sulfate radical ($SO_4^{\bullet-}$) produced from persulfate ($S_2O_8^{2-}$) under UV irradiation was observed to be more effective for PFASs degradation over ·OH. Therefore, $S_2O_8^{2-}$ was used as an oxidant to degrade PFAS frequently [47]. For instance, Hori et al. found that the short-chain PFCAs in aqueous solution were oxidized to CO_2 and F^- by $SO_4^{\bullet-}$ when the concentration of $S_2O_8^{2-}$ was 50 mmol/L [46]. Subsequently, this group also found that Fe^{3+} was an effective catalyst for photodegradation, and the photodegradation efficiency of PFBA and PFPeA were improved to 49.9% and 64.5% due to strong light absorption from the complexation of PFASs with Fe^{3+} [47]. Water-soluble polyacid photocatalysts could also promote C-F bond cleavage of PFPrA, thereafter generated to TFA, CO_2, and F^-, but the energy consumption of the process was exorbitant [48].

Zero-valent iron (ZVI) reduction. Reductants including zero-valent iron particles or iodine salt could also enhance the photodegradation of PFASs. ZVI was used as a reductive agent for the photodegradation of PFASs due to its high reduction potential (−0.447 V) and reactive surface area [15]. For instance, Hori et al. observed that the degradation efficiency of short-chain PFSAs (C2-C6) could be up to 95% using ZVI, owing to the generation of iron oxide in the ZVI surface, which could interact with PFASs ions synergistically, thus promoting defluoridation efficiently [49].

Photocatalytic degradation. Photocatalysts were scarcely implemented in short-chain PFASs; the photodegradation efficiency could be deduced by examining the degradation intermediates and degradation efficiency of long chain PFASs. For example, Panchangam et al. adopted TiO_2 photocatalysts for 120 mmol/L of PFOA degradation under UV irradiation at 254 nm within 7 h, 97% of PFOA were converted to short-chain intermediates, including PFPrA, PFBA, PFPeA, PFHxA, and PFHpA, in which PFHpA reached the maximum at 5 h [50]. Currently, the widely used method for modifying TiO_2 was doping with precious metals (Pt, Pd, Au) or other metals (Pb, Cu, Fe). Li et al. explored TiO_2 doped with Pt to completely decompose 144.9 mmol/L PFOA within 7 h under UV irradiation at 365 nm; the defluoridation efficiency was 34.8%. It was 12.5-fold faster than unmodified TiO_2 due to the deposited Pt particles could store excessive electrons and promote the electrons transferred to PFAS availably [51]. In addition, the composite materials were used to improve the photocatalytic degradation of PFASs, such as TiO_2-MWCNT and TiO_2-rGO. It was found that the degradation efficiency of PFOA could reach 100% when using TiO_2-MWCNT after 8 h under irradiation of UV at 265 nm [52].

In addition to the above n-type photocatalysts of TiO_2, p-type photocatalysts involving In_2O_3 and Ga_2O_3 have attracted widespread attention because they could enhance the degradation capacity of PFASs. However, p-type photocatalysts showed a strong dependence on the material shape and microstructure [15]. For example, the In_2O_3 porous microsphere had the highest photocatalytic activity on the degradation of PFOA, which was 74.7 times faster than TiO_2. Similarly, short-chain intermediate PFHpA could be completely degraded within 3 h by the In_2O_3 nanoplates, whereas In_2O_3 nanocubes were much less effective. However, in the degradation process, the short-chain intermediates were still generated from PFOA degradation, and the concentration of intermediates was positively related to their carbon chain length [50,53].

New photocatalysts could decompose PFAS, but the information about the application for short-chain PFASs was still limited. According to the degradation data of long-chain PFASs, the degradation conclusions could be deduced as the following: photocatalytic degradation of PFASs was a gradual splitting decomposition of CF_2. The reaction of breaking the chain generated short-chain intermediates. The short-chain products showed strong resistance to photocatalytic degradation [15].

During the photodegradation process, the catalyst dosage and pH were imperative influencing factors. In the low concentration range between 20–100 mmol/L of persulfate, the photochemical reactivity could be improved with the concentration increase. Whereas further increasing persulfate concentrations could result in saturation of the reaction rate, Because $SO_4^{\bullet-}$ could react with persulfate or itself, this side reaction would reduce the degradation efficiency [15]. Moreover, a high concentration of hydrated hydrogen ions (H_3O^+) would quench e_{aq}^- under acidic conditions, thus contributing to the reactivity decrease due to the quantum yield of e_{aq}^- declining sharply. However, under alkaline conditions, the reactivity would be enhanced due to the quantum yield of e_{aq}^- increasing by the reaction of H· and OH^- [54].

Table 4. Photocatalytic efficiency of PFASs with different catalysts.

Catalyst	Catalyst Dose	PFASs	Experiment Condition	Removal Efficiency (%)	References
Fe^{3+}	5 mmol/L	PFBA PFPeA	UV UV	49.9 64.5	[47]
ZVI particles	960 mmol/L	Short-chain PFSAs (C2–C6)	UV, 350 °C 20 MPa	95	[49]
TiO_2	0.66 g/L	PFOA	254 nm pH < 3	100%	[50]
TiO_2-MWCNT	0.4 g/L	PFOA	300 W, 365 nm pH = 5	100%	[52]
TiO_2-rGO	0.1 g/L	PFOA	150 W, 254 nm pH = 3.8	93 ± 7%	[55]

3.3. Other Techniques

3.3.1. Plasma Technique

Plasma is the collection of positive and negative electric particles that consist of electrons, ions, radicals, and neutral particles, which are electric and electroneutral, thus called the fourth state beyond gas, liquid, and solid electrically. Diverse from most AOPs and conventional techniques, plasma techniques could convert water into highly active substances, involving ·OH, O, H·, HO_2·, $O_2^{\bullet-}$, H_2, O_2, H_2O_2, and e_{aq}^- [56]. When using the plasma technique for PFASs elimination, the degradation process was gradually reduced to intermediates, and then intermediates, perfluoroalkyl radicals, and perfluoro alcohols/ketones-perfluoroalkyl were oxidized [57]. However, this technique was hardly applied and is invalid in the degradation for short-chain PFASs. For example, Takaki et al. adopted $BaTiO_3$ iron beads as a filling medium to degrade C_2F_6; the degradation efficiency was only 20% [58]. Moreover, short-chain PFASs, fluorine ions, and CO_2 by-products were produced during the plasma treatment of PFASs.

3.3.2. Thermolytic and Sonochemical Degradation

In the recent decade, thermolytic and sonochemical degradation have attracted wide attention. Tsang et al. [59] found that 99% of CF_4 could be removed under urban incineration conditions (about 850 °C) and assumed the pyrolysis of 800–900 °C could resolve long and short-chain PFASs efficiently. Krusic et al. [60] explored the gas phase decomposition of PFOA in quartz tubes at 355–385 °C, finding that PFOA retained thermal stability below 300 °C but completely degraded at 370 °C after 360 min. However, the by-products of small molecule substances were generated continuously, which demonstrated the greater resistance of short-chain PFASs. Subsequently, Campbell et al. explored pre-concentrating PFHxA on GAC, which could improve the thermal mineralization rate from 46% to 74% [61]. Furthermore, in order to investigate the effect of ultrasonic chemistry on the degradation of PFASs with different carbon chain lengths, Campbell et al. investigated the degradation efficiency on six kinds of PFASs at 358 kHz, the order was PFOA > PFHxA > PFBA and PFOS ≈ PFHxS > PFBS [62]. Similarly, Fernandez et al. found that the degradation efficiency of ultrasonic chemicals raised with carbon chain length increased [63]. For one thing, long-chain PFASs with strong hydrophobicity were inclined to be adsorbed on the gas-liquid interface for thermal decomposition or oxidation degradation. For another, short-chain PFASs were more difficult to be defluorinated than long-chain congeners. Generally, these techniques could mineralize short-chain PFASs, but nongreen environmental factors about the generous discharge of CO_2 and high energy consumption limited the development in actual application.

3.3.3. Membrane Separation

In recent years, low carbon treatment of membrane separation techniques, including nanofiltration (NF) and reverse osmosis (RO), have made great advances. During the rejection in membrane processes, the molecular size and structure of PFASs were considered key factors. For instance, NF membranes could reject more than 96% of PFHxA (μg/L to mg/L) under neutral pH, whereas the rejection rate of shorter-chains PFASs about PFBS decreased to 69% attributable to the small molecular size [64]. Furthermore, charge, hydrophobicity, pH, and dipole moment might also affect the solute-membrane interactions and thus the rejection efficiency of PFAS. It was shown that pH reduction could increase the membrane rejection efficiency of PFHxA. Moreover, the presence of the ions generally suppressed the electrical repulsion, but the exclusion efficiency was enhanced with increasing ionic strength, which indicated that the exclusion of membrane size was the dominant factor. In the practical application, the rejection efficiency of short-chain PFASs would be easily influenced because of membrane contamination. In addition, membrane separation techniques could produce a high concentration of PFASs and still require subsequent treatment or disposal.

3.3.4. Bioremediation Techniques

Kwon et al. degraded PFOS (1400–1800 μg/L) by cultivating *P. aeruginosa* microorganism with the removal rate of 67% [65]. However, short-chain PFASs were difficult to degrade by common microbes. For example, the concentration of PFBS in the effluent of sewage treatment plants remained unchanged or increased after conventional activated sludge or biofilm bioreactors. That is, short-chain PFASs showed high resistance to various activated sludge systems [5]. While in plant tissues, both short and long-chain PFASs all have a tendency to accumulate. Recent studies have shown that the biological accumulation of PFASs followed a U-type trend. The lowest hydrophobicity (e.g., PFBA and PFPeA) and the maximum hydrophobic species (e.g., PFNA and PFDA) presented the greatest absorb efficiency [66]. Another study on the uptake and distribution of PFASs in maize showed that plant adsorption and distribution of PFASs were dependent on chain length, functional groups, and plant tissue. Generally, short-chain PFASs were transferred to the overground portion of plants, while long-chain PFASs were mainly transported to the root [67].

4. Comparisons on Different Treatment Techniques

So far, low carbon treatment techniques of short-chain PFASs have included adsorption, membrane separation, bioremediation, as well as degradation techniques relating to advanced oxidation, plasma, thermolytic, and sonochemical degradation. The above technologies could remove short-chain PFASs to a certain extent, but their treatment effects, operating conditions, removal mechanism, and applicability were quite different. The comparisons on treatment techniques for short-chain PFASs are presented in Table 5.

Adsorption was the utmost widely applied technique for short-chain PFASs, and its energy consumption could be nearly ignored besides the low energy cost of the adsorbent regeneration. The removal mechanisms of short-chain PFASs were mainly electrostatic action, hydrophobic effect, and ion exchange [4]. As shown in Table 5, this technique has the advantages of convenient operation, low carbon, low cost, and low energy consumption, as well as application with a wide concentration range of short-chain PFASs, especially trace levels. However, the technique has the drawbacks of a long adsorption period and low regeneration efficiency [27,36].

Advanced oxidation techniques, involving electrochemical oxidation and photocatalytic degradation, degraded short-chain PFASs primarily relying on active free radicals, which possessed the advantages of a short treatment period and low energy consumption [5,15]. However, the techniques were not suitable for trace levels of short-chain PFASs, and its low mineralization rate and subsequent CO_2 generation rate from mineralization were also the main problems for achieving the low-carbon goals [43]. In addition, electrochemical oxidation could produce high expenses of electrode materials and the risk of electrode contamination. The photocatalytic degradation technique generated the problem of secondary pollution by catalyst addition and might thus be limited in actual applications [15].

The plasma technique was applied to the degradation of long-chain PFASs effectively, but few studies focused on its degradation of short-chain PFASs, and the energy consumption was still higher for complete mineralization [58]. The thermolytic and sonochemical techniques could achieve complete mineralization, but the energy consumption was too high, and the operating conditions were stringent [61]. The membrane separation technique could reject short-chain PFAS pollutants effectively, but membrane pollution and membrane flux instability were the main problems [64]. Bioremediation techniques could take advantage of their low carbon and environmentally-friendly processes, but they generated the problems of long remediation period and low efficiency, as well as inefficient short-chain PFAS elimination and remaining in organisms [65].

Based upon the analysis, the most widely used and effective method could be adsorption, followed by advanced oxidation. However, there were still limitations of removal efficiency in the application of eliminating short-chain PFASs. Since short-chain PFASs were more resistant to be adsorbed and degraded than long-chain congeners, PFASs of C1–C3 were barely degradable. Combined techniques might be developed based on concentration/recycling-degradation of short-chain PFASs in water bodies rather than degradation. The combination of adsorption and gas–liquid series electrical discharge treatment has been applied in the degradation of dyestuff, which achieved excellent removal efficiency compared with the adsorption alone [68].

Table 5. Comparisons on treatment techniques for short-chain PFASs.

Technique	Materials	Advantage	Disadvantage	Removal Mechanism	Treatment Time	Removal Efficiency	Energy Consumption	References
Adsorption	ACs, Anion-exchange resin	Low carbon, cost, energy consumption, and convenient operation; No change in physicochemical properties; Wide concentrations and trace short-chain PFASs could be treated.	Long adsorption time and unfortunate regeneration capacity of sorbent; Secondary contamination of elution solvent.	Electrostatic interaction; Hydrophobic interaction; Ion exchange	10 min–10 d	10–95.6%	—	[25,27,36]
Electrochemical oxidation	Ti/SnO_2, BDD	Low energy consumption and short time; Good treatment for short-chain PFASs.	Expensive electrode materials; Not suitable for trace contamination; Prone to secondary contamination and produced intermediates.	Oxidation; Hydrophobic interaction	1–3 h	31.8–98%	45 Wh/L	[40,43]
Photocatalytic degradation	UV, $S_2O_8^{2-}$ Fe^{3+}, ZVI	Mineralizable.	Additional catalyst; Low degradation efficiency; Complex by-products.	Oxidation; Reduction.	6 h–10 d	16–95%	29–9091 Wh/L	[15,69,70]
Plasma	Grinding nickel chrome rod	Short time; Suitable for long-chain PFASs.	High energy consumption; Low mineralization efficiency; Generate intermediates.	Reduction.	1–2 h	20%	—	[57,58]
Thermolytic and sonochemical degradation	High temperature; Ultrasonic	Fully mineralized.	Long time; High energy consumption.	Disrupts molecular structure by high energy	6–65 h	46–74%	2129 Wh/L	[60,61]
Membrane separation	NF RO	Low carbon and energy consumption; Better rejection of short-chain PFASs.	Not suitable for the actual waters; Easy to occur in membrane pollution.	Rejection	—	69–96%	—	[64]
Bioremediation	Corn	Low carbon and environmentally-friendly; Effective to some degree for some short-chain PFASs by plant adsorption.	Long time and ineffective.; Need targeted training.	—	>2 d	−67%	—	[65]

5. Conclusions and Future Research Recommended

This paper showed that the adsorption, electrochemical oxidation, and photocatalytic degradation have certain removal effects on short-chain PFASs by comparisons on various treatment techniques. Considering the removal efficiency, treatment time, energy consumption, and cost, adsorption was the most widely applied technique for the effective removal of short-chain PFASs, which was suitable for a wide concentration range of pollution and to meet the low-carbon policy. Whereas, long adsorption period and unsatisfied regeneration ability were the main problems. The advanced oxidation techniques of electrochemical and photocatalytic activity could degrade short-chain PFASs, but low mineralization efficiency contributed to intermediates of short-chain PFASs, as well as abundant organic matter and CO_2 and they were especially inappropriate to eliminate trace short-chain PFASs. Therefore, it was desirable to choose suitable techniques according to PFAS properties, as well as the advantages and disadvantages of various techniques.

The contaminations of short-chain PFASs have attracted much attention at present, while most studies still focused on the laboratory-scale treatment of long-chain PFASs, including PFOA and PFOS, and data on the treatment of short-chain PFASs is still absent. Therefore, future studies need to focus on the following topics: (1) Targeting long duration period and poor material regeneration ability for the adsorption technique; the functional groups of adsorbent materials need to be modified for improving electrostatic attraction and hydrophobic effects and enhancing the adsorption efficiency and elution of short-chain PFASs. (2) Innovative design for short-chain PFAS treatment by electrode and catalyst modification to develop advanced oxidation techniques, with high degradation efficiency in low-carbon and low-energy consumption, as well as adaptability for low-concentration PFASs. (3) With the increasing pollution of short-chain PFASs in drinking water, extensive emphasis should be placed on the development of advanced treatment techniques for actual groundwater and surface water, along with exploration of new adsorption materials, electrode materials, and catalysts, which could remove low concentrations of short-chain PFAS under the background of multiple substances co-existing in the actual waters. (4) Short-chain PFASs were more resistant to degradation than long-chain congeners, and PFASs of C1-C3 were barely degradable. Therefore, concentration/recycling of short-chain PFASs in water bodies should be considered rather than degradation. In addition, combined techniques might be developed based on concentration/recycling-degradation, such as adsorption and advanced oxidation, for the efficient removal of short-chain PFASs from actual waters.

Author Contributions: Conceptualization, Y.L. and J.B.; methodology, Y.L. and T.L.; software, M.L.; validation, X.Z., C.L. and L.S.; formal analysis, T.L.; investigation, T.L; resources, Y.L. and J.B.; data curation, T.L. and X.H.; writing—original draft preparation, Y.L. and J.B.; writing—review and editing, Y.L. and X.H.; visualization, T.L.; supervision, Y.L. and T.L.; project administration, Y.L. and J.B.; funding acquisition, J.B. All authors have read and agreed to the published version of the manuscript.

Funding: This research was funded by the National Natural Science Foundation of China (No. 21976124 and No. 21507092), the Natural Science Foundation of Liaoning Province of China (No. 2019-ZD-0217), and the Liaoning Revitalization Talents Program (No. XLYC2007195).

Acknowledgments: Thanks for the financial support from the National Natural Science Foundation of China (No. 21976124 and No. 21507092), the Natural Science Foundation of Liaoning Province of China (No. 2019-ZD-0217), and the Liaoning Revitalization Talents Program (No. XLYC2007195).

Conflicts of Interest: The authors declare no conflict of interest.

References

1. Rahman, M.F.; Peldszus, S.; Anderson, W.B. Behaviour and fate of perfluoroalkyl and polyfluoroalkyl substances (PFASs) in drinking water treatment: A review. *Water Res.* **2013**, *50*, 318–340. [CrossRef] [PubMed]
2. Jian, J.; Chen, D.; Han, F.; Guo, Y.; Zeng, L.; Lu, X.; Wang, F. A short review on human exposure to and tissue distribution of per- and polyfluoroalkyl substances (PFASs). *Sci. Total Environ.* **2018**, *636*, 1058–1069. [CrossRef] [PubMed]
3. Park, M.; Wu, S.; Lopez, I.J.; Chang, J.Y.; Karanfil, T.; Snyder, S.A. Adsorption of perfluoroalkyl substances (PFAS) in groundwater by granular activated carbons: Roles of hydrophobicity of PFAS and carbon characteristics. *Water Res.* **2020**, *170*, 115364. [CrossRef] [PubMed]
4. Liu, Y.; Hu, X.; Zhao, Y.; Wang, J.; Bao, J. Treatment techniques for perfluorinated compounds and their alternatives. *Environ. Chem.* **2018**, *37*, 1860–1868. [CrossRef]
5. Ateia, M.; Maroli, A.; Tharayil, N.; Karanfil, T. The overlooked short- and ultrashort–chain poly- and perfluorinated substances: A review. *Chemosphere* **2019**, *220*, 866–882. [CrossRef] [PubMed]
6. Reemtsma, T.; Berger, U.; Arp, H.P.H.; Gallard, H.; Knepper, T.P.; Neumann, M.; Quintana, J.B.; Voogt, P.D. Mind the gap: Persistent and mobile organic compounds-water contaminants that slip through. *Environ. Sci. Technol.* **2016**, *50*, 10308–10315. [CrossRef]
7. Arp, H.P.H.; Brown, T.N.; Berger, U.; Hale, S.E. Ranking REACH registered neutral, ionizable and ionic organic chemicals based on their aquatic persistency and mobility. *Environ. Sci-Proc. Imp.* **2017**, *19*, 939–955. [CrossRef]
8. Kotthoff, M.; Bücking, M. Four Chemical trends will shape the next decade's directions in perfluoroalkyl and polyfluoroalkyl substances research. *Front. Chem.* **2018**, *6*, 1–6. [CrossRef]
9. Gellrich, V.; Stahl, T.; Knepper, T.P. Behavior of perfluorinated compounds in soils during leaching experiments. *Chemosphere* **2012**, *87*, 1052–1056. [CrossRef]
10. Brendel, S.; Fetter, É.; Staude, C.; Vierke, L.; Biegel-Engler, A. Short-chain perfluoroalkyl acids: Environmental concerns and a regulatory strategy under REACH. *Environ. Sci. Eur.* **2018**, *30*, 9. [CrossRef]
11. Zhou, Z.; Liang, Y.; Shi, Y.; Xu, L.; Cai, Y. Occurrence and transport of perfluoroalkyl acids (PFAAs), including short–chain PFAAs in Tangxun Lake, China. *Environ. Sci. Technol.* **2013**, *47*, 9249–9257. [CrossRef] [PubMed]
12. Bao, J.; Li, C.; Liu, Y.; Wang, X.; Yu, W.; Liu, Z.; Shao, L.; Jin, Y. Bioaccumulation of perfluoroalkyl substances in greenhouse vegetables with long-term groundwater irrigation near fluorochemical plants in Fuxin, China. *Environ. Res.* **2020**, *188*, 109751. [CrossRef] [PubMed]
13. Eschauzier, C.; Beerendonk, E.; Scholte-Veenendaal, P.; Voogt, P.D. Impact of treatment processes on the removal of perfluoroalkyl acids from the drinking water production chain. *Environ. Sci. Technol.* **2012**, *46*, 1708–1715. [CrossRef] [PubMed]
14. Schmidt, C.K.; Brauch, H.J. N, N-dimethylsulfamide as precursor for N-nitrosodimethylamine (NDMA) formation upon ozonation and its fate during drinking water treatment. *Environ. Sci. Technol.* **2008**, *42*, 6340–6346. [CrossRef] [PubMed]
15. Li, F.; Duan, J.; Tian, S.; Ji, H.; Zhu, Y.; Wei, Z.; Zhao, D. Short–chain per- and polyfluoroalkyl substances in aquatic systems: Occurrence, impacts and treatment. *Chem. Eng. J.* **2020**, *380*, 122506. [CrossRef]
16. Wang, H.; Lu, X.; Deng, Y.; Sun, Y.; Nielsen, C.P.; Liu, Y.; Zhu, G.; Bu, M.; Bi, J.; McElroy, M. China's CO_2 peak before 2030 implied from characteristics and growth of cities. *Nat. Sustain.* **2019**, *2*, 748–754. [CrossRef]
17. Watts, M. Commentary: Cities spearhead climate action. *Nat. Clim. Chang.* **2017**, *7*, 537–538. [CrossRef]
18. Wang, Y.; Xu, Z.; Zhang, Y. Influencing factors and combined scenario prediction of carbon emission peaks in megacities in China: Based on Threshold-STIRPAT model. *Acta Sci. Circumst.* **2019**, *39*, 4284–4292. [CrossRef]
19. Song, Q.; Liu, T.; Qi, Y. Policy innovation in low carbon pilot cities: Lessons learned from China. *Urban Clim.* **2021**, *39*, 100936. [CrossRef]
20. Liu, Q.; Zhang, W.; Yao, M.; Yuan, J. Carbon emissions performance regulation for China's top generation groups by 2020: Too challenging to realize? *Resour. Conserv. Recy.* **2017**, *122*, 326–334. [CrossRef]
21. Ma, W.; De, J.M.; De, B.M.; Mu, R. Mix and match: Configuring different types of policy instruments to develop successful low carbon cities in China. *J. Clean. Prod.* **2021**, *282*, 125399. [CrossRef]
22. Glüge, J.; Scheringer, M.; Cousins, I.T.; DeWitt, J.C.; Goldenman, G.; Herzke, D.; Lohmann, R.; Ng, C.A.; Trier, X.; Wang, Z. An overview of the uses of per- and polyfluoroalkyl substances (PFAS). *Environ. Sci. Process Impacts* **2020**, *22*, 2345–2373. [CrossRef] [PubMed]
23. Ochoa-Herrera, V.; Sierra-Alvarez, R. Removal of perfluorinated surfactants by sorption on to granular activated carbon, zeolite and sludge. *Chemosphere* **2008**, *72*, 1588–1593. [CrossRef] [PubMed]
24. Hansen, M.C.; Borresen, M.H.; Schlabach, M.; Cornelissen, G. Sorption of perfluorinated compounds from contaminated water to activated carbon. *J. Soils Sediments* **2010**, *10*, 179–185. [CrossRef]
25. Du, Z.; Deng, S.; Chen, Y.; Wang, B.; Huang, J.; Wang, Y.; Yu, G. Removal of perfluorinated carboxylates from washing wastewater of perfluorooctanesulfonyl fluoride using activated carbons and resins. *J. Hazard. Mater.* **2015**, *286*, 136–143. [CrossRef] [PubMed]
26. Wang, B.; Lee, L.S.; Wei, C.; Fu, H.; Zheng, S.; Xu, Z.; Zhu, D. Covalent triazine-based framework: A promising adsorbent for removal of perfluoroalkyl acids from aqueous solution. *Environ. Pollut.* **2016**, *216*, 884–892. [CrossRef]
27. McCleaf, P.; Englund, S.; Östlund, A.; Lindegren, K.; Wiberg, K.; Ahrens, L. Removal efficiency of multiple poly- and perfluoroalkyl substances (PFASs) in drinking water using granular activated carbon (GAC) and anion exchange (AE) column tests. *Water Res.* **2017**, *120*, 77–87. [CrossRef]

28. Wang, M.; Zhou, Q. Environmental effects and their mechanisms of biochar applied to soils. *Environ. Chem.* **2013**, *32*, 768–780. [CrossRef]
29. Inyang, M.; Dickenson, E.R.V. The use of carbon adsorbents for the removal of perfluoroalkyl acids from potable reuse systems. *Chemosphere* **2017**, *184*, 168–175. [CrossRef]
30. Deng, S.; Zhang, Q.; Nie, Y.; Wei, H.; Wang, B.; Huang, J.; Yu, G.; Xing, B. Sorption mechanisms of perfluorinated compounds on carbon nanotubes. *Environ. Pollut.* **2012**, *168*, 138–144. [CrossRef]
31. Gagliano, E.; Sgroi, M.; Falciglia, P.P.; Vagliasindi, F.G.A.; Roccaro, P. Removal of poly- and perfluoroalkyl substances (PFAS) from water by adsorption: Role of PFAS chain length, effect of organic matter and challenges in adsorbent regeneration. *Water Res.* **2020**, *171*, 115381. [CrossRef] [PubMed]
32. Gao, Y.; Deng, S.; Du, Z.; Liu, K.; Yu, G. Adsorptive removal of emerging polyfluoroalky substances F-53B and PFOS by anion-exchange resin: A comparative study. *J. Hazard. Mater.* **2017**, *323*, 550–557. [CrossRef] [PubMed]
33. Zaggia, A.; Conte, L.; Falletti, L.; Fant, M.; Chiorboli, A. Use of strong anion exchange resins for the removal of perfluoroalkylated substances from contaminated drinking water in batch and continuous pilot plants. *Water Res.* **2016**, *91*, 137–146. [CrossRef] [PubMed]
34. Maimaiti, A.; Deng, S.; Meng, P.; Wang, W.; Wang, B.; Huang, J.; Wang, Y.; Yu, G. Competitive adsorption of perfluoroalkyl substances on anion exchange resins in simulated AFFF-impacted groundwater. *Chem. Eng. J.* **2018**, *348*, 494–502. [CrossRef]
35. Deng, S.; Zhou, Q.; Yu, G.; Huang, J.; Fan, Q. Removal of perfluorooctanoate from surface water by polyaluminium chloride coagulation. *Water Res.* **2011**, *45*, 1774–1780. [CrossRef]
36. Liu, Y.; Hu, X.; Zhao, Y.; Wang, J.; Lu, M.; Peng, F.; Bao, J. Removal of perfluorooctanoic acid in simulated and natural waters with different electrode materials by electrocoagulation. *Chemosphere* **2018**, *201*, 303–309. [CrossRef]
37. Bao, J.; Yu, W.; Liu, Y.; Wang, X.; Liu, Z.; Duan, Y. Removal of perfluoroalkanesulfonic acids (PFSAs) from synthetic and natural groundwater by electrocoagulation. *Chemosphere* **2020**, *248*, 125951. [CrossRef]
38. Niu, J.; Wang, C.; Shang, E. Removal of perfluorinated compounds from wastewaters by electrochemical methods: A general review. *Sci. Sin. Technol.* **2017**, *47*, 1233–1255. [CrossRef]
39. Llorca, M.; Schirinzi, G.; Martínez, M.; Barceló, D.; Farré, M. Adsorption of perfluoroalkyl substances on microplastics under environmental conditions. *Environ. Pollut.* **2018**, *235*, 680–691. [CrossRef]
40. Niu, J.; Lin, H.; Xu, J.; Wu, H.; Li, Y. Electrochemical mineralization of per-fluorocarboxylic acids (PFCAs) by Ce-doped modified porous nanocrystalline PbO_2 film electrode. *Environ. Sci. Technol.* **2012**, *46*, 10191–10198. [CrossRef]
41. Soriano, A.; Gorri, D.; Urtiaga, A. Efficient treatment of perfluorohexanoic acid by nanofiltration followed by electrochemical degradation of the NF concentrate. *Water Res.* **2017**, *112*, 147–156. [CrossRef] [PubMed]
42. Liao, Z.; Farrell, J. Electrochemical oxidation of perfluorobutane sulfonate using borondoped diamond film electrodes. *J. Appl. Electrochem.* **2009**, *39*, 1993–1999. [CrossRef]
43. Niu, J.; Li, Y.; Shang, E.; Xu, Z.; Liu, J. Electrochemical oxidation of perfluorinated compounds in water. *Chemosphere* **2016**, *146*, 526–538. [CrossRef] [PubMed]
44. Lin, H.; Niu, J.; Liang, S.; Wang, C.; Wang, C.; Wang, Y.; Jin, F.; Luo, Q.; Chiang, S.Y.D.; Huang, Q. Development of macroporous magneli phase Ti_4O_7 ceramic materials: As an efficient anode for mineralization of poly- and perfluoroalkyl substances. *Chem. Eng. J.* **2018**, *354*, 1058–1067. [CrossRef]
45. Niu, J.; Lin, H.; Gong, C.; Sun, X. Theoretical and experimental insights into the electrochemical mineralization mechanism of perfluorooctanoic acid. *Environ. Sci. Technol.* **2013**, *47*, 14341–14349. [CrossRef]
46. Hori, H.; Yamamoto, A.; Hayakawa, E.; Taniyasu, S.; Yamashita, N.; Kutsuna, S.; Kiatagawa, H.; Arakawa, R. Efficient decomposition of environmentally persistent perfluorocarboxylic acids by use of persulfate as a photochemical oxidant. *Environ. Sci. Technol.* **2005**, *39*, 2383–2388. [CrossRef]
47. Hori, H.; Yamamoto, A.; Koike, K.; Kutsuna, S.; Osaka, I.; Arakawa, R. Photochemical decomposition of environmentally persistent short–chain perfluorocarboxylic acids in water mediated by iron (II)/(III) redox reactions. *Chemosphere* **2007**, *68*, 572–578. [CrossRef]
48. Hori, H.; Takano, Y.; Koike, K.; Kutsuna, S.; Einaga, H.; Ibusuki, T. Photochemical decomposition of pentafluoropropionic acid to fluoride ions with a water-soluble heteropolyacid photocatalyst. *Appl. Catal. B-Environ.* **2003**, *46*, 333–340. [CrossRef]
49. Hori, H.; Nagaoka, Y.; Yamamoto, A.; Sano, T.; Yamashita, N.; Taniyasu, S.; Kutsuna, S.; Osaka, I.; Arakawa, R. Efficient decomposition of environmentally persistent perfluorooctanesulfonate and related fluorochemicals using zerovalent iron in subcritical water. *Environ. Sci. Technol.* **2006**, *40*, 1049–1054. [CrossRef]
50. Panchangam, S.C.; Lin, A.; Shaik, K.L.; Lin, C. Decomposition of perfluorocarboxylic acids (PFCAs) by heterogeneous photocatalysis in acidic aqueous medium. *Chemosphere* **2009**, *77*, 242–248. [CrossRef]
51. Li, M.; Yu, Z.; Liu, Q.; Sun, L. Photocatalytic decomposition of per-fluorooctanoic acid by noble metallic nanoparticles modified TiO_2. *Chem. Eng. J.* **2016**, *286*, 232–238. [CrossRef]
52. Song, C.; Chen, P.; Wang, C.; Zhu, L. Photodegradation of perfluorooctanoic acid by synthesized TiO_2–MWCNT composites under 365nm UV irradiation. *Chemosphere* **2011**, *86*, 853–859. [CrossRef] [PubMed]
53. Chen, M.; Lo, S.L.; Lee, Y.C.; Huang, C.C. Photocatalytic decomposition of perfluorooctanoic acid by transition-metal modified titanium dioxide. *J. Hazard. Mater.* **2015**, *288*, 168–175. [CrossRef] [PubMed]

54. Jin, L.; Zhang, P. Photochemical decomposition of perfluorooctane sulfonate (PFOS) in an anoxic alkaline solution by 185nm vacuum ultraviolet. *Chem. Eng. J.* **2015**, *280*, 241–247. [CrossRef]
55. Gomez-Ruiz, B.; Ribao, P.; Diban, N.; Rivero, M.J.; Ortiz, I.; Urtiaga, A. Photocatalytic degradation and mineralization of perfluorooctanoic acid (PFOA) using a composite TiO2-rGO catalyst. *J. Hazard. Mater.* **2017**, *344*, 950–957. [CrossRef]
56. Stratton, G.R.; Dai, F.; Bellona, C.L.; Holsen, T.M.; Dickenson, E.R.V.; Thagard, S.M. Plasma-Based water treatment: Efficient transformation of perfluoroalkyl substances in prepared solutions and contaminated groundwater. *Environ. Sci. Technol.* **2017**, *51*, 1643–1648. [CrossRef]
57. Singh, R.K.; Fernando, S.; Baygi, S.F.; Multari, N.; Thagard, S.M.; Holsen, T.M. Breakdown products from perfluorinated alkyl substances (PFAS) degradation in a plasma-based water treatment process. *Environ. Sci. Technol.* **2019**, *53*, 2731–2738. [CrossRef]
58. Takaki, K.; Urashima, K.; Chang, J.S. Scale-up of ferro-electric packed bed reactor for C_2F_6 decomposition. *Thin Solid Films* **2006**, *506*, 414–417. [CrossRef]
59. Tsang, W.; Burgess, D.R.; Babushok, V. On the Incinerability of Highly Fluorinated Organic Compounds. *Combust. Sci. Technol.* **1998**, *139*, 385–402. [CrossRef]
60. Krusic, P.J.; Roe, D.C. Gas-phase NMR studies of the thermolysis of perfluorooctanoic acid. *Anal. Chem.* **2005**, *126*, 1510–1516. [CrossRef]
61. Watanabe, N.; Takata, M.; Takemine, S.; Yamamoto, K. Thermal mineralization behavior of PFOA, PFHxA, and PFOS during reactivation of granular activated carbon (GAC) in nitrogen atmosphere. *Environ. Sci. Pollut. Res. Int.* **2018**, *25*, 7200–7205. [CrossRef] [PubMed]
62. Campbell, T.Y.; Vecitis, C.D.; Mader, B.T.; Hoffmann, M.R. Perfluorinated surfactant chain-length effects on sonochemical kinetics. *J. Phys. Chem. A* **2009**, *113*, 9834–9842. [CrossRef] [PubMed]
63. Fernandez, N.A.; Rodriguez-Freire, L.; Keswani, M.; Sierra-Alvarez, R. Effect of chemical structure on the sonochemical degradation of perfluoroalkyl and poly-fluoroalkyl substances (PFASs). *Water Res.* **2016**, *2*, 975–983. [CrossRef]
64. Wang, J.; Wang, L.; Xu, C.; Zhi, R.; Miao, R.; Liang, T.; Yue, X.; Lv, Y.; Liu, T. Perfluorooctane sulfonate and perfluorobutane sulfonate removal from water by nanofiltration membrane: The roles of solute concentration, ionic strength, and macromolecular organic foulants. *Chem. Eng. J.* **2018**, *332*, 787–797. [CrossRef]
65. Kwon, B.G.; Lim, H.J.; Na, S.H.; Choi, B.I.; Shin, D.S.; Chung, S.Y. Biodegradation of perfluorooctanesulfonate (PFOS) as an emerging contaminant. *Chemosphere* **2014**, *109*, 221–225. [CrossRef] [PubMed]
66. Müller, C.E.; LeFevre, G.H.; Timofte, A.E.; Hussain, F.A.; Sattely, E.S.; Luthy, R.G. Competing mechanisms for perfluoroalkyl acid accumulation in plants revealed using an arabidopsis model system. *Environ. Toxicol. Chem.* **2016**, *35*, 1138–1147. [CrossRef] [PubMed]
67. Krippner, J.; Brunn, H.; Falk, S.; Georgii, S.; Schubert, S.; Stahl, T. Effects of chain length and pH on the uptake and distribution of perfluoroalkyl substances in maize (Zea mays). *Chemosphere* **2014**, *94*, 85–90. [CrossRef]
68. Zhang, Y.Z.; Zheng, J.T.; Qu, X.F.; Chen, H.G. Effect of granular activated carbon on degradation of methyl orange when applied in combination with high-voltage pulse discharge. *J. Colloid Interface Sci.* **2007**, *316*, 523–530. [CrossRef]
69. Chen, J.; Zhang, P.; Liu, J. Photodegradation of perfluorooctanoic acid by 185 nm vacuum ultraviolet light. J. *Environ. Sci.* **2007**, *19*, 387–390. [CrossRef]
70. Hori, H.; Yamamoto, A.; Kutsuna, S. Efficient photochemical decomposition of long-chain perfluorocarboxylic acids by means of an aqueous/liquid CO_2 biphasic system. *Environ. Sci. Technol.* **2005**, *39*, 7692–7697. [CrossRef]

Article

Study on the Removal of Iron and Manganese from Groundwater Using Modified Manganese Sand Based on Response Surface Methodology

Han Kang *, Yan Liu, Dan Li and Li Xu

School of Municipal and Environmental Engineering, Shenyang Jianzhu University, Shenyang 110168, China
* Correspondence: hj_kh@sjzu.edu.cn; Tel.: +86-1389-793-1713

Abstract: This study used modified manganese sand as an adsorbent to explore its adsorption effect on iron and manganese ions from groundwater. The effects of pH, manganese sand dosage, and the initial concentration of Fe/Mn on the removal rate of iron and manganese ions were studied through single-factor experiments. Based on the above three factors, a quadratic polynomial model between the adsorption rate and the above factors was established to determine the optimal adsorption conditions. The response surface analysis showed that pH had the most significant effect on the adsorption process. The optimum conditions for the adsorption of iron and manganese ions by modified manganese sand were pH = 7.20, the dosage of manganese sand = 3.54 g/L, and the initial concentration ratio of Fe/Mn = 3.80. The analysis of variance showed that the RSM model could accurately reflect the adsorption process of manganese sand. In addition, we confirmed that the relative error between model predictions and experimental values was close to 1%, proving that the response surface model was reliable. The kinetic data of the manganese sand were described well with the pseudo-second-order model. The isothermal adsorption of iron and manganese ions by modified manganese sand was fitted well using the Langmuir equation.

Keywords: modified manganese sand; iron and manganese ions; response surface methodology

Citation: Kang, H.; Liu, Y.; Li, D.; Xu, L. Study on the Removal of Iron and Manganese from Groundwater Using Modified Manganese Sand Based on Response Surface Methodology. *Appl. Sci.* **2022**, *12*, 11798. https://doi.org/10.3390/app122211798

Academic Editors: Xin Zhao, Lili Dong and Zhaoyang Wang

Received: 24 October 2022
Accepted: 17 November 2022
Published: 20 November 2022

1. Introduction

Iron and manganese are natural components in the crust. High levels of iron and manganese in groundwater are common. In China, 20% of groundwater resources have excess iron and manganese [1]. In water supply networks, iron and manganese in tap water are oxidized to high valence during disinfection. The oxide precipitation formed in the pipeline is easily adsorbed into the water supply network, affecting the quality of the drinking water supply [2]. Although iron and manganese are necessary trace elements for the human body, drinking high-iron and -manganese surface water or groundwater for a long time will lead to chronic poisoning and damage to human health [3–5]. Physiologically, a large amount of iron ingested by the human body cannot be removed through metabolism. Excessive iron accumulation will induce diabetes, skin diseases, and other diseases, while excessive manganese can cause pathological changes in human organs and even cause neurotoxicity [6–8].

The coexistence of iron and manganese is common in groundwater. In recent years, how to efficiently and stably remove iron and manganese has become the focus of research. Iron usually exists in a soluble ferrous (Fe(II)) state. The most stable oxidation state of manganese is +2 valence [9,10]. The traditional removal methods of Fe(II) and Mn(II) in groundwater include natural oxidation, biological, and adsorption [11–14]. The natural oxidation method is to oxidize Fe^{2+} to Fe^{3+} via aeration and then generate $Fe(OH)_3$ precipitation. The removal of manganese requires adding alkali based on aeration to improve pH. The procedure flow of this method is complex, and the high pH in effluent needs acidification treatment, which increases the treatment cost and management difficulty.

The biological method mainly depends on microorganisms to reduce iron and manganese concentration in groundwater. However, the metabolism of microorganisms is influenced by the oxygen content, and insufficient oxygen slows down the metabolism of microorganisms [15]. Although oxidation and biological methods are widely used, their disadvantages are unstable effluent quality and difficulty to control reaction conditions. In contrast, the adsorption method has the advantages of large capacity, less energy consumption, and less pollution [16,17]. It is widely used in removing iron and manganese from groundwater and is considered one of the most effective methods to remove iron and manganese [18,19]. Manganese sand is the most widely used adsorbent for water adsorption of iron and manganese ions [20].

This study aimed to determine the best conditions for the adsorption effect of modified manganese sand. Response surface methodology can effectively optimize the optimal process parameters and evaluate the interactions between various influencing factors [21,22]. We established a response surface model (RSM) based on single-factor experiments to determine the optimal adsorption factors. The model helped analyze various factors affecting the experiment in a limited number of experiments. Meanwhile, we analyzed the interaction between the factors and proved the rationality of the response surface model through experiments.

2. Materials and Methods

2.1. Experiment Material and Equipment

A standard solution of iron and manganese ions, hydrochloric acid (Sinopharm Chemical Reagent Co., Ltd., Shanghai, China), NaOH (Tianjin Bodi Chemical Co., Ltd., Tianjin, China), and manganese sand (Shenyang Keer Automation Instrument Co., Ltd., Shenyang, China) were used in this study. MnO_2 content was 40%, iron content was 15%, SiO_2 content was 18%, MnC2 content was 27%, the solubility of hydrochloric acid was <3.5%, and the particle size was 0.9–1.7 mm. A p611 pH tester (Shanghai Youke Instrument Co., Ltd., Shanghai, China) and a UV-5500 ultraviolet-visible spectrophotometer (Shanghai Yuan Xi Instrument Co., Ltd., Shanghai, China) were used for the analyses.

2.2. Preparation of High-Efficiency Manganese Sand and Determination of Removal Rate

Studies have shown that Mn^{2+} is very stable under acidic conditions and difficult to remove through oxidation. Nevertheless, modified manganese sand can reduce the lower limit of pH in the manganese removal. In order to obtain high-efficiency manganese sand, we adopted the impregnation method to modify manganese sand [23]. Untreated manganese sand (0.9–1.7 mm) was cleaned with deionized water to remove the impurities on the surface and then placed in a drying oven at a constant temperature of 100 °C. The dried manganese sand was put into a 500 mL beaker and mixed with 5% potassium permanganate solution, and the mixture of manganese sand and the modifier was then heated at 50 °C for 16 h in a constant-temperature water bath. Then, it was dried again for 12 h in an oven at 100 °C to obtain the modified manganese sand.

To confirm the effect of adsorption, we studied the removal rate of the adsorption process and configured several water samples with different iron and manganese ion concentrations. Briefly, 2 mg/L iron together with the manganese ion water sample and the manganese sand adsorbent were added to a 250 mL conical flask. The pH of the solution was adjusted with 0.1 mol/L NaOH and HCl. After sealing, it was placed in a constant temperature shaker at 25 °C with a rotating speed of 120 r/min. For the sampling after adsorption, a 0.45 μm filter membrane was used. The absorbance was measured at 510 nm via o-phenanthroline spectrophotometry (detection limit: 0.03 mg/L; the relative standard deviation of the laboratory was 0.44%). The absorbance was measured at 525 nm through potassium periodate spectrophotometry (detection limit: 0.05 mg/L; the relative standard deviation was 3.94%). The iron and manganese ion removal rate is calculated according to Formula 1.

$$R = \left(\frac{C_0 - C_e}{C_0}\right) \times 100\% \tag{1}$$

where C_0 is the initial concentration of adsorbate, mg/L. C_e is the final concentration of adsorbate, mg/L.

2.3. Characterization of Properties

The sample pore size distribution and other parameters were determined using an ASAP 2020 physical adsorption meter manufactured by an American microphone instrument company. Nitrogen adsorption was determined at 77 K and within the range of 10^{-3}~1.0 relative pressure (p/p_0), using nitrogen as the adsorbing medium. The sample was degassed for 2 h at 300 °C before the test. The micromorphology of manganese sands before and after modification was observed using an S-4800 scanning electron microscope (Hitachi Limited, Tokyo, Japan).

2.4. Single-Factor Experimental Design

In order to obtain reasonable experimental factors, we designed single-factor experiments, which provided guidance for the design of the response surface experiments. The single-factor experiments were based on three influencing factors: pH (4, 5, 6, 7, 8, 9); manganese sand dosage (1.5, 2, 2.5, 3, 3.5, 4, 4.5 g/L); and initial concentration of Fe/Mn (2, 4, 6, 8, 10). The concentrations of iron and manganese ions were 1 and 0.5 mg/L, 2 and 0.5 mg/L, 3 and 0.5 mg/L, 4 and 0.5 mg/L, and 5 and 0.5 mg/L, respectively. Three groups of parallel tests were set up to detect the concentration of iron and manganese in the water sample and calculate the removal rate. The appropriate value range of each factor was determined.

2.5. Response Surface Experimental Design

The interaction of pH, the amount of adsorbent, and the initial concentration ratio of iron and manganese ions on the adsorption of manganese sand was studied. The adsorption conditions were optimized within the experimental range, and the experimental design was optimized using the Box–Behnken response surface. Based on the single-factor experiments described in Section 2.4, the range of three input variables was determined: pH (A), manganese sand dosage (B), and the Fe/Mn initial concentration ratio (C). A central composite design was carried out for the adsorption experiment at three levels: low (−1), medium (0), and high (1). There were 17 groups of experiments; each group of experiments was repeated three times, and the average value was taken as the corresponding response value. Table 1 shows the coding values for each level.

Table 1. Variables and experimental design levels for the Box–Behnken design.

Factors	Code	Levels		
		−1	0	1
pH	A	6.5	7	7.5
Manganese sand dosage (g/L)	B	3.2	3.5	3.8
Initial concentration of Fe/Mn	C	3	4	5

2.6. Adsorption Kinetic Experiment

Several solutions containing 2 mg/L iron and manganese ions were prepared. Briefly, 2 g of manganese sand was weighed for each sample and put into 250 mL conical flasks in an incubator. The temperature of the incubator was set to 25 °C, and the oscillation intensity was set to 120 r/min to ensure full contact between the modified manganese sand and the water sample. The samples were taken at 15, 30, 60, 120, 240, 360, 480, 600, and 720 min, respectively, to determine the concentration of iron and manganese ions in the filtrate. An adsorption kinetic model of manganese sand was established to explore the mechanism of iron and manganese removal.

2.7. Adsorption Isotherm Experiment

Briefly, 2 g of manganese sand was weighed in a 250 mL conical flask to prepare iron and manganese solutions with different initial concentrations. The iron and manganese solutions with different initial mass concentrations were poured into the water samples, and the pH value was adjusted to 7.2. The solutions were placed in a 120 r/min constant temperature oscillator, shaken, and adsorbed for 12 h, and then samples were taken to determine the concentration of iron and manganese ions in the filtrate. The Langmuir and Freundlich models were used to explore the mechanism of iron and manganese removal.

3. Results and Discussion

3.1. SEM Results of Manganese Sand

The unmodified (a) and modified (b) manganese sand were scanned with a scanning electron microscope. From Figure 1b, it can be seen that the manganese sand modified by potassium permanganate had a loose cluster distribution, and the roughness of the surface increased. There were many different sizes of pores on the surface of manganese sand, which may be one of the reasons for the improvement in its adsorption performance.

Figure 1. SEM of unmodified manganese sand (**a**) and modified manganese sand (**b**).

3.2. Surface Area and Porosity Analysis

Compared with unmodified manganese sands, the specific surface area, the pore volume, and the average pore size of the modified manganese sand samples in Table 2 increased by 26.9%, 41.7%, and 26.9%, respectively. During the modification process, the influence of the potassium permanganate solution on the internal thin layer and impurities of manganese sand samples changed the micropore structure of the manganese sand samples themselves, and through the connection of small pores, intermediate pores were formed. With the increase in the specific surface area and pore diameter, modified manganese sand was more beneficial to the diffusion and adsorption of iron and manganese ions.

Table 2. BET test results of manganese sand.

Sample	BET Surface Area m^2/g	Total Pore Volume cm^3/g	Average Pore Size nm
Unmodified manganese sand	19.281	0.060	12.451
Modified manganese sand	24.459	0.085	15.797

The adsorption volume of manganese sand increased with the increase in relative pressure. When the relative pressure was about 0.5 in Figure 2, the curve was divided into two smooth curves. The hysteresis loop occurred at high pressure, and capillary condensation occurred. According to the figure, the nitrogen adsorption–desorption isotherms of manganese sand and modified manganese sand belonged to type IV isotherms in the BDDT classification system. The adsorption–desorption curve revealed an obvious hysteresis phenomenon, which was the result of nitrogen capillary condensation in the mesopore. The adsorption effect of the modified manganese sand was improved, which showed that the pore expanded after being corroded by potassium permanganate.

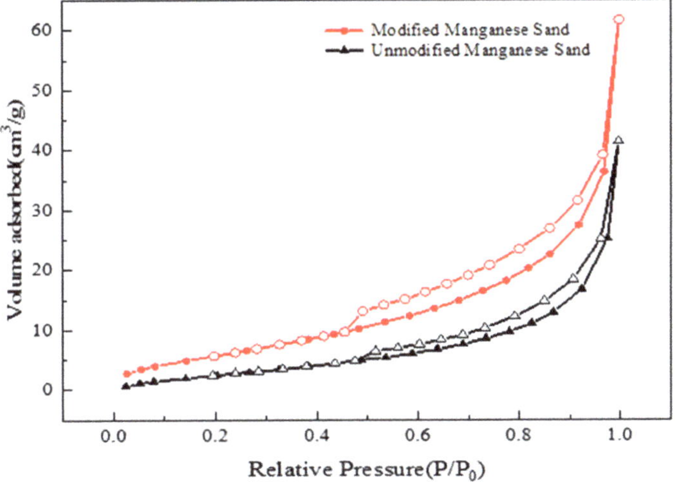

Figure 2. Nitrogen adsorption–desorption curve.

3.3. Single-Factor Experimental Results and Analysis

3.3.1. Effect of pH on the Iron and Manganese Ion Removal

PH was one of the essential factors for iron and manganese ion removal. We studied the effect of the initial pH value on the adsorption process. As shown in Figure 3, the iron and manganese ions' removal rate showed an upward trend with the increase in pH. When the pH rose from 6.0 to 7.0, the iron ion removal rate rapidly increased from 69.8% to 80.4%, and the manganese ion removal rate rapidly increased from 62.3% to 75.5%. However, when the pH was greater than 7, the iron and manganese ion removal rate increased slowly, and the change was not significant.

The hydrated ion radius of H^+ in water was much smaller than Fe^{2+} and Mn^{2+}. Under acidic conditions, a large number of H^+ competed with iron and manganese ions for adsorption sites [24], resulting in the low removal efficiency of iron and manganese. At the same time, when the pH was too low, an iron filter membrane was easily formed on the surface of manganese sand, and iron infiltrated the filter layer and interfered with the formation of the manganese active filter membrane, thus affecting the manganese removal effect [25,26]. However, if the pH was too high, on the one hand, the iron ions in the solution precipitated in the form of hydroxide, which reduced the catalytic capacity, and on the other hand, it would inhibit the production of OH^-. The above results showed that when the pH value was close to neutral, the removal rate of iron and manganese ions was significantly improved. Based on the consideration of economic factors and manganese sand adsorption conditions [27], the optimal pH value of manganese sand adsorption was 7.0.

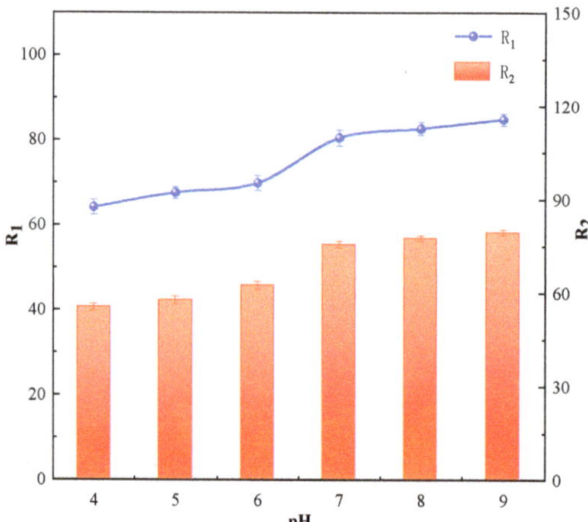

Figure 3. Effect of pH on the iron and manganese ions' removal. R_1 is the iron ion removal rate; R_2 is the manganese ion removal rate.

3.3.2. Effect of Manganese Sand Dosage on the Iron and Manganese Ion Removal

It can be seen from Figure 4 that, with the increase in manganese sand consumption, the system had more adsorption sites, and the removal rate of iron and manganese ions gradually increased. When the amount of manganese sand increased from 1.5 g/L to 3.5 g/L, the iron removal rate increased from 61.6% to 84.5%, and the manganese removal rate increased from 61.1% to 74.7%. When the dosage continued to increase from 3.5 to 4.5 g/L, the iron ion removal rate slowly increased from 84.5% to 85.2%, and the manganese ion removal rate slowly increased from 74.7% to 75.8%. This showed that increasing the dosage of manganese sand has little effect on adsorption in this range.

With the increase in manganese sand dosage and adsorption sites, the removal rate of iron and manganese ions increased. However, an excessive amount of manganese sand also aggravated the collision between the particles, which was not conducive to the adsorption of ions, and the maximum amount of the adsorbed substances when the unit mass of the adsorbent reached equilibrium was reduced. Considering process costs and the regeneration cost of manganese sand, we considered the amount of manganese sand used in this test to be controlled at about 3.5 g/L. Therefore, the optimal dosage of manganese sand was determined to be 3.5 g/L.

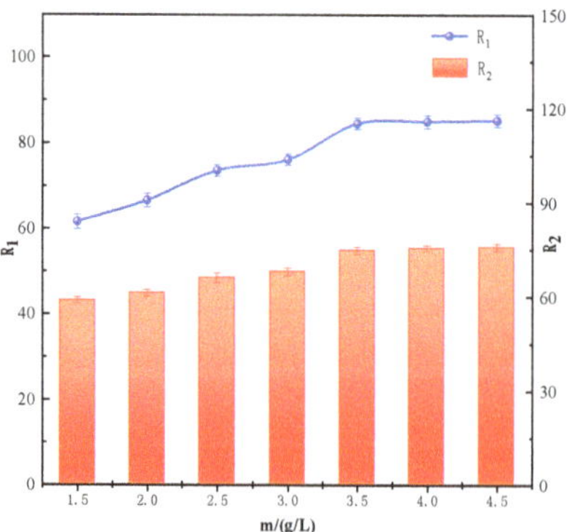

Figure 4. Effect of manganese sand dosage on the iron and manganese ion removal. R_1 is the iron ion removal rate; R_2 is the manganese ion removal rate.

3.3.3. Effect of Initial Concentration of Fe/Mn on the Iron and Manganese Ion Removal

As can be seen from Figure 5, with the increase in the initial concentration ratio of Fe/Mn from 2 to 4, the iron ion removal rate in the solution increased from 78.4% to a maximum of 82.9%. The manganese ion removal rate showed a decreasing trend, and the degree of decrease became obvious after the Fe/Mn ion concentration ratio was greater than 4.

The above phenomenon indicated that when the concentration of iron in the solution was small, the generated iron hydroxide acted as a coagulant and adsorbed iron, so the removal rate of iron was slightly increased. With the increase in the iron ion concentration, the system had a dynamic adsorption equilibrium. At the same time, there was a competitive adsorption relationship between iron and manganese ions. Excess iron ions dissociated in the solution and affected the removal of manganese ions. Therefore, a higher Fe/Mn concentration ratio was not favorable for the removal of Fe and Mn ions. Based on the above analysis, the reasonable initial Fe/Mn concentration ratio in this experiment was about 4.

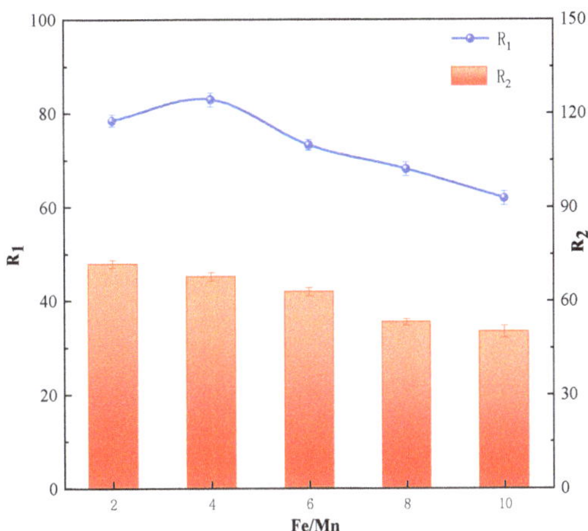

Figure 5. Effect of the initial concentration ratio of iron and manganese ions on the iron and manganese ions' removal. R_1 is the iron ion removal rate; R_2 is the manganese ion removal rate.

3.4. Analysis of Response Surface Experimental Results

Experimental Results

Based on the single-factor experimental results, 17 groups of comparative experiments were designed by using the response surface method. The experimental results are shown in Table 3.

Design-Expert 10.0 analysis was performed, and the experimental results are shown in Table 4. The quadratic polynomial regression model for the interaction between pH (A), manganese sand dosage (B), and the Fe/Mn initial concentration ratio (C) was obtained as follows:

$$Y_1 = 83.14 + 1.49A - 0.057B - 0.65C + 0.62AB - 0.41AC + 0.16BC - 1.42A^2 - 0.97B^2 - 1.18C^2 \quad (2)$$

$$Y_2 = 76.04 + 1.36A + 1.15B + 0.13C - 0.75AB - 0.70AC - 2.39BC - 3.16A^2 - 3.53B^2 - 2.89C^2 \quad (3)$$

where Y_1 is the predicted Fe removal efficiency (%), Y_2 is the predicted Mn removal efficiency (%), A is the pH value, B is the manganese sand dosage, and C is the initial concentration ratio of Fe/Mn.

Table 3. Experimental design and results.

No.	Independent Factors			Removal Rate %		No.	Independent Factors			Removal Rate %	
	A	B	C	Fe^{2+}	Mn^{2+}		A	B	C	Fe^{2+}	Mn^{2+}
1	−1	−1	0	79.92	65.24	10	0	1	−1	81.34	72.43
2	1	−1	0	81.60	70.24	11	0	−1	1	80.32	71.58
3	−1	1	0	78.57	69.95	12	0	1	1	80.52	68.17
4	1	1	0	82.81	71.94	13	0	0	0	83.15	75.23
5	−1	0	−1	79.37	68.32	14	0	0	0	83.24	76.37
6	1	0	−1	83.15	71.65	15	0	0	0	83.29	76.95
7	−1	0	1	78.74	69.72	16	0	0	0	82.95	75.38
8	1	0	1	80.89	70.24	17	0	0	0	83.06	76.28
9	0	−1	−1	81.78	66.27						

The analysis of variance is an essential statistical method to test the significance and suitability of regression models. The results of the analysis of variance for the regression models of Y_1 and Y_2 obtained from the response surface are shown in Table 4. The F-value was used for statistical saliency detection, and the p-value was used for each regression coefficient. If the p-value in the model was smaller, the experimental results were more significant. In this experiment, the p-values of the two models were <0.01, indicating that the model had high significance and was suitable for the optimization study of this parameter. Similarly, the F-value in Table 4 showed that the quadratic model had certain sufficiency and significance. The p-values of the lack of fit for both models were greater than 0.05, showing that the model was consistent in the regression study. As can be seen from the data in Table 4, the correlation coefficients of the predicted value of the model's removal rate and the experimental value were 0.9972 and 0.9747, respectively. This could better reflect the removal effect of iron and manganese ions. The values of C.V. were 0.16 < 10% and 1.20 < 10%, respectively, proving that the experiment had good accuracy and reliability. In the model, the values of Adeq precision referred to the effective signal-to-noise ratio and were considered reasonable if they were greater than 4. The Adeq precision values of the two models, as shown in the table, were 45.006 > 4 and 15.392 > 4, which means the model had high accuracy [28]. In conclusion, the results of this analysis revealed that the model could replace the real point of the test to analyze the results.

Table 4. Analysis of variance of regression models.

Source	Sum of Squares		df	Mean Square		F-Value		p-Value		
	Fe^{2+}	Mn^{2+}		Fe^{2+}	Mn^{2+}	Fe^{2+}	Mn^{2+}	Fe^{2+}	Mn^{2+}	
Model	43.87	197.39	9	4.84	21.93	279.12	30.02	<0.0001	<0.0001	significant
A-pH	17.79	14.69	1	17.79	14.69	1018.64	20.01	<0.0001	0.0029	
B-Dosage	0.026	10.49	1	0.026	10.49	1.51	14.35	0.2582	0.0068	
C-Fe/Mn	3.34	0.14	1	3.34	0.14	191.30	0.219	<0.0001	0.6800	
AB	1.54	2.27	1	1.54	2.27	88.04	3.10	<0.0001	0.1217	
AC	0.66	1.97	1	0.66	1.97	38.03	2.70	0.0005	0.1442	
BC	0.10	22.90	1	0.10	22.90	5.86	31.34	0.0460	0.0008	
A^2	8.52	42.09	1	8.52	42.09	488.01	57.61	<0.0001	0.0001	
B^2	3.96	52.52	1	3.96	52.52	226.95	71.88	<0.0001	<0.0001	
C^2	5.84	35.21	1	5.84	35.21	334.41	48.19	<0.0001	0.0002	
Residual error	0.12	5.11	7	0.017	0.73					
Lack of fit	0.047	3.05	3	0.016	1.02	1.07	1.97	0.5395	0.2613	not significant
Pure error	0.075	2.07	4	0.019	0.52					
summation	44	202.51	16							
C.V.%		0.16 1.20	Adeq Precision		45.006 15.392	R^2		0.9972 0.9747		

Figure 6 illustrates the correlation between the predicted value and the actual value of the removal efficiency in the response surface model. The experimental data points in the figure were more distributed on the straight line or on both sides of the straight line, indicating a good fit of the model. The summary of the fitted output report showed that the quadratic response model was suitable for explaining the relationship between the pollutant variables, so it could be used to analyze and optimize the effect of iron and manganese ions in water on the adsorption system.

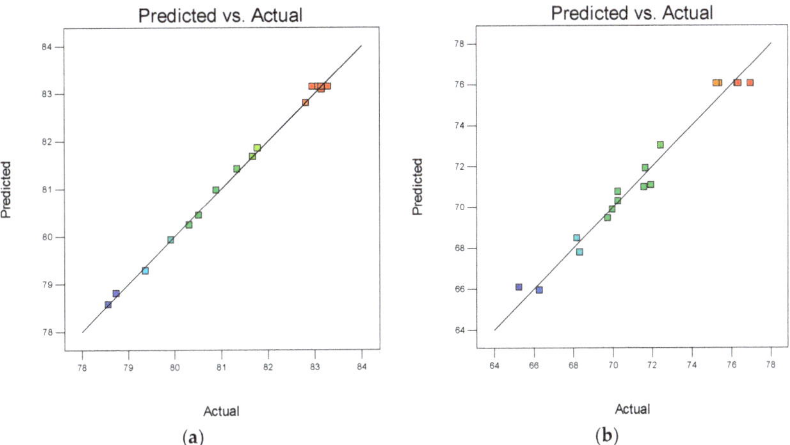

Figure 6. Residual plots of Fe and Mn ions (**a**) and plots of predicted vs. actual values (**b**).

3.5. Analysis of Interaction between Factors

The response surface method overcomes the problem with orthogonal experiments, namely that they cannot give intuitive graphics. According to the fitted quadratic equation model, the response surfaces between different test factors can be plotted. Two-dimensional and three-dimensional response surface maps can better explain independent variables and interaction effects [29]. The shape of the contour can reflect the size of the interaction. A circle indicates that the interaction between the two factors is not apparent, and an ellipse indicates that the interaction is obvious. In other words, the greater the flattening degree of the ellipse, the more significant the interaction between the two factors [30]. Using this method, we analyzed and evaluated the effect of manganese sand on ion removal according to the interaction between any two factors. An additional factor was controlled at the intermediate level when discussing the influence of the interaction on the removal rate. The 3D surface plots of the independent and dependent variables are illustrated in Figures 7 and 8.

3.5.1. Interaction of Three Factors in Iron Ion Removal

It can be seen from Figure 7a–c that when the pH was 7.1–7.3, the dosage of manganese sand was 3.4–3.6 g/L; that is, the initial concentration ratio of iron and manganese ions was between 3.5 and 4, the 3D surface had the deepest chromaticity, and the iron ion removal effect was better. It can be seen from the figure that the removal rate of iron ions could be significantly improved by increasing pH, which indicated that the change in pH significantly affected the whole adsorption system during the process of the adsorption of iron ions by manganese sand. The figure shows that with the increase in pH, the dosage of manganese sand, or the concentration of iron and manganese ions, the degradation rate of iron ions first increased and then decreased. Therefore, there was an optimal matching value between the different influencing factors, and appropriate experimental conditions can maximize the economic benefits of the model and effectively improve the iron ion removal rate.

Figure 7a shows the interaction between the pH value and the manganese sand dosage and their effect on the iron ion removal rate when the initial concentration ratio of iron to manganese was 4. It can be seen from the figure that the two-dimensional contour line was oval, indicating that the interaction between the pH and the dosage was significant. Figure 7b shows the interaction between the pH and the initial concentration ratio of iron and manganese when the dosage was fixed at 3.5 g/L. The steepness of the response surface in the figure was not as significant as the above combination, indicating that the interaction between the pH value and the initial concentration ratio of iron and manganese was not as significant as the interaction between the above pH value and the dosage of manganese sand. The slope of the pH in the response surface graph was steep, indicating that the pH value had a greater impact on the iron removal rate than these two factors. The interaction between the manganese sand dosage and the initial concentration ratio of iron and manganese is shown in Figure 7c. The steepness of the response surface in the figure was not obvious, indicating that the interaction between the dosage of manganese sand and the initial concentration ratio of iron and manganese was not significant.

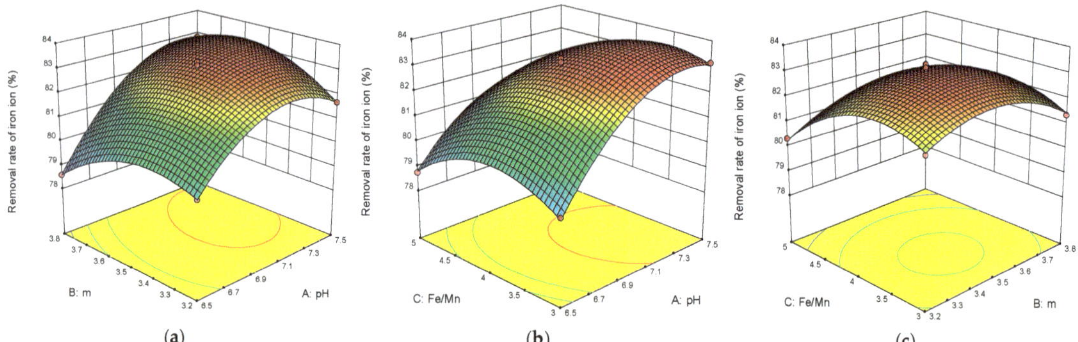

Figure 7. The 3D response surface diagram of manganese sand dosage and pH (**a**), initial concentration ratio of Fe/Mn and pH (**b**), initial concentration ratio of Fe/Mn and manganese sand dosage (**c**) influence iron ion removal.

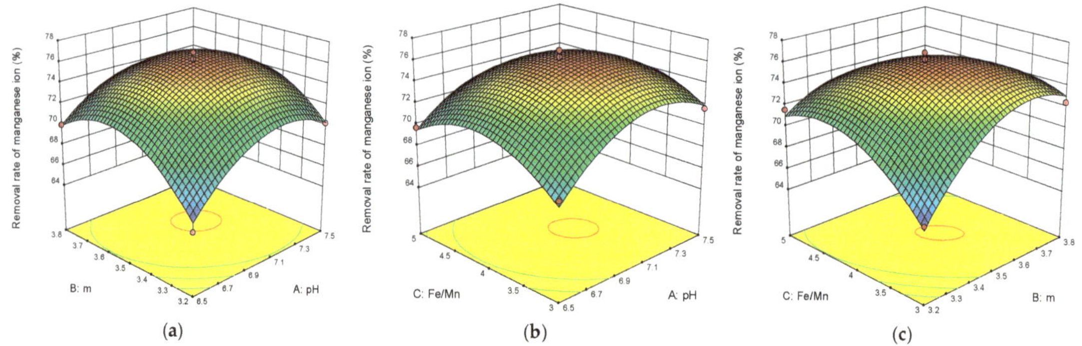

Figure 8. The 3D response surface diagram of manganese sand dosage and pH (**a**), initial concentration ratio of Fe/Mn and pH (**b**), initial concentration ratio of Fe/Mn and manganese sand dosage (**c**) influence manganese ion removal.

3.5.2. Interaction of the Three Factors in Manganese Removal

The insignificant ellipticity of the contours in Figure 8a indicated that the interaction between the pH and the dosage had no significant effect on the manganese ion removal rate. When the initial concentration ratio of Fe/Mn was fixed at 4, the slope of the pH on the response surface was very steep, indicating that the pH value had a more obvious influence on manganese ion removal in these two factors. Under the condition of a fixed dosage, the removal efficiency of manganese ions changed with the increase in the pH value. It can be seen from the response surface model that, with the increase in the pH value and the dosage of manganese sand, the removal of Mn^{2+} by manganese sand first increased and then decreased. When the dosage of manganese sand was 3.4–3.6 g/L, the pH value remained between 6.9 and 7.3, and the removal rate of manganese ions was within the maximum range.

As can be seen from the contour lines in Figure 8b, there was an interaction between the pH value and the initial concentration ratio of Fe/Mn. The slope of the pH value in the response surface was large, which means that the influence of the pH value on the experimental adsorption process was more significant. With the increase in the pH value from 6.5 to 7.5, the removal of Mn^{2+} by manganese sand first increased and then decreased. When the concentration ratio of Fe/Mn was 3.5–4.5, and the pH value was 6.9–7.3, the removal rate of manganese ions by manganese sand was within the maximum range.

Figure 8c shows the interaction model of the manganese sand dosage and the concentration ratio of iron and manganese ions at pH = 7. It can be seen that the response surface had an "arch shape", indicating that the interaction between these two factors in the manganese ion removal process was very significant. When the initial concentration ratio of Fe/Mn was constant, the removal of manganese ions first increased and then decreased with the increase in dosage. The slope reflecting the dosing amount in the response surface plot was larger, indicating that when considering the interaction between these two factors, the dosage significantly affected the removal of manganese. When the amount of manganese sand was controlled at 3.5–3.6 g/L, and the initial concentration ratio of Fe/Mn was controlled at 3.5–4, the removal effect of manganese ions was the best.

3.6. Validation Experiment

From the response surface analysis, it can be seen that there was an efficient combination of operating parameters for the complex interaction among the three influencing factors. Through the optimization function of the Design-Expert software, the optimal parameters of the reaction system were predicted as follows: pH = 7.20, the dosage of manganese sand was 3.54 g/L, and the initial concentration ratio of iron and manganese was 3.80. Under this optimal condition, the predicted iron ion removal rate was 83.62%, and the manganese ion removal rate was 76.10%. In order to verify the prediction results, three groups of parallel experiments were carried out in the same reaction system. The average removal rate of iron ions was 82.78%, and the average removal rate of manganese ions was 75.89%, both of which were close to the prediction values, and the relative deviation was less than 1%. These results show that the model can truly reflect the influence of the various analyzed factors on iron and manganese ion removal rate.

3.7. Analysis of Adsorption Kinetic Model

Figure 9 shows the adsorption kinetic curves reflecting the adsorption of iron and manganese ions by manganese sand. The trends of the adsorption rate curves of the two models were basically similar. During the initial 240 min of the adsorption process, manganese sand had many adsorption sites, a large curve slope, and a fast adsorption rate. With the progress in adsorption, the adsorption sites on the surface of manganese sand were gradually occupied by iron and manganese ions, and the adsorption rate slowly decreased until the adsorption equilibrium was reached at 480 min.

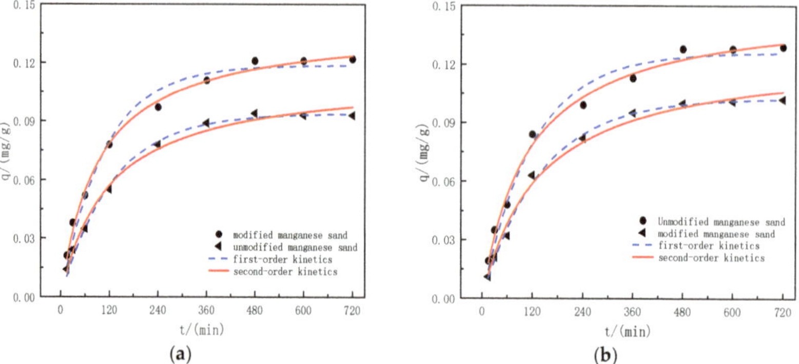

Figure 9. The adsorption kinetic model of iron (**a**) and manganese (**b**) ions.

The quasi-primary and quasi-secondary kinetic fitting parameters are shown in Tables 5 and 6. In the adsorption of iron and manganese ions by the unmodified manganese sand and modified manganese sand, the fitting regression coefficients of the quasi-first-order kinetic model were less than those of the quasi-second-order kinetic model, so it can be described by the quasi-second-order kinetic model. Compared with the unmodified manganese sand, the theoretical adsorption capacity of the modified manganese sand was improved. This was because the specific surface area of the modified manganese sand increased, and the adsorption point increased, which was conducive to increasing the reaction residence time of the adsorbent and the adsorbate, and at the same time, increasing the equilibrium adsorption capacity.

Table 5. Kinetic fitting parameters of the adsorption of iron ions by manganese sand.

	First-Order Kinetic Equation			Second-Order Kinetic Equation		
	Q_e	k_1	R^2	Q_e	k_1	R^2
Unmodified manganese sand	0.0941	0.4701	0.9839	0.110	5.059	0.995
Modified manganese sand	0.1189	0.5564	0.9772	0.139	4.828	0.998

Table 6. Kinetic fitting parameters of the adsorption of manganese ions by manganese sand.

	First-Order Kinetic Equation			Second-Order Kinetic Equation		
	Q_e	k_1	R^2	Q_e	k_1	R^2
Unmodified manganese sand	0.1026	0.4298	0.9819	0.126	3.307	0.992
Modified manganese sand	0.1261	0.4961	0.9790	0.150	3.687	0.996

3.8. Analysis of Adsorption Isotherm Model

We used the Langmuir and Freundlich models to fit the adsorption process. The thermodynamic model parameters are shown in Table 7. The Langmuir and Freundlich isotherm models showed a good linear relationship in Figure 10. It can be seen from Table 7 that R^2_L was greater than R^2_F. Obviously, the determination coefficient of the Langmuir isotherm equation was closer to 1, and the fitting effect was good. The adsorption of the ions by manganese sand was a single-layer surface chemical adsorption. In the quasi-second-order kinetic model, the adsorption reaction was assumed to be a single-layer adsorption system, and the adsorption mechanism was chemical adsorption, which was consistent with the fitting result of the adsorption isotherm.

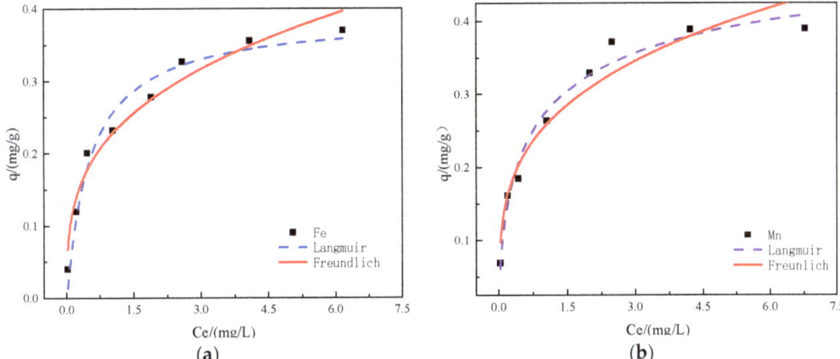

Figure 10. Adsorption isotherm models of iron (**a**) and manganese (**b**) ions.

Table 7. Constants of Langmuir and Freundlich isotherms.

	Langmuir			Freundlich		
	K_L	q_m	R^2	K_F	1/n	R^2
Fe	1.97	0.395	0.9908	0.2148	0.395	0.9670
Mn	2.05	0.444	0.9918	0.2505	0.337	0.9759

4. Conclusions

In this study, we investigated the interaction of three variables (pH, the manganese sand dosage, and the initial concentration ratio of Fe/Mn) on the removal of iron and manganese ions using RSM. We first determined the optimal experimental range of the three factors through single-factor experiments. ANOVA explained the significance of the factors, and the results proved the accuracy of the model. To determine the interactions between the input variables, we developed a three-dimensional response surface. The isothermal model of the adsorption process and the adsorption mechanism were also investigated.

(i) The Box–Behnken experimental design results showed that the interaction between the pH value and the dosage was the most obvious for iron removal. The interaction between the dosage and the initial concentration ratio of iron to manganese was the most obvious for manganese removal;

(ii) The results of response surface optimization showed that when pH was 7.20, the amount of adsorbent was 3.54 g/L, and when the initial concentration ratio of iron and manganese ions was 3.80, the adsorption rates of iron and manganese ions by manganese sand was higher, reaching 83.62% and 76.10%;

(iii) The adsorption of iron and manganese ions by manganese sand followed the Langmuir isotherm adsorption model and the quasi-second-order kinetic model. The modified manganese sand had a remarkable adsorption effect on iron and manganese ions and thus can be used as a good material for a new type of ion adsorbent.

Author Contributions: Conceptualization, H.K. and Y.L.; methodology, D.L.; software, D.L.; validation, H.K. and L.X.; formal analysis, L.X.; investigation, H.K.; resources, Y.L.; data curation, L.X.; writing—original draft preparation, D.L. and Y.L.; writing—review and editing, H.K.; visualization, Y.L.; supervision, H.K.; project administration, H.K.; funding acquisition, H.K. and L.X. All authors have read and agreed to the published version of the manuscript.

Funding: This research was funded by the Department of Science and Technology of Liaoning province (Grant No. 20170520224).

Institutional Review Board Statement: Not applicable.

Informed Consent Statement: Not applicable.

Data Availability Statement: Not applicable.

Conflicts of Interest: The authors declare no conflict of interest.

References

1. Zhao, L.; Zhou, L.; Li, G.; Zhang, F.; Li, T. Distribution and genetic analysis of iron and manganese microbial community in soil and groundwater of mining area. *Environ. Chem.* **2021**, *40*, 1464–1479.
2. Zhang, W.; Zhu, J. Iron and Manganese Pollution in Groundwater and Its Treatment Methods. *Guangdong Chem. Ind.* **2018**, *45*, 163–164.
3. Yu, D.; Zhou, J.; Chen, J. Spatial distribution characteristics and genesis of groundwater with high iron and manganese content in Kashi Prefecture, Xinjiang. *Environ. Chem.* **2020**, *39*, 3235–3245.
4. Adeyeye, O.; Xiao, C.; Zhang, Z.; Liang, X. State, source and triggering mechanism of iron and manganese pollution in groundwater of Changchun, Northeastern China. *Environ. Monit. Assess* **2020**, *192*, 619. [CrossRef] [PubMed]
5. Tan, W.; Wang, Y.; Yu, C.; Pan, Z. Treatment of groundwater with high iron and manganese by modified zeolite. *Chin. J. Environ. Eng.* **2013**, *7*, 2203–2207.
6. Fleming, R.E.; PONK, A.P. Iron overload in human disease. *N. Engl. J. Med.* **2012**, *366*, 348–359. [CrossRef]
7. Santra, S.; Agrawal, D.; Kumar, S.; Mishra, S.S. Incidence and prevalence of chronic iron poisoning and it's management: A review. *Int. J. Pharma Bio Sci.* **2014**, *5*, 722–737.
8. Ferreira, D.C.; Graziele, I.; Marques, R.C.; Gonçalves, J. Investment in drinking water and sanitation infrastructure and its impact on waterborne diseases dissemination: The Brazilian case. *Sci. Total Environ.* **2021**, *779*, 146–279. [CrossRef]
9. Ghazi, M.M.; Qomi, M.H. Removal of manganese from an aqueous solution using micellar-enhanced ultrafiltration (MEUF) with SDS surfactants. *Adv. Environ. Technol.* **2015**, *1*, 17–23.
10. Bright, K.A.; Baah, S.N.; Elizabeth, V.K.; Nkrumah, I.; Williams, C. Adsorptive Removal of Iron and Manganese from Groundwater Samples in Ghana by Zeolite Y Synthesized from Bauxite and Kaolin. *Water* **2019**, *11*, 1912.
11. Bruins, J.H. *Manganese Removal from Groundwater: Role of Biological and Physico-Chemical Autocatalytic Processes*; CRC Press: Boca Raton, FL, USA, 2017.
12. Li, G.; Liang, H.; Yu, H.; Du, X.; Yang, H. Research on manganese removal by chemical auto-catalytic oxidation mechanism involved in active manganese oxides film. *Water Wastewater Eng.* **2019**, *45*, 1–5.
13. Diaz-Alarcón, J.A.; Alfonso-Pérez, M.P.; Vergara-Gómez, I.; Díaz-Lagos, M.; Martínez-Ovalle, S.A. Removal of iron and manganese in groundwater through magnetotacticbacteria. *J. Environ. Manag.* **2019**, *249*, 109381. [CrossRef] [PubMed]
14. Elsheikh, M.A.; Guirguis, H.S.; Fathy, A. Removal of Iron and Manganese from Groundwater: A Study of Using Potassium Permanganate and Sedimentation. *Mansoura Eng. J.* **2017**, *42*, 7–12.
15. Marsidi, N.; Hasan, H.A.; Abdullah, S.R.S. A review of biological aerated filters for iron and manganese ions removal in water treatment. *J. Water Process Eng.* **2018**, *23*, 1–12. [CrossRef]
16. Shrestha, S.; Dhami, A.K.; Nyachhyon, A.R. Adsorptive Removal of Fe (II) By NaOH Treated Rice Husk: Adsorption Equilibrium And Kinetics. *Sci. World* **2021**, *14*, 75–82. [CrossRef]
17. Arafat, M.; Marzouk, S.Y.; El Monayeri, O.D. Hybrid system for iron and manganese reduction from polluted water using adsorption and filtration. *Ain Shams Eng. J.* **2021**, *12*, 2465–2470. [CrossRef]
18. Li, X.K.; Chu, Z.R.; Liu, Y.J.; Zhu, M.T.; Yang, L.; Zhang, J. Molecular characterization of microbial populations in full-scale biofilters treating iron, manganese and ammonia containing groundwater in Harbin. *Bioresour. Technol.* **2013**, *147*, 234–239. [CrossRef]
19. Guo, F.; Li, X.; Yang, Y. Effect of Filter Material Characteristics on Iron and Manganese Removal Efficiency during Start-up Period of Manganese Sand Filter. *China Water* **2018**, *34*, 16–25.
20. Ye, M.X.; Pan, J.; Chen, P. Study on optimum operation parameters of modified manganese sand filter for treatment of high iron and manganese water containing ammonia nitrogen. *Jiangsu Water Resour.* **2018**, *9*, 11–15.
21. Liu, J.; Zhang, H.L.; You, K.; Yuan, Y.S. Different Filter Material on the Northeast Biological Iron Manganese Removal Effect of Groundwater in the Countryside. *Appl. Mech. Mater.* **2014**, *522*, 465–468. [CrossRef]
22. Eri, I.R.; Hadi, W.; Slamet, A. Clarification of pharmaceutical wastewater with Moringa oleifera: Optimization through response surface methodology. *J. Ecol. Eng.* **2018**, *19*, 126–134. [CrossRef]
23. Dong, X.; Jin, B.; Sun, Y.; Yu, L. Urban gas production from low H2/CO biogas using Re-promoted Ni catalysts supported on modified manganese sand. *Fuel* **2018**, *220*, 60–71. [CrossRef]
24. Agbovi, H.K.; Wilson, L.D. Optimisation of orthophosphate and turbidity removal using an amphoteric chitosan-based flocculant-ferric chloride coagulant system. *Environ. Chem.* **2019**, *16*, 599–612. [CrossRef]
25. Wang, Y.H.; Lv, W.Y.; Zou, X.G.; Shu, R.J.; Huang, J.L.; Yao, K.; Liu, G.G. The study of adsorption mechanism of Cu(II) from aqueous solutions by humin with response surface methodology. *Acta Sci. Circumstantiae* **2017**, *37*, 624–632.
26. Chen, T.Y.; Chen, Z.H.; Jin, S.F.; Li, H.S.; Li, G.B.; Liang, H. Effect of pH Value on Treatment of Groundwater Containing High Concentrations of Iron, Manganese and Ammonia Nitrogen. *China Water* **2015**, *31*, 1–9.

27. Wang, Y.Y.; Li, J.Y.; Lu, Y.W. Optimization of Cu^{2+} adsorption on corncob by response surface methodology. *J. Shanghai Ocean Univ.* **2020**, *29*, 355–363.
28. Agarwal, M.; Patel, D.; Dinker, A. Optimization of Manganese Removal from Water Using Response Surface Methodology. *Iran. Technol.* **2016**, *40*, 63–73. [CrossRef]
29. Agbovi, H.K.; Wilson, L.D. Flocculation optimization of orthophosphate with $FeCl_3$ and alginate using the Box-Behnken response surface methodology. *Ind. Eng. Chem. Res.* **2017**, *56*, 3145–3155. [CrossRef]
30. Yan, Y.P.; Wang, G.; Jiang, S.J.; Wang, L.L. Response surface methodology for optimizing Cu^{2+} removal by heavy metal flocculant DTAPAM. *Acta Sci. Circumstantiae* **2021**, *41*, 2156–2161.

applied sciences

Article

Water Quality Evaluation and Prediction Based on a Combined Model

Guimei Jiao [1,*], Shaokang Chen [1], Fei Wang [1], Zhaoyang Wang [2], Fanjuan Wang [1], Hao Li [1], Fangjie Zhang [1], Jiali Cai [1] and Jing Jin [1]

1 School of Mathematics and Statistics, Lanzhou University, Lanzhou 730000, China
2 College of Earth and Environmental Sciences, Lanzhou University, Lanzhou 730000, China
* Correspondence: gmjiao@lzu.edu.cn

Abstract: Along with increasingly serious water pollution, water environmental problems have become major factors that hinder the sustainable development of our economy and society. Reliable evaluation of water quality and accurate prediction of water pollution indicators are the key links in water resource management and water pollution control. In this paper, the water quality data of Lanzhou Xincheng Bridge section in the Yellow River Basin and Sichuan Panzhihua Longdong section in the Yangtze River Basin were used to establish a water quality evaluation model and a prediction model. For the water quality evaluation model, we constructed the research samples by means of equal intervals and uniform distribution of interpolated water quality index data according to *Environmental Quality Standards for Surface Water*. The training samples were determined by a stratified sampling method, and the water quality evaluation model was established using a T-S fuzzy neural network. The experimental results show that the highest accuracy achieved by the evaluation model in water quality classification was 94.12%. With respect to the water quality prediction model, we propose ARIMA-WNN, which combines the autoregressive integrated moving average model (ARIMA) and a wavelet neural network (WNN) with the bat algorithm (BA) to determine the optimal weight of each individual model. The experimental results show that the highest prediction accuracy of ARIMA-WNN is 68.06% higher than that of the original model.

Keywords: water quality evaluation; water quality forecasting; T-S fuzzy neural network; combined model

Citation: Jiao, G.; Chen, S.; Wang, F.; Wang, Z.; Wang, F.; Li, H.; Zhang, F.; Cai, J.; Jin, J. Water Quality Evaluation and Prediction Based on a Combined Model. *Appl. Sci.* **2023**, *13*, 1286. https://doi.org/10.3390/app13031286

Academic Editor: Chang-Gu Lee

Received: 10 December 2022
Revised: 30 December 2022
Accepted: 14 January 2023
Published: 18 January 2023

1. Introduction

Water is the foundation of human existence and the driving force for social stability and a nation's prosperity. However, water resource management has been ignored and forgotten for a long time. It was not until the mid-19th century that, due to the rapid development of industry, water pollution became increasingly serious and water resource management became increasingly prominent. [1]. Since then, the declining water quality of rivers, lakes and groundwater has become a global problem. Although an increasing number of countries has begun to attach importance to water resources and implement a series of protection measures for the sustainable development of water resources, the water resources environment is still deteriorating, with increasing pollution and waste caused by economic development, the acceleration of urbanization and population growth.

The river pollution situation is serious. The water quality level has fallen to IV or worse in 31.4% of the more than 208,000 km of managed river sections in China and below class V in 14.9% of managed sections, indicating that water resources have completely lost their potential for daily use [2]. Of the ten major river basins in China, only some in the southwest and northwest have moderate water quality (categories I to III), and the major river systems in the north, such as the Yellow River, Liao River and Huai River, are rated IV or V. The declining self-purification ability of rivers and deteriorating industrial wastewater management have further worsened the water quality of small tributaries flowing into the major rivers of our country.

Lakes are also heavily polluted. The water quality of nearly half of the 62 key lakes in the country is of grades IV and V or inferior. The three major lakes in China, Taihu Lake, Chaohu Lake and Dianchi Lake, are polluted to varying degrees, in states of mild, moderate and severe pollution, respectively, with total phosphorus and chemical oxygen demand representing the main pollutants.

The groundwater quality situation is also worrying. In China's major cities, 27 percent of centralized drinking water sources do not meet official standards. Among the 5118 groundwater monitoring points in various provinces and cities across the country, the proportion of poor and extremely poor water quality is more than half, threatening people's daily water use [3,4].

Water environmental problems have become a major factor hindering the sustainable development of China's economy and society, and the effective treatment of water pollution and the rational management of water resources are urgent problems to be solved. The accurate prediction of water quality indicators and the reliable evaluation of water quality grades are the basis for understanding the current water quality status and taking corresponding protection measures, so water quality prediction and evaluation have great practical significance.

In this paper, we take the water quality of Lanzhou Xincheng Bridge section in Yellow River Basin and Longdong section in Yangtze River Basin as the research object, establish a water quality evaluation model and propose a new water quality prediction model.

A T-S fuzzy neural network was used to establish the evaluation model combined with the relevant water quality information of the two basins. In the process of model training, an innovative method of interpolating water quality index data with equal intervals and uniform distribution was adopted to construct research samples, and the method of stratified sampling was used to construct training samples. The trained model was applied to water quality evaluation of Lanzhou Xincheng Bridge section in the Yellow River Basin and Longdong section in the Yangtze River Basin. A total of 52 groups were randomly selected from the real water quality index data from 2004 to 2015, and the results of water quality status were output and compared with the real water quality grade to prove the effectiveness and generalizability of the evaluation model.

Furthermore, a new model is proposed for water quality prediction, which combines the autoregressive integrated moving average (ARIMA) model and the wavelet neural network (WNN) mode with the bat algorithm to determine the optimal weight of each individual model. The combined model was used to predict the water quality indices of Lanzhou Xincheng Bridge section in the Yellow River Basin and Longdong section in the Yangtze River Basin. First, 624 weekly monitoring data points of each indicator from 2004 to 2015 were used as the training set, and 52 data points from 2016 were used as the validation set. ARIMA and WNN were used for prediction. The empirical mode decomposition (EMD) algorithm was used to denoise the data before WNN prediction [5]. Secondly, the bat algorithm was used to determine the optimal weight; the final prediction result was the weighted sum of the prediction results of ARIMA and WNN. Then, the prediction results of the combined model were compared and analyzed relative to the prediction results of three individual models (backpropagation (BP), neural network and least squares support vector machine (LSSVM)) to prove the ability of the proposed combined model for water quality prediction [6,7]. Finally, the prediction results of each index in 2016 were substituted into the previously established water quality evaluation model, and the output results had a high coincidence rate with the real water quality grade, which further verified the effectiveness of the evaluation model.

2. Literature Review

2.1. Research Status of Water Quality Evaluation

At present, multivariate statistical methods are widely used in water quality evaluation and analysis abroad. Singh K P et al. (2005) applied multivariate statistical methods to the water quality evaluation of eight monitoring sites of the Gomti River in India from 1999 to 2001, demonstrating the advantages of multivariate statistical methods in processing and evaluating a large number of complex water quality datasets and obtaining effective water quality evaluation results [8]. Shrestha S et al. (2007) collected a total of 14,976 data points measuring 12 water quality indicators at 13 different monitoring points from 1995 to 2002 using multivariate statistical tools for spatiotemporal variable analysis of a large set of complex water quality data of the Fuji River. The water quality status of the 13 observation points was divided into three categories by stratified cluster analysis: mild, moderate and severe pollution [9]. Zhang X et al. (2011) used the monthly data of 23 indicators from 16 different monitoring points in southwest Kowloon, Hong Kong, from 2000 to 2007, employing hierarchical cluster analysis to divide the 12 months into two periods and the 16 monitoring points into three categories. Discriminant analysis provides analysis results from both spatial and temporal aspects. Among the 23 indicators, 4 are the main factors affecting the temporal distribution, and 8 indicators are the main factors affecting the spatial distribution [10]. Ogwueleka T C (2015) used principal component analysis, cluster analysis and factor analysis to study the water quality of the Kaduna River and analyze the potential pollution factors of the river [11]. The multivariate statistical method is a classical method for water quality assessment and management, but it cannot provide comprehensive information about water quality. On this basis, in this study, the selected neural network constantly updates and iterates the model parameters, adjusts the weights and thresholds to the state that can output the optimal results and applies the trained model to the evaluation task.

2.2. Research Status of Water Quality Prediction

Previously proposed water quality prediction models are based on qualitative analysis. Water quality prediction was first studied in 1925, when Phelps and Streeter proposed the S-P model to track BOD-DO changes in water quality. Since then, with the worsening global water pollution and drinking water crises, an increasing number of water quality prediction models have been proposed. However, due to the complexity of water environments, it is difficult to obtain accurate prediction results with traditional mathematical models, so scholars began to use neural networks to predict water quality. Singh KP et al. (2009) used the monthly data of 11 water quality indicators from 8 different monitoring points over 10 years to establish two BP neural network models, 11-23-1 and 11-11-1, to calculate the levels of dissolved oxygen and biochemical oxygen demand of Gomti River in India and indirectly judge the water quality [12]. Seo IW et al. (2016) used an artificial neural network model to predict eight water quality indicators downstream of Cheongpyeong Dam [13]. Zhang ran et al. (2013) established a GM(1,1) model to predict the water quality of the Yellow River estuary from 2012 to 2015 [14]. Zhang Ying et al. (2015) took the section of Shanghai Qingpu Urgent Water Port in Taihu Lake Basin as an example and applied the gray model after residual correction of an extreme learning machine regression model for the prediction of water quality indicators. They used the data of the first 100 days of 2013 of six water quality indicators, including dissolved oxygen and chemical oxygen demand, to predict the data of the 101st to 110th days [15]. Xu Hongmin et al. (2007) proposed a weighted support vector regression model to predict the concentration of permanganate in Taihu Lake Basin using the same method; the results showed that the prediction accuracy of this algorithm was higher than that of SVM and RBF neural network alone [16]. According to the idea of neural network weighting introduced in the abovementioned literature, in this study, we designed and implemented a mixed model to predict water quality.

3. Materials and Methods

3.1. T-S Fuzzy Neural Network

T-S fuzzy neural network is a new fuzzy neural network proposed by Takagi and Sugeno in 1985 [17]. Fuzzy reasoning rules are adopted in '*if then*' form:

$$R^i : If\ x_1\ is\ A_1^i,\ x_2\ is\ A_2^i, \ldots,\ x_k\ is\ A_k^i\ then\ y_i = p_0^i + p_1^i x_1 + \ldots + p_k^i x_k \tag{1}$$

where '*if then*' are the front part and back part of fuzzy rules, respectively (the former is the input part of fuzzy rules, and the latter is the determined output); A_j^i is a fuzzy set of fuzzy models; $p_j^i (j = 1, 2, \ldots, k)$ is a fuzzy model parameter; and y_i is the output of the fuzzy rules. The process shows that the output is a linear combination of inputs, as shown in Figure 1.

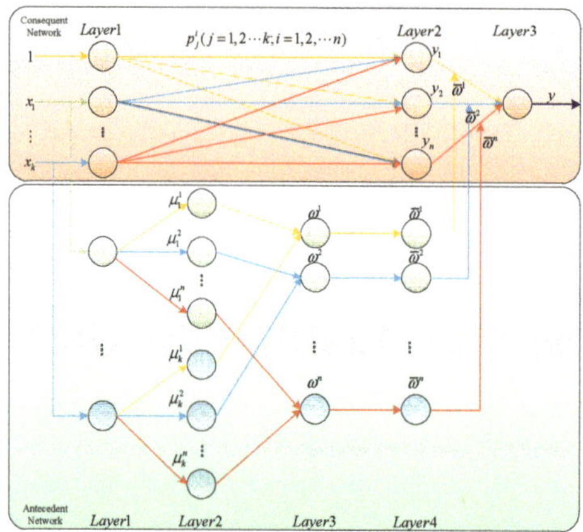

Figure 1. Structural diagram of T-S fuzzy neural network.

3.2. ARIMA

The autoregressive integrated moving average (ARIMA) model was proposed by Box and Jenkins in the 1970s [18]. It is based on the autoregressive model (AR) proposed by Yule in 1927 and the combination of the moving average model (MA) and autoregressive moving average model (ARMA) with AR and MA proposed by Walker in 1931 [19–21]. In this model, the future value of a variable is considered a linear combination of past values and past errors:

$$y_t = \theta_0 + \varphi_1 y_{t-1} + \varphi_2 y_{t-2} + \ldots + \varphi_p y_{t-p} + \varepsilon_t - \theta_1 \varepsilon_{t-1} - \theta_2 \varepsilon_{t-2} - \ldots - \theta_q \varepsilon_{t-q} \tag{2}$$

$$\varphi(B) \nabla^d y_t = \theta(B) \varepsilon_t \tag{3}$$

where y_t is the actual value; ε_t is the random error at time t; φ_i and θ_j are the coefficients; p and q are the orders of autoregressive and sliding average polynomials, respectively; B represents the lag operator, where $\Delta_d = (1 - B)^d$; d is the number of differences; and $\varphi(B)$ and $\theta(B)$ are defined as:

$$\varphi(B) = 1 - \varphi_1 B - \varphi_2 B^2 - \ldots - \varphi_p B^p \tag{4}$$

$$\theta(B) = 1 - \theta_1 B - \theta_2 B^2 - \ldots - \theta_q B^q \tag{5}$$

3.3. Wavelet Neural Network

Wavelet neural network (WNN) is a neural network model that combines wavelet transform and artificial neural network, which replaces the excitation function of the traditional neural network with the wavelet basis function. It was first proposed in 1992 by Zhang Q and Benveniste A of LRISA, a famous French information science institute [22]. The combination of wavelet transform and neural network provides unique advantages. In recent years, it has been widely used in nonlinear function approximation, dynamic modeling and non-stationary time series prediction. $X_1, X_2, \ldots, X_k$ are input parameters, ω_{ij} and ω_{jk} are connection weights and $Y_1, Y_2, \ldots, Y_m$ are predicted outputs. The formula for calculating the output of the hidden layer is:

$$h(j) = h_j\left[\frac{\sum_{i=1}^{k}\omega_{ij}x_i - b_j}{a_j}\right], j = 1, 2, \ldots, l \tag{6}$$

where h_j is the wavelet basis function; a_j and b_j are the scale factor and translation factor of the wavelet basis function, respectively; ω_{ij} is the connection weight between the input layer and the hidden layer; and $h(j)$ is the output value of the node of the seventh hidden layer. The calculation formula of the output layer is:

$$y(k) = \sum_{j=1}^{l} \omega_{jk}h(j) \quad k = 1, 2, \ldots, m \tag{7}$$

where ω_{jk} is the connection weight between the hidden layer and the output layer, and $y(k)$ is the output value.

3.4. ARIMA-WNN

Actual time series data often have both linear and nonlinear characteristics, and ARIMA or WNN alone cannot reflect the dual linear and nonlinear characteristics of time series. In order to simultaneously utilize the good linear fitting ability of the differential autoregressive moving average model and the powerful nonlinear relationship mapping ability of the wavelet neural network model, we combine ARIMA and WNN methods.

Assuming $y_t(t = 1, 2, \ldots, L)$ is the actual value of the time series, L is the number of sample points, and $\hat{y}_t$ and $\hat{y}_{it}(i = 1, 2, \ldots, N, t = 1, 2, \ldots, L)$ is the predicted value of the combined model and the first single method, respectively; then:

$$\hat{y}_t = \sum_{i=1}^{N} \lambda_i \hat{y}_{it} \tag{8}$$

where λ_i is the weight of the prediction method, and $\sum_{i=1}^{N} \lambda_i = 1$. The weight coefficients of each of the component models in the combined model are determined by solving the following optimization problems:

$$Min \sum_{t=1}^{L} (y_t - \hat{y}_t)^2, \quad s.t. \sum_{i=1}^{N} \lambda_i = 1, \ 0 \leq \lambda_i \leq 1 \tag{9}$$

On this basis, we propose a new combination model composed of the ARIMA model and WNN [23]. The optimal weight of a single model is obtained by the bat algorithm, and the predicted value of the combination model is expressed as follows:

$$P_\Sigma = \lambda_1 P_{WNN} + \lambda_2 P_{ARIMA} \tag{10}$$

where P_Σ is the final predicted value; $\lambda_1, \lambda_2, P_{WNN}, P_{ARIMA}$ are the weight coefficients and predicted values of WNN and ARIMA models, respectively; and $\lambda_1 + \lambda_2 = 1$, as shown in Figure 2.

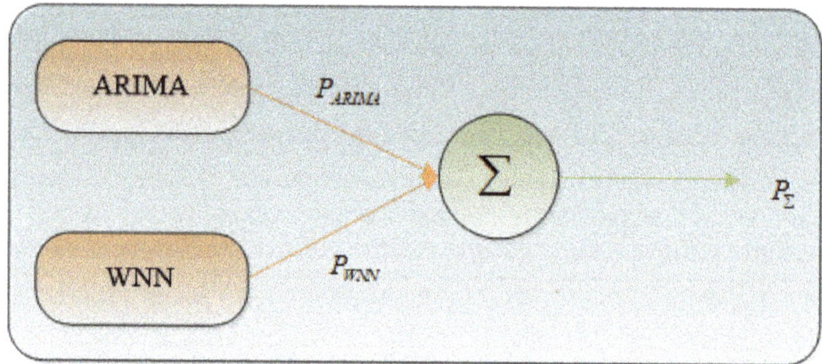

Figure 2. ARIMA—WNN structure.

3.5. Bat Algorithm

The bat algorithm (BA) is a swarm intelligence optimization algorithm that simulates the behavior of bats to hunt down prey, which was proposed by Yang X S in 2010 [24]. Let the bat search for prey in the n dimensional space at the moment of $t-1$; the flight speed and position of bat i are v_i^{t-1}, x_i^{t-1}, respectively; then, the update rules of the bat's flight speed (v_i^t) and location at time t are:

$$f_i = f_{min} + (f_{max} - f_{min}) * \beta \tag{11}$$

$$v_i^t = v_i^{t-1} + \left(x_i^{t-1} - x^*\right) * f_i \tag{12}$$

$$x_i^t = x_i^{t-1} + v_i^t \tag{13}$$

Among them, $\beta \in [0,1]$ is a uniformly distributed random number; f_{max} and f_{min} are the maximum and minimum search frequency, respectively; the bat's pulse search frequency is f_i; and $f_i \in [f_{min}, f_{max}]$. The bat algorithm controls the prey hunting range of bats by adjusting f_i and controls the global search in the whole updating process, which is the optimal solution for the current bat population.

For a local search, the bat algorithm is completed by random disturbance. Each bat randomly selects one solution from the current optimal solution set as the current optimal solution (x_{old}); then, we use the following formula to update the position to obtain a new solution:

$$x_{new} = x_{old} + \varepsilon A^t \tag{14}$$

where $\varepsilon \in [-1,1]$ is a random number, and A^t is the average loudness of all bats at time t.

3.6. Materials

(1) ADF

The ADF test determines whether the series is stationary by checking whether the sum of the autoregressive coefficients is 1 [25]. Compared with the DF test, which can only be used to determine whether the AR(1) model is stationary, the ADF test can be used to determine the stationarity of the AR(P) model. The hypothesis test is established as:

$$H_0 : \rho = 0$$

$$H_1 : \rho < 0$$

The test statistic is:

$$\tau = \frac{\hat{\rho}}{S(\hat{\rho})} \tag{15}$$

where $\rho = \varphi_1 + \Lambda + \varphi_p - 1$, and $S(\hat{\rho})$ is the sample standard deviation of parameter ρ.

(2) AF

The autocorrelation function describes the correlation between the random sequence at time t and the value at time $t-k$ [26].

$$\rho_k = \frac{Cov(y_t, y_{t-k})}{\sqrt{Var(y_t)}\sqrt{Var(y_{t-k})}} = \frac{E[(y_t-\mu)(y_{t-k}-\mu)]}{\sqrt{Var(y_t)}\sqrt{Var(y_{t-k})}} = \frac{\gamma_k}{\sigma^2} \tag{16}$$

where $E(y_t) = \mu, \sigma^2 = E(y_t - \mu)^2$, and γ_k denotes the covariance of y_t and y_{t-k}.

(3) PAF

The partial autocorrelation function shows that for sequence y_t in determining y_{t-1}, $y_{t-2}, \ldots, y_{t-k+1}$, the correlation between y_t and y_{t-k} is denoted by φ_{kk} [26]:

$$\varphi_{kk} = \begin{cases} \gamma_1 & k = 1 \\ \frac{\gamma_k - \sum_{j=1}^{k-1} \varphi_{k-1,j}\gamma_{k-j}}{1-\sum_{j=1}^{k-1} \varphi_{k-1,j}\gamma_j} & k = 2, 3, \ldots \end{cases} \tag{17}$$

(4) The AIC criterion

The AIC criterion was proposed by Japanese statistician Hiroji Akike as a measure of model fit excellence [27].

$$AIC = n \log \sigma^2 + 2(p+q) \tag{18}$$

where n is the number of samples; σ^2 is the sum of squares of the fitted residuals; and p and q are the orders of the AR and MA models, respectively. Models are established from low-order to high-order according to the ARIMA values, and the AIC value of each model is calculated. According to the criterion, the model with the lowest AIC value is the optimal model.

(5) MAE

Mean absolute error (MAE) is a measure of accuracy for regression [28]. It sums up absolute values of errors and divides them by the total number of values. It gives equal weight to each error value. The formula for calculating MAE is shown in Equation (19).

$$MAE = \frac{\sum(|\hat{y}_i - y_i|)}{n} \tag{19}$$

(6) MAPE

The reason why mean absolute percentage error can describe accuracy is that it is often used as a statistical index to measure the accuracy of prediction [29].

$$MAPE = \frac{1}{n}\sum\left(\left|\frac{\hat{y}_i - y_i}{y_i}\right|\right) * 100\% \tag{20}$$

(7) RMSE

Root mean squared error (RMSE) is the square root of MSE and scales the values of MSE to the ranges of observed values [28]. It is estimated according to Equation (21).

$$RMSE = \sqrt{\frac{\sum(|\hat{y}_i - y_i|)^2}{n}} \tag{21}$$

(8) AI

In this paper, a new index (AI, accuracy improvement) is introduced to show the improvement of prediction ability of the combined model compared with the single model [30]. S_i and S_c are the MAPE of the single model and combined model, respectively.

$$RI = \frac{|S_i - S_c|}{S_i} \times 100\% \tag{22}$$

4. Results and Discussion

4.1. Water Quality Evaluation Model

4.1.1. Data Preprocessing

The three water quality monitoring indicators used in this paper are from the data center of the Ministry of Environmental Protection (http://datacenter.mep.gov.cn/index (accessed on 1 September 2021)): dissolved oxygen (DO), chemical oxygen demand permanganate (COD_{Mn}) and ammonia nitrogen (NH_3-N). The corresponding water quality grades of each index value are shown in Table 1:

Table 1. Environmental quality standards for surface water (GB3838-2002).

Classification	Class I	Class II	Class III	Class IV	Class V	Class Inferior V
DO/$(mg \cdot L^{-1}) \geq$	7.5	6.0	5.0	3.0	2.0	rest
$COD_{Mn}/(mg \cdot L^{-1}) \leq$	2.0	4.0	6.0	10	15	rest
NH_3-N/$(mg \cdot L^{-1}) \leq$	0.15	0.50	1.0	1.5	2.0	rest

In order to solve the problem of inadequate sample size due to only taking the water quality evaluation grading standard as the research sample, the water quality index data are interpolated by the method of equal interval and uniform distribution [31]. For the convenience of modeling, the output value is continuous, and its value range is (0.5,6.5). The relationship between output value and water quality grade is shown in Table 2.

Table 2. Corresponding water quality grade of output values.

Output Value	Water Quality Grade
$0.5 < y \leq 1.5$	Class I
$1.5 < y \leq 2.5$	Class II
$2.5 < y \leq 3.5$	Class III
$3.5 < y \leq 4.5$	Class IV
$4.5 < y \leq 5.5$	Class V
$5.5 < y \leq 6.5$	Class inferior V

4.1.2. Model Building Process

In this study, T-S fuzzy neural network is used to evaluate water quality. The number of input and output nodes of the model is determined by the input and output dimensions of training samples. According to the index data considered in this paper, the input and output dimensions are determined to be three and one, respectively, so the number of input and output nodes is three and one, respectively. Through trial-and-error method, the number of hidden layer nodes is determined to be six, so the structure of water quality evaluation model is as follows: 3–6–1 [32]. The four coefficients (P_0, P_1, P_2 and P_3), the width (b) and center (c) of the membership function were randomly initialized.

After normalizing the input data, training of each parameter in the model was started [33]. After 100 iterations, the network finally converged, and the training error was $3.46 * 10^{-4}$.

4.1.3. Water Quality Evaluation of Lanzhou Xincheng Bridge Section

The process of water quality evaluation is realized in MATLAB, and the confusion matrix and water quality evaluation chart are output, as shown in Figure 3 [34]:

$$\begin{bmatrix} 0 & 0 & 0 & 0 & 0 & 0 \\ 1 & 41 & 3 & 0 & 0 & 0 \\ 0 & 1 & 6 & 0 & 0 & 0 \\ 0 & 0 & 0 & 0 & 0 & 0 \\ 0 & 0 & 0 & 0 & 0 & 0 \\ 0 & 0 & 0 & 0 & 0 & 0 \end{bmatrix}$$

Figure 3. Water quality grade confusion matrix of Lanzhou Xincheng Bridge section.

In classification models, confusion matrices are often used to observe and judge the number of correct and incorrect classifications. As shown in Figure 3, among the 52 weeks of random sampling, correct in water quality evaluation was achieved for 47 weeks, and the total correct judgment rate reached 90.38%. The correct prediction rate of class II and class III water quality was 91.11% and 85.71%, respectively. Class II water quality was misjudged as class I for one week, class II water quality was misjudged class as III for three weeks, and class III water quality was misjudged class II for one week.

The total number of correct and incorrect judgments and the corresponding water quality grade can be determined according to the above analysis, but it is not clear in which week the incorrect judgment occurred. This information can be obtained from the water quality evaluation chart presented below.

In Figure 4, the x-axis indicates the number of weeks; the y-axis indicates the water quality evaluation; and the green dot and the red dots represent the real water quality grade and the water quality grade determined by the model, respectively. Overlapping of the red and green dots indicates that the water quality level is correctly judged for that week, whereas divergence indicates an incorrect judgment. As shown in Figure 4, the water quality grade is concentrated in class II and class III, which is not accidental because during the whole time period from 2004 to 2015, the water quality grade is mainly class II and class III. Five of the fifty-two water quality grades were incorrectly judged; details are shown in Table 3.

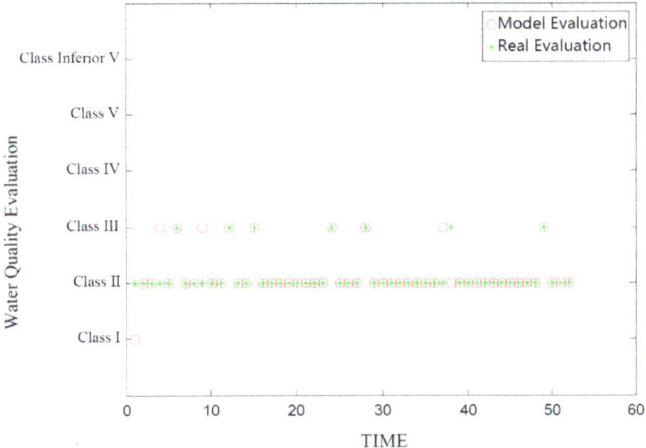

Figure 4. Water quality evaluation of Lanzhou Xincheng Bridge section.

Table 3. Water quality grade misjudgment of Lanzhou Xincheng Bridge section.

Week	True Water Quality	Misjudged Quality
1	II	I
4	II	III
9	II	III
37	II	III
38	III	II

4.1.4. Water Quality Evaluation of Longdong Section

Similarly, the trained T-S fuzzy neural network is applied to the water quality evaluation of Longdong section in Panzhihua, Sichuan, in the Yangtze River Basin, with a confusion matrix and water quality evaluation map as output, as shown in Figures 5 and 6, respectively.

$$\begin{bmatrix} 32 & 2 & 0 & 0 & 0 & 0 \\ 3 & 14 & 0 & 0 & 0 & 0 \\ 0 & 1 & 0 & 0 & 0 & 0 \\ 0 & 0 & 0 & 0 & 0 & 0 \\ 0 & 0 & 0 & 0 & 0 & 0 \\ 0 & 0 & 0 & 0 & 0 & 0 \end{bmatrix}$$

Figure 5. Confusion matrix of water quality grades in the Longdong section.

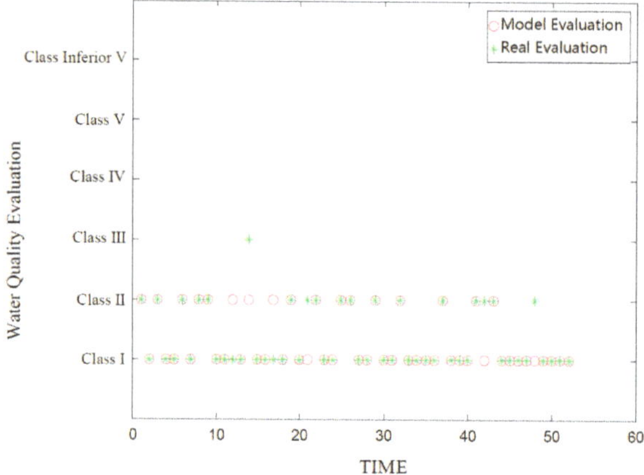

Figure 6. Water quality evaluation of Longdong section.

According to the confusion matrix, the water quality grade was correctly judged for 46 weeks and incorrectly judged for 6 weeks; the total correct judgment rate reached 88.46%. The correct judgment rates of class I and II water quality were 94.12% and 82.35%, respectively. Class I water quality was misjudged as class II for two weeks, class II was misjudged as class I for three weeks and class III was misjudged as class II for one week.

As shown in Figure 6, the water quality was mainly classified as class I and class II for the 52-week investigation period. In the 12th and 17th weeks, water quality in class I was incorrectly classified as class II; in the 21st, 42nd and 48th weeks, the water quality in class II was incorrectly classified as class I; and in the 14th week, the water quality in class III was incorrectly classified as class II. Results are shown in Table 4:

Table 4. Misjudgment of water quality grade in Longdong section.

Week	True Water Quality	Misjudged Quality
12	I	II
14	III	II
17	I	II
21	II	I
42	II	I
48	II	I

The water quality evaluation results presented above indicate that an ideal state was achieved in all investigated river basins. In the Yellow River in Lanzhou Xincheng Bridge section and Panzhihua, Sichuan province, in the Yangtze River Basin Longdong section, the water levels were correctly classified at rates of 90.38% and 88.46%, respectively, indicating that the trained T-S fuzzy neural network achieved satisfactory quality evaluation with good generalization ability.

4.2. Water Quality Prediction Model

4.2.1. Water Quality Prediction of Lanzhou Xincheng Bridge Section

For prediction with the ARIMA model, we used E-views10 software, which is created by IHS Global Inc in the State of California, United States. An original sequence diagram was generated to intuitively judge whether the data sequence was stable, as shown in Figure 7.

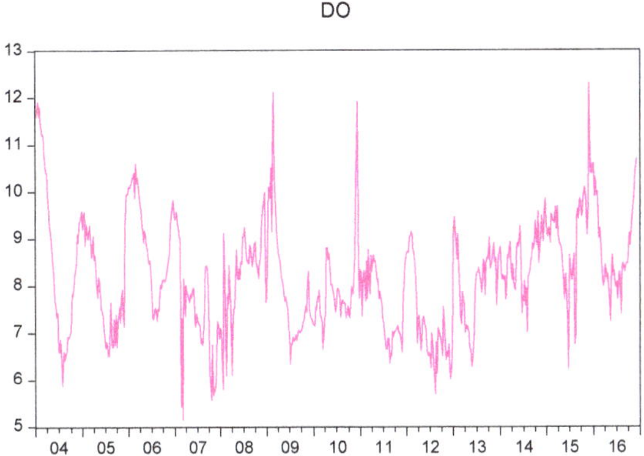

Figure 7. Original dissolved oxygen values.

In Figure 7, the *x*-axis shows the time in months, and the *y*-axis shows dissolved oxygen in milligrams per liter. The series in the figure does not indicate an obvious trend or seasonality, so it was preliminarily judged to be stationary series [35]. For further confirmation, an ADF unit root test was performed. The results show that the ADF test statistic is less than the critical value at levels of 1%, 5% and 10%, so the dissolved oxygen series is stationary and does not require differential processing [36]. An autocorrelation diagram and partial autocorrelation diagram were generated for model order determination, as shown in Figure 8.

As shown in Figure 8, the trailing and censoring characteristics of the autocorrelation function and partial autocorrelation function are very obvious in the dissolved oxygen data series. The decay of the autocorrelation function is very slow, which is a typical characteristic of trailing. However, the partial autocorrelation function rapidly decays to within two times the standard deviation after two steps, so the AR(2) model is determined. The estimation results of the model are shown in Table 5.

The R^2 of the model reached 79.34%, and an adaptability test was conducted on the AR(2) model was conducted, i.e., a white noise test of the residuals, as shown in Figure 9 [37]:

As shown in Figure 9, the autocorrelation and partial autocorrelation functions of the residual sequence both fall within two standard deviations, and the P value is significantly greater than 0.05. Therefore, it is a white noise sequence, indicating that the useful information in the original sequence has been extracted and the model has passed the adaptability test. Then, the AR(2) model established in this paper is used to predict the test data. A comparison between the predicted value and the real value is shown in Figure 10.

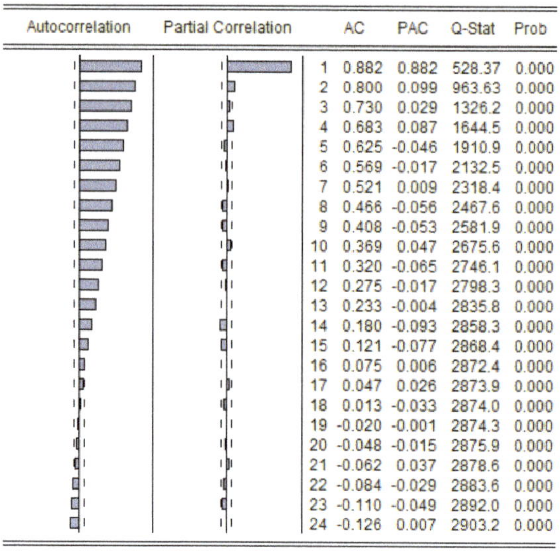
Autocorrelation Partial Correlation AC PAC Q-Stat Prob
1 0.882 0.882 528.37 0.000
2 0.800 0.099 963.63 0.000
3 0.730 0.029 1326.2 0.000
4 0.683 0.087 1644.5 0.000
5 0.625 -0.046 1910.9 0.000
6 0.569 -0.017 2132.5 0.000
7 0.521 0.009 2318.4 0.000
8 0.466 -0.056 2467.6 0.000
9 0.408 -0.053 2581.9 0.000
10 0.369 0.047 2675.6 0.000
11 0.320 -0.065 2746.1 0.000
12 0.275 -0.017 2798.3 0.000
13 0.233 -0.004 2835.8 0.000
14 0.180 -0.093 2858.3 0.000
15 0.121 -0.077 2868.4 0.000
16 0.075 0.006 2872.4 0.000
17 0.047 0.026 2873.9 0.000
18 0.013 -0.033 2874.0 0.000
19 -0.020 -0.001 2874.3 0.000
20 -0.048 -0.015 2875.9 0.000
21 -0.062 0.037 2878.6 0.000
22 -0.084 -0.029 2883.6 0.000
23 -0.110 -0.049 2892.0 0.000
24 -0.126 0.007 2903.2 0.000

Figure 8. Autocorrelation and partial autocorrelation of dissolved oxygen.

Table 5. Parameter estimation of the AR(2) model.

Variable	Coefficient	Standard Error	T Statistic	p Value
C	8.16	0.20	40.13	<0.001
AR(1)	0.78	0.04	20.46	<0.001
AR(2)	0.11	0.04	2.99	<0.001

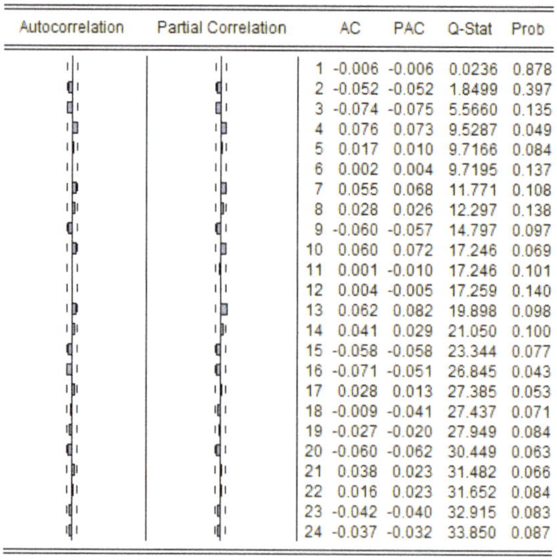
Autocorrelation Partial Correlation AC PAC Q-Stat Prob
1 -0.006 -0.006 0.0236 0.878
2 -0.052 -0.052 1.8499 0.397
3 -0.074 -0.075 5.5660 0.135
4 0.076 0.073 9.5287 0.049
5 0.017 0.010 9.7166 0.084
6 0.002 0.004 9.7195 0.137
7 0.055 0.068 11.771 0.108
8 0.028 0.026 12.297 0.138
9 -0.060 -0.057 14.797 0.097
10 0.060 0.072 17.246 0.069
11 0.001 -0.010 17.246 0.101
12 0.004 -0.005 17.259 0.140
13 0.062 0.082 19.898 0.098
14 0.041 0.029 21.050 0.100
15 -0.058 -0.058 23.344 0.077
16 -0.071 -0.051 26.845 0.043
17 0.028 0.013 27.385 0.053
18 -0.009 -0.041 27.437 0.071
19 -0.027 -0.020 27.949 0.084
20 -0.060 -0.062 30.449 0.063
21 0.038 0.023 31.482 0.066
22 0.016 0.023 31.652 0.084
23 -0.042 -0.040 32.915 0.083
24 -0.037 -0.032 33.850 0.087

Figure 9. White noise test of residual value.

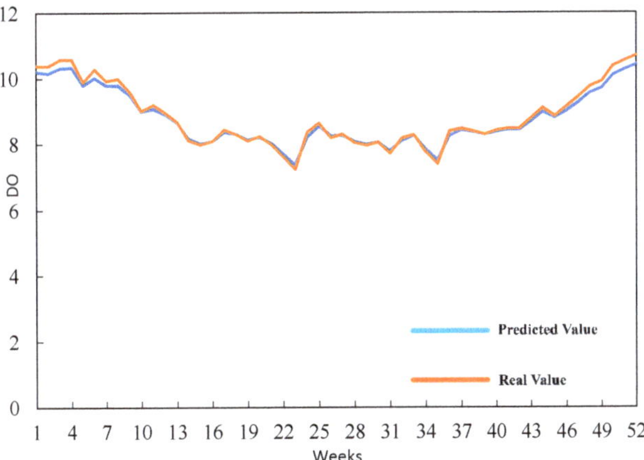

Figure 10. ARIMA Prediction Diagram.

In Figure 10, the x-axis represents time, the y-axis represents the concentration of DO, the blue line represents the predicted value of the ARIMA model and the orange line represents the real value. We can intuitively observe that the coincidence degree of the two lines is relatively high, which indicates that the prediction effect of the ARIMA model is strong. Before using wavelet neural network for prediction, the irrelevant noise in the data is removed. The empirical mode decomposition (EMD) of the dissolved oxygen sequence was carried out by MATLAB, as shown in Figure 11:

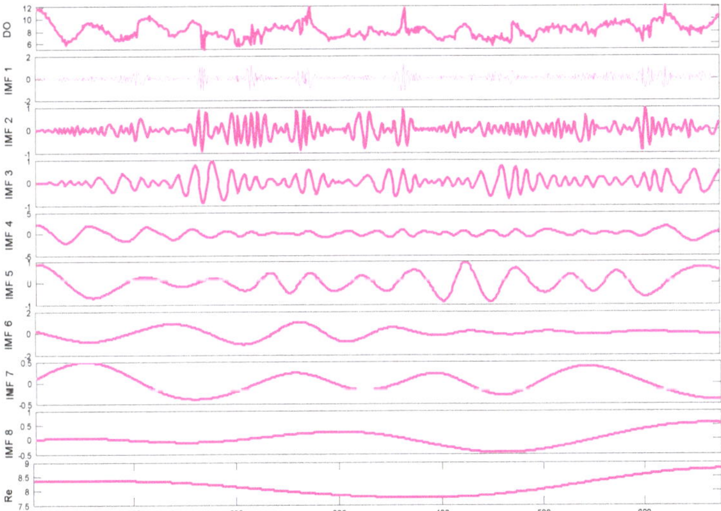

Figure 11. EMD diagram of dissolved oxygen.

The original data series were decomposed into eight intrinsic mode functions (IMFs) and a residual sequence. The frequency decreases from IMF1 to IMF8, and each IMF has its own unique frequency and cycle [38]. Due to the high-frequency property of IMF1 and its chaotic fluctuation trend, it was removed on the basis of the original data sequence too obtain a new sequence after denoising. The original sequence was denoised as shown in Figure 12:

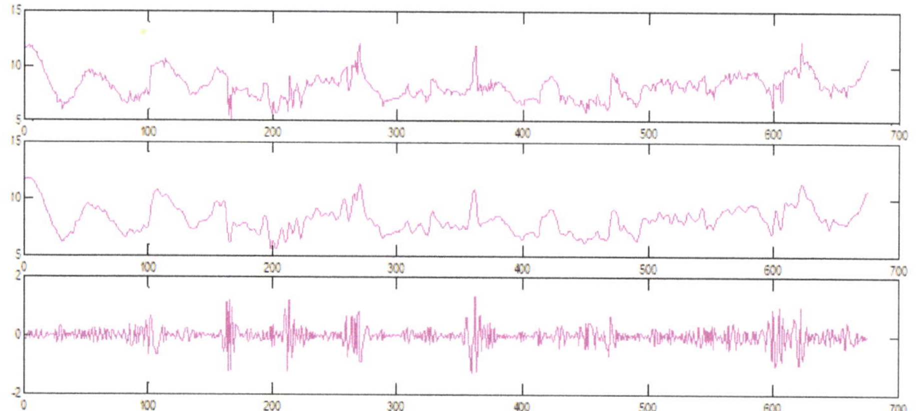

Figure 12. Comparison of dissolved oxygen sequence before and after denoising.

The denoised sequence is smoother than the original sequence, showing its original fluctuation trend more clearly. Therefore, the denoised sequence can be used for subsequent experimental demonstration and analysis.

First, a three-layer forward neural network is created, the parameters of each network are initialized and a training set is constructed to train the wavelet neural network. The data for weeks T-1, T-2, T-3 and T-4 are used to predict the number of neurons in week T. Therefore, the number of neurons in the input layer and output layer is four and one, respectively, and the number of neurons in the hidden layer is nine, as determined by the trial-and-error method; therefore, the structure of the wavelet neural network proposed in this paper is 4–9–1. The Morlet wavelet function is selected as the wavelet basis function, the number of iterations is set as 100, the learning probability is 0.001 and the training target is 10^{-6}. The weight and parameters of the network are modified by gradient correction method so that the predicted output is close to the desired output [39]. The predicted result is output and compared with the real value, as shown in Figure 13.

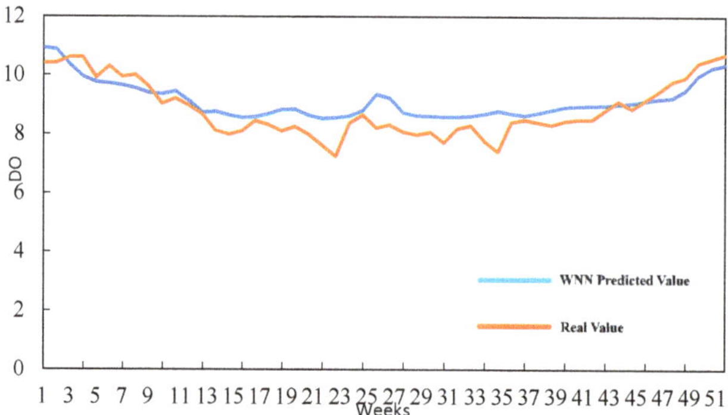

Figure 13. WNN prediction diagram.

The bat algorithm is used to determine the coefficients of each model, and the process of determining the optimal weight is transformed into the process of hunting prey by bats; therefore, the combined prediction model of dissolved oxygen in Lanzhou Xincheng Bridge section of the Yellow River Basin is as follows:

$$P_{DO} = 0.1495P_{WNN} + 0.8505P_{ARIMA}$$

The combined prediction model of permanganate and ammonia nitrogen was obtained according to the same steps:

$$P_{COD_{Mn}} = 0.7852P_{WNN} + 0.2148P_{ARIMA}$$

$$P_{NH_3-N} = 0.3274P_{WNN} + 0.6726P_{ARIMA}$$

4.2.2. Forecast Results and Analysis

The MAPE, MAE, RMSE of DO, COD_{Mn} and NH_3-N in 2016 under the single model and combined model prediction, as well as the degree of improvement of the combined model compared with the single model prediction ability were obtained according to the modeling process described above, as shown in Tables 6 and 7, respectively.

Table 6. Comparison of water quality index predictions.

Water Quality Index	Prediction Accuracy Index	ARIMA	WNN	Combined Model
DO	MAPE	2.73%	5.57%	2.58%
	MAE	0.2395	0.4716	0.2248
	RMSE	0.2867	0.5587	0.2758
COD_{Mn}	MAPE	19.64%	12.97%	11.96%
	MAE	0.4885	0.3517	0.3160
	RMSE	0.6853	0.4611	0.4079
NH_3-N	MAPE	16.18%	17.40%	13.85%
	MAE	0.0320	0.0307	0.0268
	RMSE	0.0427	0.0324	0.0324

Table 7. Improvement in the predictive ability of the combined model relative to the single model.

	DO	COD_{Mn}	NH_3-N
AI_{ARIMA}	5.49%	39.10%	14.40%
AI_{WNN}	53.68%	7.79%	20.40%

For the three water quality indicators, the MAPE, MAE and RMSE values of the combined model are lower than those of the single model, and the prediction effect is better. In the prediction of DO, COD_{Mn} and NH_3-N, the prediction accuracy of the combined model is improved by 5.49% and 53.68%, 39.10% and 7.79%, and 14.40% and 20.40% compared with the ARIMA model and WNN, respectively.

For a particular set of data, a higher weight is assigned to a single model of the combined model, which indicates that the method has a better predictive ability. Taking DO as an example, the prediction accuracy of the ARIMA model is higher than that of WNN, and it has a better prediction ability. In the ARIMA-WNN model, the bat algorithm was used to calculate the weights of the two individual models as 0.8505 and 0.1495, respectively, meaning that the individual model with better prediction ability had more weight and proving the reliability of to the bat algorithm to determine the weight coefficient in the combined model. The combined model of water quality prediction proposed in this paper was compared with the prediction results of each individual model, the BP neural network and LSSVM. The results are shown in Figures 14–17:

As shown in the figures presented above, the MAPE, MAE and RMSE of the combined model are significantly lower than those of the comparison models. The fitting figures of the real and predicted values also show that the water quality prediction model proposed in this paper has a higher fitting degree to the real data.

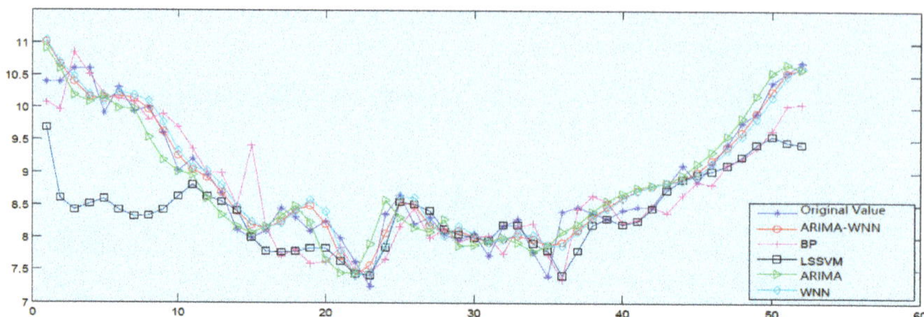

Figure 14. Comparison of each model fitting for dissolved oxygen.

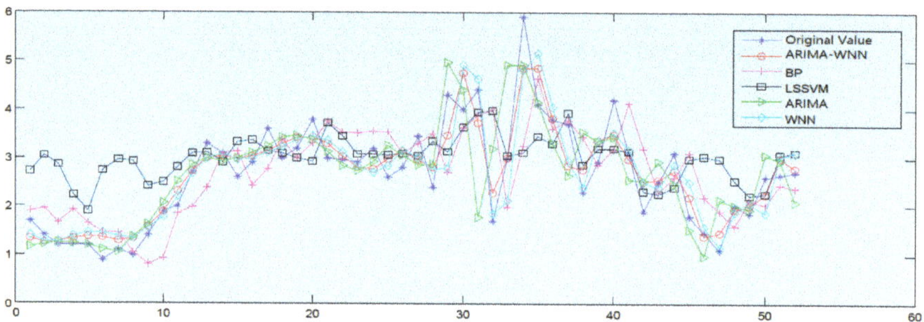

Figure 15. Comparison of each model fitting for CODMn.

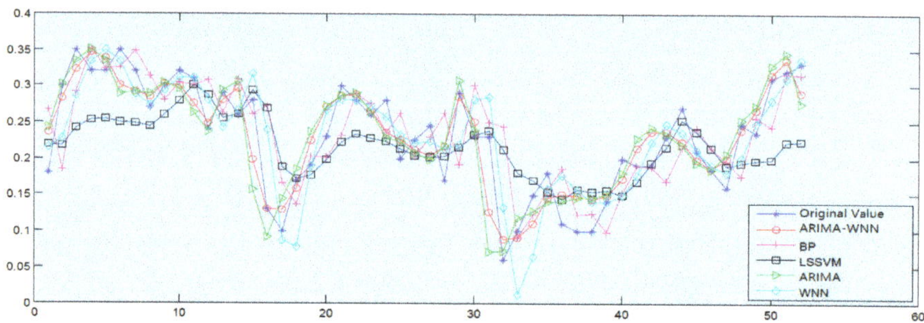

Figure 16. Comparison of each model fitting for NH_3-N.

4.2.3. Water Quality Prediction of Longdong Section

In order to verify the predictive ability of the combined model proposed in this paper for different water systems, the water quality index data of the Longdong section of Panzhihua in Sichuan Province of the Yangtze River Basin are selected for prediction and analysis. The modeling results are as follows:

$$P_{DO} = 0.1044P_{WNN} + 0.8956P_{ARIMA}$$

$$P_{COD_{Mn}} = 0.0432P_{WNN} + 0.9568P_{ARIMA}$$

$$P_{NH_3-N} = 0.8646P_{WNN} + 0.1354P_{ARIMA}$$

The improvement in the prediction accuracy and combined model prediction ability of the single and combined models for each indicator are compared in Tables 8 and 9.

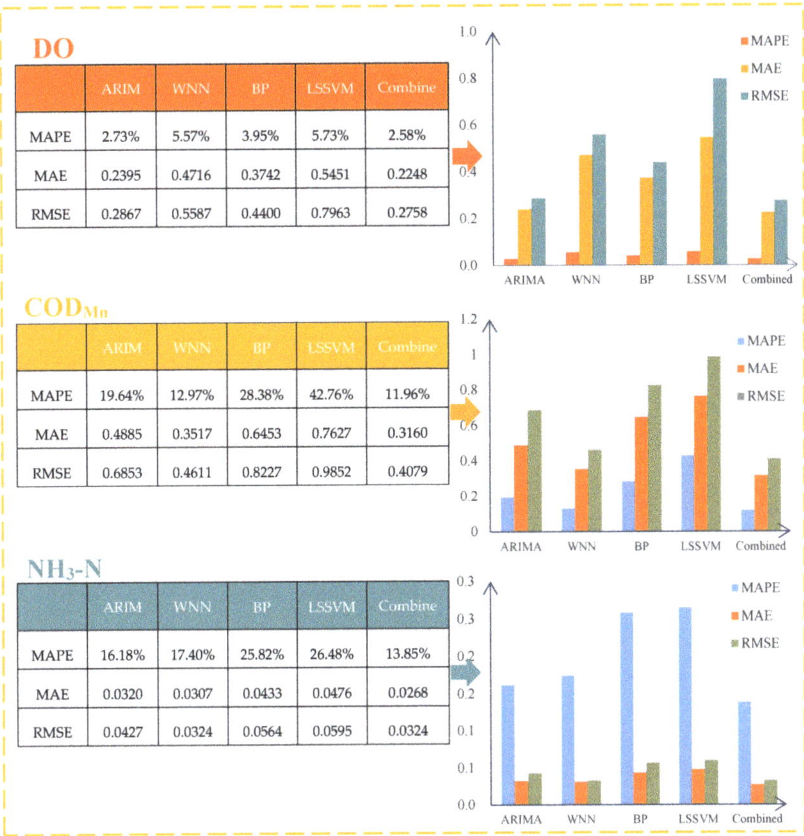

DO

	ARIM	WNN	BP	LSSVM	Combine
MAPE	2.73%	5.57%	3.95%	5.73%	2.58%
MAE	0.2395	0.4716	0.3742	0.5451	0.2248
RMSE	0.2867	0.5587	0.4400	0.7963	0.2758

COD_{Mn}

	ARIM	WNN	BP	LSSVM	Combine
MAPE	19.64%	12.97%	28.38%	42.76%	11.96%
MAE	0.4885	0.3517	0.6453	0.7627	0.3160
RMSE	0.6853	0.4611	0.8227	0.9852	0.4079

NH_3-N

	ARIM	WNN	BP	LSSVM	Combine
MAPE	16.18%	17.40%	25.82%	26.48%	13.85%
MAE	0.0320	0.0307	0.0433	0.0476	0.0268
RMSE	0.0427	0.0324	0.0564	0.0595	0.0324

Figure 17. Comparison of the combined model with other models.

As shown in Table 8, the prediction effect of the combined model is better than that of each single model. For DO, COD_{Mn} and NH_3-N, the combinatorial model outperformed the ARIMA model by 2.99%, 6.73% and 68.06%, respectively, in terms of prediction ability and outperformed the prediction of the WNN model by, 58.33%, 66.17% and 11.67%, respectively.

In the prediction of various indicators of the Longdong section of Panzhihua in Sichuan in the Yangtze River Basin, the bat algorithm still assigns more weight to the single model with better prediction ability and less weight to the single model with poor prediction ability. With respect to DO, the MAPE value predicted by the wavelet neural network is 4.68%, compared with 2.01% for the ARIMA model, which suggests that the ARIMA model achieves better predictive performed; therefore, in the combinatorial model, the bat algorithm assigns it a weight of 0.8956, whereas the WNN only assigns it a weight of 0.1044. In comparison with a BP neural network and LSSVM, the prediction accuracy of the combined model proposed in this paper is higher; the comparison results are shown in Figures 18–21.

In the prediction of various water quality indices in the Longdong section of Panzhihua, Sichuan, in the Yangtze River Basin, the prediction effect of the combined model is better and the accuracy is higher than that of the single model, the BP neural network and the LSSVM, indicating that the combined model proposed in this paper is suitable for water quality prediction in different river basins, with satisfactory generalization performance [40].

Table 8. Comparison of single and combined model prediction accuracy of each indicator.

Water Quality Indicator	Prediction Accuracy Indicator	ARIMA	WNN	Combined Model
DO	MAPE	2.01%	4.68%	1.95%
	MAE	0.1860	0.4116	0.1815
	RMSE	0.2481	0.5201	0.2446
COD_{Mn}	MAPE	6.39%	17.62%	5.96%
	MAE	0.1169	0.3220	0.1045
	RMSE	0.1578	0.5055	0.1498
NH_3-N	MAPE	27.02%	9.77%	8.63%
	MAE	0.0236	0.0092	0.0088
	RMSE	0.0315	0.0144	0.0148

Table 9. Improvement in the predictive power of the combined model compared to a single model.

	DO	COD_{Mn}	NH_3-N
AI_{ARIMA}	2.99%	6.73%	68.06%
AI_{WNN}	58.33%	66.17%	11.67%

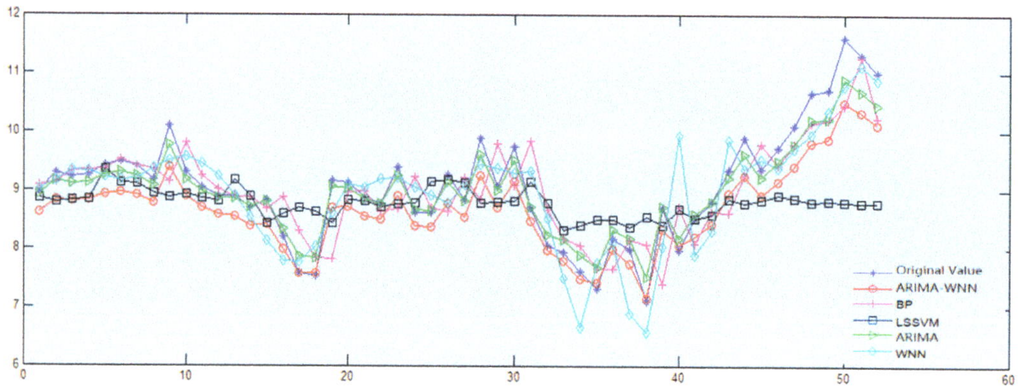

Figure 18. Comparison of model prediction of dissolved oxygen.

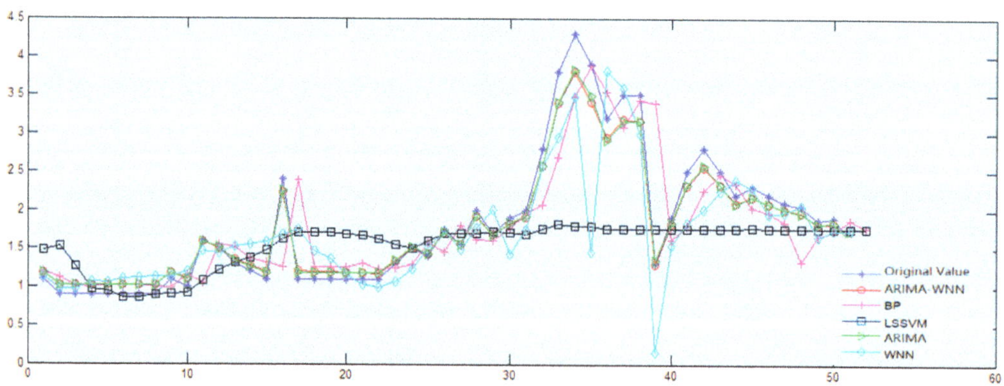

Figure 19. Comparison of model prediction of CODMn.

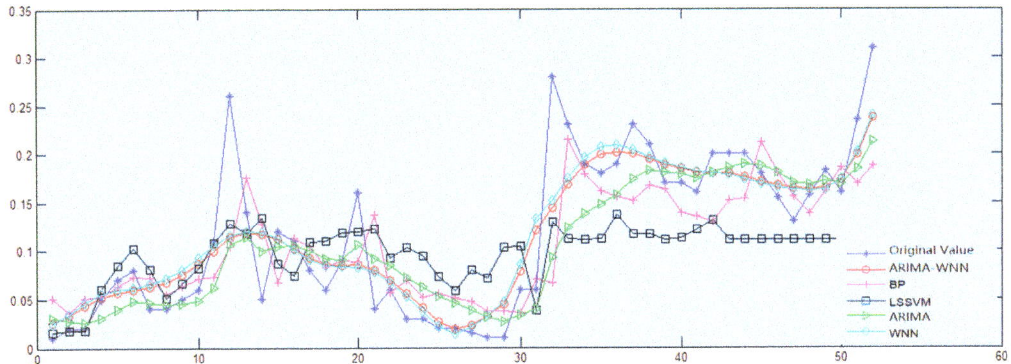

Figure 20. Comparison of model prediction of NH_3-N.

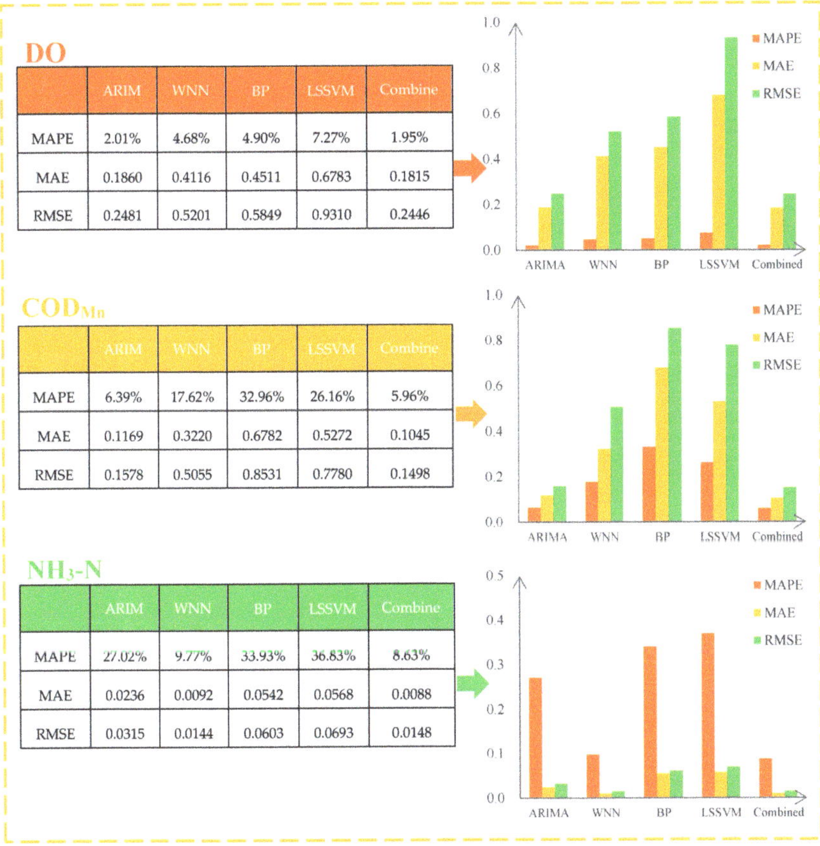

DO

	ARIM	WNN	BP	LSSVM	Combine
MAPE	2.01%	4.68%	4.90%	7.27%	1.95%
MAE	0.1860	0.4116	0.4511	0.6783	0.1815
RMSE	0.2481	0.5201	0.5849	0.9310	0.2446

COD_{Mn}

	ARIM	WNN	BP	LSSVM	Combine
MAPE	6.39%	17.62%	32.96%	26.16%	5.96%
MAE	0.1169	0.3220	0.6782	0.5272	0.1045
RMSE	0.1578	0.5055	0.8531	0.7780	0.1498

NH_3-N

	ARIM	WNN	BP	LSSVM	Combine
MAPE	27.02%	9.77%	33.93%	36.83%	8.63%
MAE	0.0236	0.0092	0.0542	0.0568	0.0088
RMSE	0.0315	0.0144	0.0603	0.0693	0.0148

Figure 21. Combined model compared to other models.

4.2.4. Water Quality Evaluation in 2016

The results in presented in Section 4.1.3 prove that the trained T-S fuzzy neural network described in this paper has excellent ability in the determination of water quality grade. On this basis, the water quality index data of two sections predicted for 2016 were substituted for water quality evaluation. The results are shown in Tables 10 and 11 and Figures 22 and 23.

Table 10. Water quality evaluation of Lanzhou Xincheng Bridge Section in 2016.

Week	Evaluation Results	True Water Quality Rating	Week	Evaluation Results	True Water Quality Rating	Week	Evaluation Results	True Water Quality Rating	Week	Evaluation Results	True Water Quality Rating
1	II	II	14	II	II	27	II	II	40	II	III
2	II	II	15	II	II	28	II	II	41	II	II
3	II	II	16	II	II	29	III	III	42	II	II
4	III	II	17	II	II	30	II	II	43	I	II
5	II	II	18	II	II	31	III	III	44	II	II
6	II	II	19	II	II	32	I	I	45	II	II
7	II	II	20	II	II	33	II	II	46	II	II
8	II	II	21	II	II	34	III	III	47	II	II
9	II	II	22	III	II	35	II	III	48	II	II
10	II	II	23	II	II	36	II	II	49	III	II
11	II	II	24	II	II	37	II	II	50	II	II
12	II	II	25	II	II	38	II	II	51	II	II
13	I	II	26	II	II	39	II	II	52	II	II

Table 11. Water quality evaluation of Longdong section in 2016.

Week	Evaluation Results	True Water Quality Rating	Week	Evaluation Results	True Water Quality Rating	Week	Evaluation Results	True Water Quality Rating	Week	Evaluation Results	True Water Quality Rating
1	I	I	14	I	I	27	I	I	40	II	II
2	I	I	15	I	I	28	I	I	41	II	II
3	I	I	16	I	I	29	I	I	42	II	II
4	II	I	17	I	I	30	I	I	43	II	II
5	I	I	18	II	I	31	II	I	44	II	II
6	I	I	19	I	I	32	II	II	45	II	II
7	I	I	20	I	II	33	II	II	46	I	II
8	I	I	21	I	I	34	III	III	47	II	II
9	I	I	22	I	I	35	II	II	48	II	II
10	I	I	23	I	I	36	II	II	49	I	II
11	I	I	24	I	I	37	II	II	50	II	II
12	I	II	25	I	I	38	II	II	51	I	II
13	I	I	26	II	I	39	I	II	52	II	II

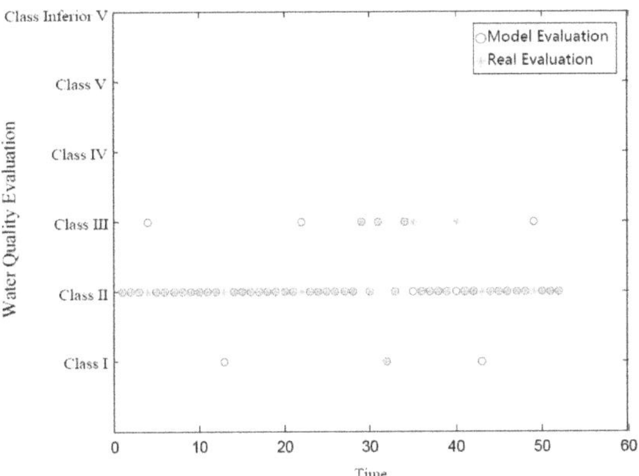

Figure 22. Water quality evaluation of Lanzhou Xincheng Bridge section in 2016.

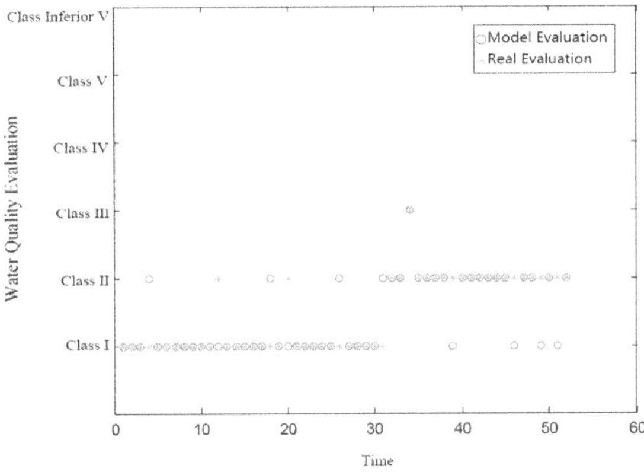

Figure 23. Water quality evaluation of Longdong section in 2016.

Water quality evaluation using the T-S fuzzy neural network trained as described in this paper revealed that among the 52 weeks of 2016, water quality was misjudged in the Lanzhou Xincheng Bridge section of the Yellow River Basin and the Longdong section of Panzhihua, Sichuan, the Yangtze River Basin for 7 and 10 weeks, respectively, with total correct judgment rates of 86.54% and 80.77%, respectively. Because the input data contain errors, this result is acceptable, verifying the reliability of the water quality evaluation model trained as described in this paper.

5. Conclusions

Water quality evaluation and prediction are not two completely independent procedures. On the contrary, they form a mutually dependent system and complement each other. Accordingly, in this study, we established a water quality evaluation–prediction system, established a water quality evaluation model using a T-S fuzzy neural network and constructed research samples by interpolating water quality index data evenly distributed on the basis of each index grading standard stipulated in the *Environmental Quality Standards for Surface Water*. A stratified sampling method is used to construct training samples. The trained T-S fuzzy neural network was applied to the water quality evaluation of the Lanzhou Xincheng Bridge section in the Yellow River Basin and the Longdong section in the Sichuan Panzhihua section in the Yangtze River Basin, with total positive water quality grade evaluation rates of 90.38% and 88.46%, respectively, indicating the positive water quality evaluation effect and generalizability of the model.

For the prediction of water quality, in this paper, we proposed a new combined model, ARIMA-WNN, which establishes the combined prediction model for each water quality index of the two basins and compares the prediction results with the combined model. The results show that compared with the single model, the combined model (ARIMA-WNN) has a higher prediction accuracy, and the prediction ability can be improved by up to 68.06%. Compared with commonly used water quality prediction models (BP neural network and LSSVM), we found that the MAPE, MAE and RMSE of the combined model are significantly lower, which demonstrates the excellent water quality prediction ability of the combined model.

Determining reasonable weight coefficients for each single model in the combined model is the basis for obtaining accurate prediction results, and in this study, we used the bat algorithm to achieve this process. Swarm intelligence optimization algorithms have developed rapidly in recent years, and a variety of new methods have emerged in succession [41]. Determining the optimal weight is a subject that can be studied in depth, and additional methods should be proposed and tested in subsequent work [42].

Author Contributions: Conceptualization, G.J.; Methodology, G.J.; Writing—original draft, F.W. (Fei Wang), F.W. (Fanjuan Wang), H.L., F.Z., J.C. and J.J.; Writing—review & editing, S.C.; Supervision, G.J. and Z.W. All authors have read and agreed to the published version of the manuscript.

Funding: The research was supported by National Natural Science Foundation of China Under Grant No.41271038.

Institutional Review Board Statement: Not applicable.

Informed Consent Statement: Not applicable.

Data Availability Statement: The data presented in the present study are available on request from the corresponding author.

Conflicts of Interest: The authors declare no conflict of interest.

References

1. Fielding, J.J.; Croudace, I.W.; Kemp, A.E.; Pearce, R.B.; Cotterill, C.J.; Langdon, P.; Avery, R. Tracing lake pollution, eutrophication and partial recovery from the sediments of Windermere, UK, using geochemistry and sediment microfabrics. *Sci. Total Environ.* **2020**, *722*, 137745. [CrossRef] [PubMed]
2. Cao, H.; Guo, Z.; Wang, S.; Cheng, H.; Zhan, C. Intelligent wide-area water quality monitoring and analysis system exploiting unmanned surface vehicles and ensemble learning. *Water* **2020**, *12*, 681. [CrossRef]
3. Li, Z.; Sun, Z.; Liu, J.; Dong, H.; Xiong, W.; Sun, L.; Zhou, H. Prediction of river sediment transport based on wavelet transform and neural network model. *Appl. Sci.* **2022**, *12*, 647. [CrossRef]
4. Jiao, G.; Guo, T.; Ding, Y. A new hybrid forecasting approach applied to hydrological data: A case study on precipitation in Northwestern China. *Water* **2016**, *8*, 367. [CrossRef]
5. Colominas, M.A.; Schlotthauer, G.; Torres, M.E. Improved complete ensemble EMD: A suitable tool for biomedical signal processing. *Biomed. Signal Process. Control* **2014**, *14*, 19–29. [CrossRef]
6. Werbos, P.J. Backpropagation through time: What it does and how to do it. *Proc. IEEE* **1990**, *78*, 1550–1560. [CrossRef]
7. Wang, H.; Hu, D. Comparison of SVM and LS-SVM for regression. In Proceedings of the 2005 International Conference on Neural Networks and Brain, Beijing, China, 13–15 October 2005; Volume 1, pp. 279–283.
8. Singh, K.P.; Malik, A.; Sinha, S. Water quality assessment and apportionment of pollution sources of Gomti river (India) using multivariate statistical techniques—A case study. *Anal. Chim. Acta* **2005**, *538*, 355–374. [CrossRef]
9. Shrestha, S.; Kazama, F. Assessment of surface water quality using multivariate statistical techniques: A case study of the Fuji river basin, Japan. *Environ. Model. Softw.* **2007**, *22*, 464–475. [CrossRef]
10. Zhang, X.; Wang, Q.; Liu, Y.; Wu, J.; Yu, M. Application of multivariate statistical techniques in the assessment of water quality in the Southwest New Territories and Kowloon, Hong Kong. *Environ. Monit. Assess.* **2010**, *173*, 17–27. [CrossRef]
11. Ogwueleka, T.C. Use of multivariate statistical techniques for the evaluation of temporal and spatial variations in water quality of the Kaduna River, Nigeria. *Environ. Monit. Assess.* **2015**, *187*, 137. [CrossRef]
12. Singh, K.P.; Basant, A.; Malik, A.; Jain, G. Artificial neural network modeling of the river water quality—A case study. *Ecol. Model.* **2009**, *220*, 888–895.
13. Seo, I.W.; Yun, S.H.; Choi, S.Y. Forecasting Water Quality Parameters by ANN Model Using Pre-processing Technique at the Downstream of Cheongpyeong Dam. *Procedia Eng.* **2016**, *154*, 1110–1115. [CrossRef]

14. Zhang, R.; Sun, B.; Wang, Y.; Si, Z.; Li, X.; Yan, T. Water quality evaluation and forecast of Yellow River estuary. *Chin. J. Environ. Eng.* **2013**, *7*, 3089–3093.
15. Zhang, Y.; Gao, Q. Comprehensive prediction model of water quality based on grey model and fuzzy neural network. *Chin. J. Environ. Eng.* **2015**, *9*, 537–545.
16. Xu, H. Weighted support vector machine for regression and its application for prediction of water quality. *Glob. Geol.* **2007**, *26*, 58–61.
17. Lee, M.A.; Takagi, H. Integrating design stage of fuzzy systems using genetic algorithms. In Proceedings of the Second IEEE International Conference on Fuzzy Systems, San Francisco, CA, USA, 28 March–1 April 1993; pp. 612–617.
18. Box, G.E.P.; Jenkins, G.M.; Reinsel, G.C. *Time Series Analysis*; North-Holland Publishing, Co.: Amsterdam, The Netherlands, 1970; Volume 61, pp. 285–293.
19. Kelejian, H.H.; Prucha, I.R. A generalized spatial two-stage least squares procedure for estimating a spatial autoregressive model with autoregressive disturbances. *J. Real Estate Financ. Econ.* **1998**, *17*, 99–121.
20. Said, S.E.; Dickey, D.A. Testing for unit roots in autoregressive-moving average models of unknown order. *Biometrika* **1984**, *71*, 599–607. [CrossRef]
21. Akaike, H. Maximum likelihood identification of Gaussian autoregressive moving average models. *Biometrika* **1973**, *60*, 255–265. [CrossRef]
22. Zhang, Q.; Benveniste, A. Wavelet networks. *IEEE Trans. Neural Netw.* **1992**, *3*, 889–898. [CrossRef]
23. Zhang, Y.; Yang, H.; Cui, H.; Chen, Q. Comparison of the ability of ARIMA, WNN and SVM models for drought forecasting in the Sanjiang Plain, China. *Natural Resour. Res.* **2020**, *29*, 1447–1464.
24. Yang, X.S.; Gandomi, A.H. Bat algorithm: A novel approach for global engineering optimization. *Eng. Comput.* **2012**, *29*, 464–483. [CrossRef]
25. Curran, P.J.; West, S.G.; Finch, J.F. The robustness of test statistics to nonnormality and specification error in confirmatory factor analysis. *Psychol. Methods* **1996**, *1*, 16. [CrossRef]
26. Alder, B.J.; Wainwright, T.E. Decay of the velocity autocorrelation function. *Phys. Rev. A* **1970**, *1*, 18. [CrossRef]
27. Akaike, H. *Factor analysis and AIC. Selected Papers of Hirotugu Akaike*; Springer: New York, NY, USA, 1987; pp. 371–386.
28. Willmott, C.J.; Matsuura, K. Advantages of the mean absolute error (MAE) over the root mean square error (RMSE) in assessing average model performance. *Clim. Res.* **2005**, *30*, 79–82. [CrossRef]
29. De Myttenaere, A.; Golden, B.; Le Grand, B.; Rossi, F. Mean absolute percentage error for regression models. *Neurocomputing* **2016**, *192*, 38–48. [CrossRef]
30. Fu, Z.; Wu, X.; Guan, C.; Sun, X.; Ren, K. Toward efficient multi-keyword fuzzy search over encrypted outsourced data with accuracy improvement. *IEEE Trans. Inf. Forensics Secur.* **2016**, *11*, 2706–2716. [CrossRef]
31. Fleshler, M.; Hoffman, H.S. A progression for generating variable-interval schedules. *J. Exp. Anal. Behav.* **1962**, *5*, 529. [CrossRef]
32. Evans, J.J.; Wilson, B.A.; Schuri, U.; Andrade, J.; Baddeley, A.; Bruna, O.; Canavan, T.; Del Sala, S.; Green, R.; Laaksonen, R.; et al. A Comparison of "errorless" and "trial-and-error" learning methods for teaching individuals with acquired memory deficits. *Neuropsychol. Rehabil.* **2000**, *10*, 67–101. [CrossRef]
33. Quackenbush, J. Microarray data normalization and transformation. *Nat. Genet.* **2002**, *32*, 496–501. [CrossRef] [PubMed]
34. Townsend, J.T. Theoretical analysis of an alphabetic confusion matrix. *Percept. Psychophys.* **1971**, *9*, 40–50. [CrossRef]
35. DeJong, D.N.; Nankervis, J.C.; Savin, N.E.; Whiteman, C.H. Integration versus trend stationary in time series. *Econom. J. Econom. Soc.* **1992**, *60*, 423–433. [CrossRef]
36. Epley, N.; Gilovich, T. Putting adjustment back in the anchoring and adjustment heuristic: Differential processing of self-generated and experimenter-provided anchors. *Psychol. Sci.* **2001**, *12*, 391–396. [CrossRef] [PubMed]
37. Wu, Z.; Huang, N.E. A study of the characteristics of white noise using the empirical mode decomposition method. *Proc. R. Soc. London Ser. A Math. Phys. Eng. Sci.* **2004**, *460*, 1597–1611. [CrossRef]
38. Rai, V.K.; Mohanty, A.R. Bearing fault diagnosis using FFT of intrinsic mode functions in Hilbert–Huang transform. *Mech. Syst. Signal Process.* **2007**, *21*, 2607–2615. [CrossRef]
39. Teodorof, L.; Ene, A.; Burada, A.; Despina, C.; Seceleanu-Odor, D.; Trifanov, C.; Ibram, O.; Bratfanof, E.; Tudor, M.I.; Tudor, M.; et al. Integrated Assessment of Surface Water Quality in Danube River Chilia Branch. *Appl. Sci.* **2021**, *11*, 9172. [CrossRef]
40. Schaffer, C. A conservation law for generalization performance. In *Machine Learning Proceedings 1994*; Morgan Kaufmann: Burlington, MA, USA, 1994; pp. 259–265.
41. Mavrovouniotis, M.; Li, C.; Yang, S. A survey of swarm intelligence for dynamic optimization: Algorithms and applications. *Swarm Evol. Comput.* **2017**, *33*, 1–17. [CrossRef]
42. Ahmed, M.; Mumtaz, R.; Anwar, Z. An Enhanced Water Quality Index for Water Quality Monitoring Using Remote Sensing and Machine Learning. *Appl. Sci.* **2022**, *12*, 12787. [CrossRef]

Article

Evaluation of Non-Biodegradable Organic Matter and Microbial Community's Effects on Achievement of Partial Nitrification Coupled with ANAMMOX for Treating Low-Carbon Livestock Wastewater

Mingchuan Zhang *, Xi Chen, Xinyang Xu, Zhongtian Fu and Xin Zhao

College of Resources and Civil Engineering, Northeastern University, Shenyang 110819, China; chenxineu@mail.neu.edu.cn (X.C.); xuxinyang@mail.neu.edu.cn (X.X.); fuzhongtian@mail.neu.edu.cn (Z.F.); zhaoxin@mail.neu.edu.cn (X.Z.)
* Correspondence: zhangmingchuan@mail.neu.edu.cn

Abstract: After the anaerobic digestion of livestock manure, high concentrations of nutrients still remain. Treatment of livestock wastewater through partial nitrification coupled with anaerobic ammonium oxidation (ANAMMOX) could be a useful technology depending on the investigation of microorganism enrichment and partial nitrification coupled with achievement of the ANAMMOX process. The results show 78.4% and 64.7% nitrite accumulation efficiency was successfully obtained in an intermittent aeration sequencing batch reactor and a continuous aeration sequencing batch reactor, respectively, at a loading rate of 0.93 kg ammonium/(m^3·d). The main reason for the high nitrite accumulation efficiency was the intermittent aeration strategy which generated a 20–30 min lag reaction for nitrite oxidation and promoted the growth of the dominant ammonium oxidation bacteria (*Nitrosomonas*). Non-biodegradable organic matter in the effluents of partial nitrification did not have obvious influence on ANAMMOX activity at low loading rates (118 ± 13 mg COD/L and 168 ± 9 mg COD/L), and up to 87.4% average nitrite removal rate was observed. However, with the influent COD concentration increasing to 242 ± 17 mg/L, the potential inhibition of ANAMMOX activity was exerted by non-biodegradable organic matter.

Keywords: livestock wastewater; partial nitrification; ANAMMOX; non-biodegradable organic matter; nitrogen removal; microbial community

Citation: Zhang, M.; Chen, X.; Xu, X.; Fu, Z.; Zhao, X. Evaluation of Non-Biodegradable Organic Matter and Microbial Community's Effects on Achievement of Partial Nitrification Coupled with ANAMMOX for Treating Low-Carbon Livestock Wastewater. *Appl. Sci.* **2022**, *12*, 3626. https://doi.org/10.3390/app12073626

Academic Editor: Dino Musmarra

Received: 25 February 2022
Accepted: 31 March 2022
Published: 2 April 2022

1. Introduction

Every year, a large amount of livestock manure containing high concentrations of nutrients and solids is produced from centralized livestock farms [1,2]. Anaerobic digestion is one of the sustainable technologies for livestock manure management. However, after anaerobic digestion treatment, most ammonium and non-biodegradable organic matter (mainly cellulose and hemicelluloses) in livestock manure still remains in the digestate wastewater [3,4]. The lack of a biodegradable carbon source in the livestock wastewater could lead to difficulties in the operation of the conventional nitrification–denitrification process. In this case, further studies of the livestock wastewater treatment biotechnology with low carbon cost and the effects of non-biodegradable organic matter on nitrogen removal are necessary.

Partial nitrification is an efficient technology for treating low-carbon wastewater by achieving biological nitrogen removal via nitrite rather than nitrate. This process has two advantages over the conventional nitrification–denitrification process: (I) reduction of the carbon demand of denitrifying bacteria by 40%; and (II) reduction of the aeration energy by 25% [5]. In the partial nitrification process, nitrite accumulation efficiency is related to nitrogen levels, pH, dissolved oxygen (DO) and loading rate, which are suggested to be the important factors for the washing out of nitrite-oxidizing bacteria (NOB) [6–8]. The high concentrations of ammonium and non-biodegradable organic matter in the livestock wastewater were unfavorable for partial nitrification. Thus, effective control strategies should be developed and investigated to achieve stable nitritation.

Anaerobic ammonium oxidation (ANAMMOX) has been considered as an effective technology for the reduction of accumulated nitrite by oxidation of ammonium rather than organic carbon [9]. In the ANAMMOX process, bacteria grow with bicarbonate as the sole carbon source. Thus, without the need for organic matter, this technology has been exploited to treat nitrogen-rich wastewater, such as landfill leachate, animal manure and supernatant of digested sludge [10,11]. ANAMMOX bacteria are considered to be difficult to cultivate because of their low specific growth rate (μ_{max} = 0.065/d) and sensitivity to organic matter, which could induce the inhibition of ANAMMOX enrichment [12]. Thus, the non-biodegradable organic matter withdrawn from partial nitrification reactors should be an important issue on the successful use of ANAMMOX for treating livestock wastewater.

In livestock wastewater, there is not enough biodegradable organic matter for the conventional biological nitrogen removal process. Thus, partial nitrification coupled with ANAMMOX was considered to be an efficient combined technology for nitrogen removal. In this research, in order to evaluate the influences of non-biodegradable organic matter and control strategies on livestock wastewater treatment efficiency, long-term partial nitrification coupled with ANAMMOX reactors were established via microbial community analysis to investigate the performance of nitrogen removal from the livestock wastewater.

2. Materials and Methods

2.1. Experimental Setup

Two identical laboratory-scale sequencing batch reactors (SBRs) and an ANAMMOX reactor were constructed (Figure 1). The reactors were made from transparent Plexiglas, each having an effective volume of 10 L.

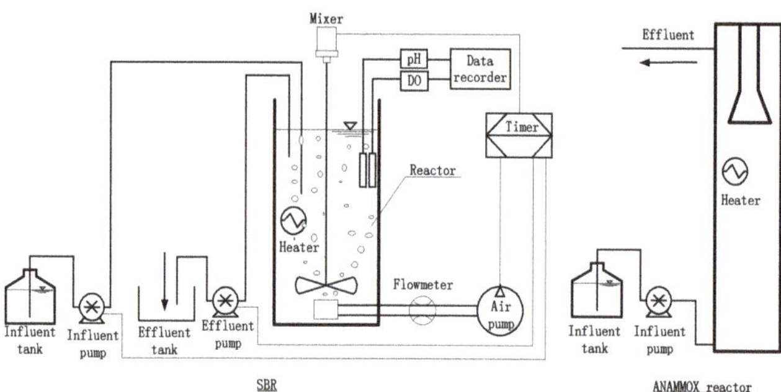

Figure 1. Schematic diagram of SBRs and ANAMMOX reactor systems.

One SBR was operated with an intermittent aeration (IASBR) strategy and the other was operated with continuous aeration (CASBR) in an 8 h cycle. For the IASBR, after a 60 min non-aeration period, the reactor was intermittently aerated with successive 50 min aeration/30 min non-aeration periods. For the CASBR, 250 min continuous aeration was adopted after a 180 min non-aeration period. Then, the settle phase lasted 40 min before the effluent was withdrawn in the last 10 min. Each SBR was stirred with a rectangular mixing paddle. The air was supplied using aquarium air pumps at an air flow rate of 0.8–0.9 L air/min.

The characteristics of the livestock wastewater taken from a mesophilic manure digester were: 10,260 ± 780 mg/L of COD; 3112 ± 94 mg/L of NH_4^+-N; 3478 ± 75 mg/L of total nitrogen (TN); 1.25 ± 0.18 g/L of suspended solids (SS); and pH of 8.26 ± 0.05. The wastewater had a COD to TN ratio of 2.96, and the BOD_5 to COD ratio was only 0.38, indicating a low proportion of biodegradable organic matter in the wastewater. The livestock wastewater was fed to SBRs with peristaltic pumps at loading rates of 3.08 kg COD/(m^3·d) and 0.93 kg NH_4^+-N/(m^3·d). The temperature of two SBRs was controlled at 26 ± 1 °C, simulating the digestate liquid temperature after anaerobic digestion.

An upflow anaerobic sludge blanket (UASB) reactor was built for cultivating ANAMMOX biomass (Figure 1). The effluents of IASBR were collected and adjusted with NH_4Cl to obtain a desirable nitrogen ratio (NH_4^+-N: NO_2^--N = 1: 1.32) for the influent of the ANAMMOX reactor [9]. The seed sludge fed to the UASB was a mixture of anaerobic denitrification sludge, anaerobic composting sludge, river sediment sludge, local soil and partial nitrification activated sludge. Before treating the raw effluents of IASBR, influent synthetic wastewater (30 mg/L NH_4^+-N and 40 mg/L NO_2^--N from Day 1 to Day 100; 60 mg/L NH_4^+-N and 80 mg/L NO_2^--N from Day 101 to Day 223; and 90 mg/L NH_4^+-N and 120 mg/L NO_2^--N from Day 223 to Day 289) was continuously fed into the UASB at a hydraulic retention time (HRT) of 8 h for ANAMMOX biomass cultivation. The temperature of the UASB was controlled at 35 ± 2 °C, and an optimum pH of 8.0 ± 0.4 was controlled through adding $NaHCO_3$ into the influent.

2.2. Analytical Methods

COD, BOD_5 and SS were tested in accordance with the standard APHA methods [13]. NH_4^+-N, NO_2^--N and NO_3^--N were measured with Hach kits (Hach, Ames, IA, USA). A pH probe (pH 320, WTW, Weilheim, Germany) and a DO sensor (Unisense, Aarhus, Denmark) were used for pH and DO measurement, respectively.

The nitrite accumulation efficiency (η) is a term used to describe the performance of partial nitrification as per Equation (1):

$$\eta = \frac{S_{NO_2^- - N}}{S_{NO_2^- - N} + S_{NO_3^- - N}} \times 100\% \quad (1)$$

where $S_{NO_2^- - N}$ and $S_{NO_3^- - N}$ are the concentrations of NO_2^--N and NO_3^--N in the effluent, respectively.

2.3. Batch Experiments

All batch experiments were carried out in 1 L beakers. Air diffusers were placed at the bottom of the beakers. Washed activated sludge (Day 120) from two SBRs was added to the beakers, and the sludge concentrations in the beakers were adjusted to close to those inside the SBRs. The pH was controlled at 8.20 ± 0.05.

Specific oxygen uptake rate (SOUR) measurements: (I) NH_4Cl or $NaNO_2$ was fed into the beaker to obtain specified 30 mg/L NH_4^+-N or NO_2^--N, respectively; (II) after feeding, the aeration commenced until DO became saturated in the beakers, (III) aeration was turned off, and DO concentrations consumed by the oxidation of NH_4^+-N or NO_2^--N were recorded.

Influence of intermittent aeration on partial nitrification: (I) 0, 10, 20, 30 and 40 min non-aeration periods were set in 5 beakers, where argon gas was striped at the bottom of the beakers in order to continuously remove DO; (II) specified 30 mg/L NH_4^+-N and NO_2^--N were added into the beakers through adjusting NH_4Cl and $NaNO_2$ dosages, (III) aeration commenced, and liquid samples were taken at intervals for the measurement of NO_2^--N and NO_3^--N concentrations.

2.4. Microbial Population Analysis by 16S rRNA Gene Sequencing

In order to assess potential population changes of the partial nitrification biomass in IASBR and CASBR, 2 mL of sludge was sampled from each reactor. DNA was extracted using the repeated bead beating and column purification extraction process [14]. Modified 16S Illumina adapter fusion primers were used to generate amplicon libraries. The primers were CaporasoNexF and CaporasoNexR [15,16]. PCR was conducted by using 20 ng of DNA from the sludge samples as a template and Kapa HiFi Hotstart ReadyMix (Kapa Biosystems, London, UK) according to the manufacturer's instructions. The thermocycling conditions were: one cycle of 95 °C for 3 min, then 26 cycles of 95 °C for 30 s, 55 °C for

30 s, 72 °C for 30 s, followed by one cycle of 72 °C for 5 min. QIAquick PCR Purification Kits (Qiagen, UK) were used to purify libraries. Two unique 8 bp indices were then added (one index at the 5′ end of the amplicon and the other at the 3′ end) to each amplicon in a second round of PCR using primers from the Illumina Nextera XT indexing kit. PCR was performed with 5 μL of each amplicon as a template and Kapa HiFi Hotstart ReadyMix. PCR conditions for this second round were: one cycle of 95 °C for 3 min, then 8 cycles of 95 °C for 30 s, 55 °C for 30 s, 72 °C for 30 s, followed by one cycle of 72 °C for 5 min. Indexed libraries were then purified, pooled, gel purified, spiked and denatured. Sequencing was performed on the Illumina MiSeq sequencer using 500 cycle MiSeq reagent kits (San Diego, CA, USA). Sequence quality control, pre-processing, amplicon sequencing and data analysis were carried out as in the descriptions in previous studies [16].

3. Results and Discussion

3.1. Overall Performance of Partial Nitrification on Treating Livestock Wastewater

The IASBR and CASBR were operated for 120 days, and the COD removal rate reached a steady state after 15 days. The average COD removal rate was 87.6 ± 1.7% and 89.7 ± 3.4% in IASBR and CASBR, respectively. In the effluent, BOD_5 concentration was only 20 ± 13 mg/L. The effluent was added into a flask with activated sludge taken from the IASBR, and then it was continuously aerated for two days. There was no evident decrease in COD, indicating that most of the remaining COD in the effluent was non-biodegradable organic matter, such as cellulose and hemicelluloses [3].

Nitrite accumulation occurred immediately in the beginning period (Figure 2) due to the activated sludge in IASBR and CASBR seeded from a partial nitrification reactor. However, during this period, it was observed that the effluent NH_4^+-N level rose with a significant pH decrease (from 7.9 in Day 1 to 6.1 in Day 10). It shows that the alkalinity in the livestock wastewater was not sufficient to sustain a stable pH. Thus, an additional 4.5 g/L of alkaline solution ($NaHCO_3$) was added to the digestate liquid from Day 13. The nitrification activity recovered immediately, and after Day 30, the effluent NH_4^+-N concentration dropped to below 10 mg/L with a stable pH of 8.1 ± 0.2.

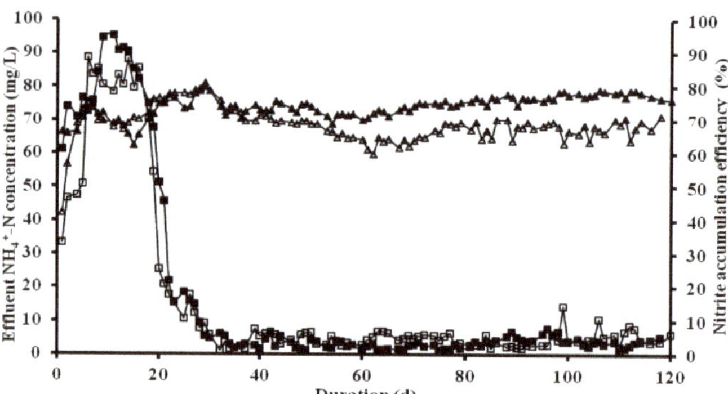

Figure 2. Performance of effluent NH_4^+-N concentration and NO_2^--N accumulation efficiency in SBRs (■: effluent NH_4^+-N in IASBR; □: effluent NH_4^+-N in CASBR; ▲: NO_2^--N accumulation efficiency in IASBR; Δ: NO_2^--N accumulation efficiency in CASBR).

In two SBRs, NO_2^--N accumulated and more than 90 mg/L of NO_2^--N appeared in the effluent as soon as the reactor operation commenced. In the steady state, effluent NO_2^--N concentration increased to 544 ± 36 mg/L and 492 ± 65 mg/L in IASBR and CASBR, and an average nitrite accumulation efficiency of 78.4 ± 1.3% and 64.7 ± 3.2% was achieved, respectively. It was observed that the accumulation efficiency of effluent NO_2^--N in CASBR was lower than that of IASBR. The nitration activity recovered when the partial nitrification sludge experienced continuous aeration. Therefore, the oxidation of NO_2^--N to NO_3^--N was encouraged in CASBR, and NOB activity partly recovered to lead to an increase in the NO_3^--N concentration in the effluent from Day 38.

Batch experiments show that in the NH_4^+-N oxidation process, the SOUR of IASBR and CASBR was 13.6 and 9.5 mg O_2/(g biomass·h), and as for the NO_2^--N oxidation, the SOUR of IASBR and CASBR was 1.4 and 2.0 mg O_2/(g biomass·h), respectively. In IASBR and CASBR, the SOUR of NH_4^+-N was over more than that of NO_2^--N. This means that the capability of nitrite accumulation was actively maintained under the intermittent aeration and continuous aeration conditions. The additional alkalinity supplement might have been the important factor promoting partial nitrification in two SBRs. It caused the pH values in two SBRs to range from 7.9–8.4. When pH is above 7.5, it may inhibit the growth and activity of NOB [17,18].

It was observed that the partial nitrification efficiency in IASBR was 21.2% higher than that in CASBR, and the SOUR ratio of NH_4^+-N oxidation to NO_2^--N oxidation reached 9.71:1 in IASBR. Under the intermittent aeration condition, the control strategy of alternating aeration and non-aeration could provide a suitable environment for partial nitrification. In order to investigate the influence of intermittent aeration on nitrite accumulation, under the condition of alternating 0, 10, 20, 30 or 40 min non-aeration periods with continuous 50 min aeration period conditions, the capability of partial nitrification was examined through batch experiments (Figure 3).

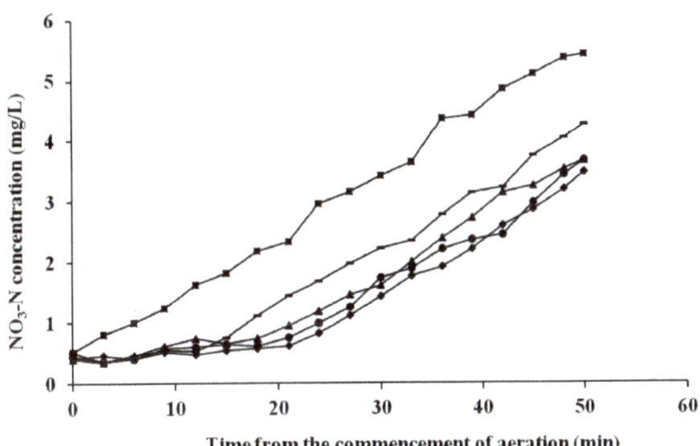

Figure 3. Influence of aeration strategy on partial nitrification (■: 0 min non-aeration duration; -: 10 min non-aeration duration; ▲: 20 min non-aeration duration; •: 30 min non-aeration duration; ♦: 40 min non-aeration duration).

Besides the 0 min non-aeration condition, after the commencement of aeration, the increase in nitrate was retarded for 12, 18, 18 and 21 min, respectively. NO_2^--N oxidation was not activated immediately after aeration started when the non-aeration period was adopted [19]. After a "lag time" of NO_2^--N oxidation, the activity of the NOB recovered, and the calculated NO_3^--N accumulation rates were 0.0972, 0.0977, 0.0967 and 0.9930 mg NO_3^--N/(L·min), respectively. Compared with 0.1030 mg NO_3^--N/(L·min) in the 0 min non-aeration condition, which could denote continuous aeration condition, there was no evident decrease of NO_3^--N accumulation rates in the intermittent aeration conditions. Thus, the intermittent aeration strategy was discovered to induce a lag reaction for NO_2^--N oxidation in a short time, but had no further effects on the inhibition of NOB activity.

The average ratios of NO_2^--N accumulation rate to NO_3^--N accumulation rate within the 50 min aeration period were investigated (Figure 4). In 10, 20, 30 and 40 min non-aeration conditions, the ratios were 1.28, 1.44, 1.51 and 1.35 times of that in 0 min condition, respectively. As for the 40 min non-aeration condition, the NH_4^+-N oxidation rate was only 0.294 mg NH_4^+-N/(L·min), compared with 0.358, 0.364, 0.356 and 0.346 mg NH_4^+-N/(L·min) in 0, 10, 20 and 30 min non-aeration conditions, respectively.

This means that a slight inhibition of ammonium-oxidizing bacteria (AOB) could be carried out in the 40 min non-aeration condition. Thus, the optimal alternating non-aeration duration was 20–30 min for partial nitrification treatment of livestock wastewater.

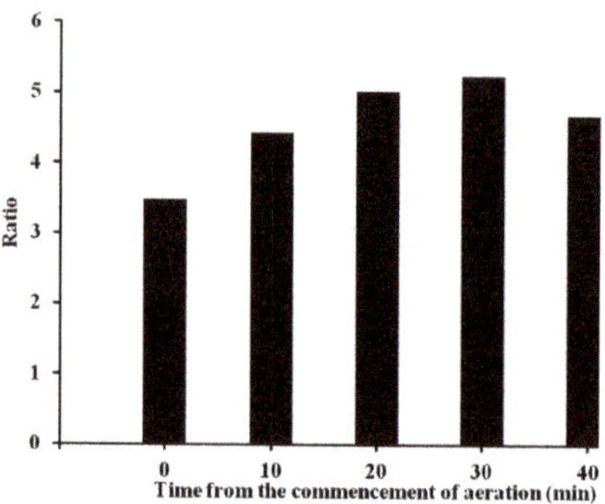

Figure 4. Ratio of NO_2^--N accumulation rate to NO_3^--N accumulation rate under the different non-aeration durations.

Sludge samples were taken from two SBRs to investigate the microbial population by 16S rRNA gene sequencing on Day 45 (partial nitrification set-up stage) and Day 109 (partial nitrification steady-state stage) compared with seed sludge (Figure 5).

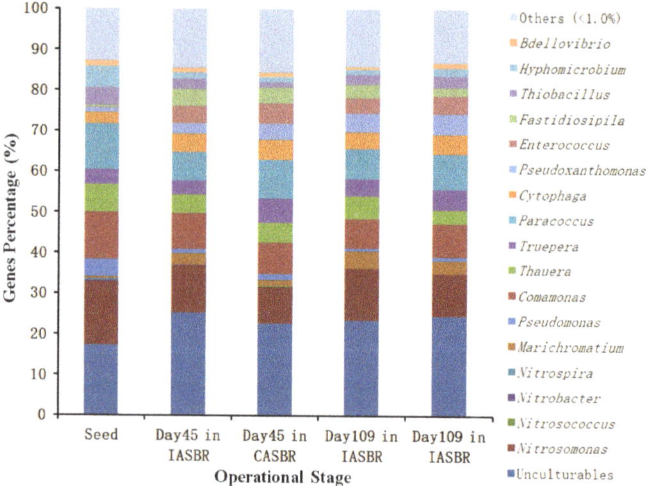

Figure 5. Microbial population analysis of partial nitrification sludge by 16S rRNA gene sequencing.

Except for unculturable bacteria, the determined microbial analysis results show that the dominant genus in the sludge samples was recognized to be capable of degrading a wide range of refractory organic pollutants; for example, *Comamonas* (7.2–11.5% of total genera), *Thauera* (3.4–6.7% of total genera), *Truepera* (3.4–5.9% of total genera), and *Pseudomonas* (0.7–4.5% of total genera), or of degrading celluloses, including *Cytophaga* (2.8–4.9% of total genera) and *Pseudoxanthomonas* (1.1–5.1% of total genera) [20–23]. *Paracoccus* (7.0–11.4% of total genera) was found to be the main denitrifier in two reactors [24]. Besides the nitrification genera (Figure 6), about 6.5–8.5% of the other genera might be species of bacteria indigenous to livestock manure, such as *Enterococcus* and *Fastidiosipila*.

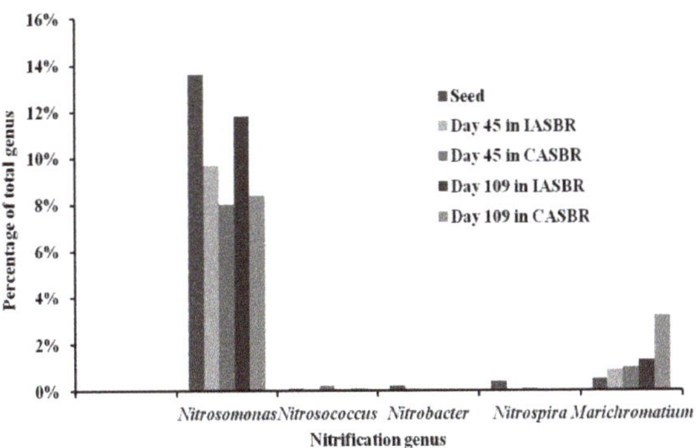

Figure 6. Microbial population analysis of nitrification genus in two SBRs.

Nitrification genus analysis (Figure 6) shows that *Nitrosomonas* was the key bacteria for NH_4^+-N oxidation in two SBRs, and was recognized as the main AOB species in IASBR and CASBR. Common NOB genera (including *Nitrobacter* and *Nitrospira*) were not detected evidently. It shows NOB was inhibited by partial nitrification conditions. However, favorable conditions for NO_2^--N oxidation actually occurred in IASBR and CASBR (Figure 2). One of the reasons might be that the concentration of common NOB genera in the SBRs was too low to be identified by 16S rRNA gene-sequencing equipment; another reason was that the special NOB species, such as *Marichromatium* which was considered as a kind of heterotrophic NO_2^--N oxidation bacteria [25], presented in two SBRs to induce the oxidization of NO_2^--N. The persistent high-COD environment in the reaction cycle was considered to be one of the reasons for the growth of *Marichromatium* bacteria.

According to Figure 6, the percentage of AOB and NOB decreased in both of IASBR and CASBR on Day 45 compared with the seed sludge. The high concentration of SS in the original livestock wastewater could dilute the density of bacteria in the reactors and inhibit the propagation of AOB and NOB in the start-up stage. In this period, the decreased pH could also affect the activity of AOB and NOB [26]. In the steady-state stage (Day 109), the quantity of *Nitrosomonas* in IASBR rapidly recovered from 9.3% to 11.8%, but only a 0.4% increase appeared in CASBR. It shows that the growth of AOB could be accelerated in the intermittent aeration condition. As for *Marichromatium*, 1.0% and 2.2% of the increase was observed in IASBR and CASBR from Day 45 to Day 109, respectively. It shows that the NOB activity recovered in the long-term livestock wastewater treatment process. However, in IASBR, due to the inhibition of the intermittent aeration strategy, the recovery of NOB was lower than that in CASBR.

3.2. Cultivation and Performance of ANAMMOX for Treating Partial Nitrification Effluents

ANAMMOX was considered as a potentially useful technology for treating NO_2^--N containing wastewater without an organic carbon source. In order to investigate the performance of NO_2^--N reduction from the effluents of the partial nitrification reactor and the effects of non-biodegradable organic matter on the activity of ANAMMOX biomass, a UASB reactor experiment was conducted.

Before introduction of IASBR effluents into the UASB, synthetic wastewater was fed to cultivate and enrich ANAMMOX bacteria. When the effluent NO_2^--N concentration in the UASB was below 15 mg/L within 15 days, the concentration of influent synthetic wastewater increased as follows: 30 mg/L NH_4^+-N and 40 mg/L NO_2^--N from Day 1 to Day 100 (Stage 1); 60 mg/L NH_4^+-N and 80 mg/L NO_2^--N from Day 101 to Day 223 (Stage 2); and 90 mg/L NH_4^+-N and 120 mg/L NO_2^--N from Day 223 to Day 289 (Stage 3). In the conventional ANAMMOX process, the theoretical ratio of consumed NO_2^--N to consumed NH_4^+-N is 1.32, and the theoretical ratio of consumed NO_3^--N to consumed NH_4^+-N is −0.26 [10]. In the cultivation period of ANAMMOX biomass, these ratios were calculated compared with the theoretical values in order to determine the ANAMMOX activity (Figure 7).

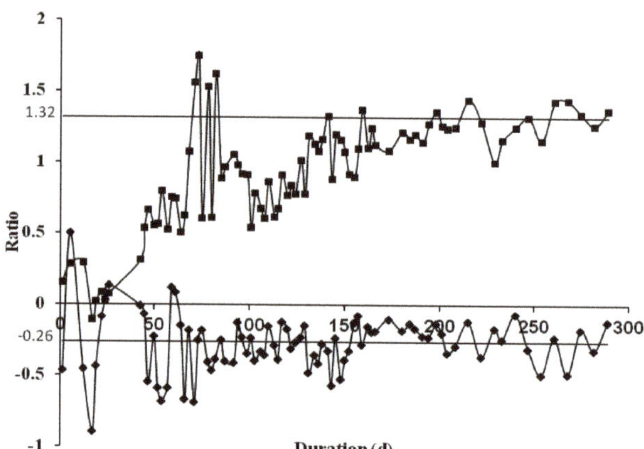

Figure 7. Performance of ANAMMOX biomass cultivation (■: ratios of consumed NO_2^--N to consumed NH_4^+-N; ♦: ratios of generated NO_3^--N to consumed NH_4^+-N).

In Stage 1, almost no NO_2^--N was removed, and NH_4^+-N was partly consumed by microbial anabolism during the start-up periods. After operation for 38 days, the phenomenon of NO_2^--N removal gradually appeared. However, it took 40–50 days for the profile of NO_2^--N to reach a pseudo-steady state. In this period, 66.2% of NO_2^--N removal was observed, and the ratio of consumed NO_2^--N to consumed NH_4^+-N reached 0.95. In Stage 2, with the increase in influent synthetic wastewater concentrations, accumulated NO_2^--N appeared again in the UASB. From Day 131, the concentration of NO_2^--N began to decrease, and satisfactory ANAMMOX performance with an average ratio (consumed NO_2^--N to consumed NH_4^+-N) of 1.24 was observed from Day 180 to Day 223. In Stage 3, no evident changes of ANAMMOX activity were obtained with higher nitrogen loading rates. This shows that a stable ANAMMOX reactor was established with an average NO_2^--N removal rate of 81.5%.

In order to investigate the effects of non-biodegradable organic matter on the activity of ANAMMOX biomass in livestock wastewater, the non-biodegradable organic matter was centrifuged from the effluent of IASBR after two days' continuous aeration in a flask with partial nitrification sludge. Influent synthetic wastewater in Stage 3 was fed into 100 mL flasks with different dosages of non-biodegradable organic matter (the initial COD concentrations were 50, 100, 150, 200, 250, 300, 350 and 400 mg/L in the flasks). Then, ANAMMOX biomass taken from the UASB was added into the flasks, and the biomass concentrations in the flasks were adjusted to close to those inside the UASB. Then, the flasks were continuously stirred with magnetic stirrers after stripping DO out for 3 min by using argon gas. Lastly, the effluents were withdrawn and influent synthetic wastewater with non-biodegradable COD was fed in again at a HRT of 8 h. After 1 day and 5 days, the effluent NH_4^+-N, NO_2^--N and NO_3^--N concentrations were measured to evaluate the effects of non-biodegradable organic matter on ANAMMOX (Figure 8).

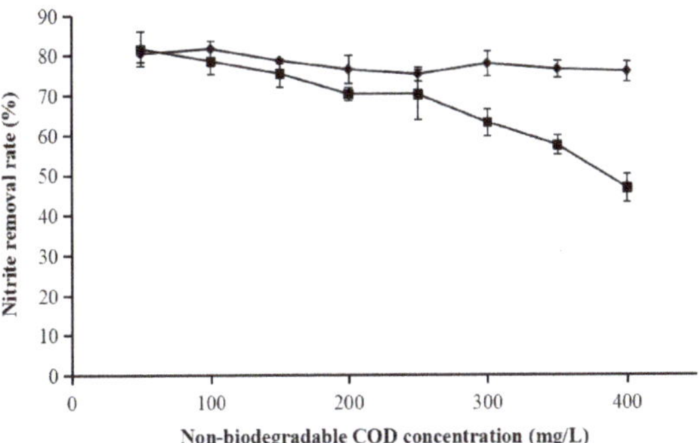

Figure 8. Influence of non-biodegradable organic matter on ANAMMOX activity (♦: nitrite removal rate after 1 day; ■: nitrite removal rate after 5 days).

The results show that short-term exposure to non-biodegradable organic matter did not affect the activity of ANAMMOX biomass evidently. After 1 day's reaction, the NO_2^--N removal rates were all kept at above 76% under the different non-biodegradable COD concentrations. However, according to the results after 5 days' reaction, high concentrations of COD could further inhibit ANAMMOX activity with the extension of exposure reaction time. When the non-biodegradable COD concentration was more than 200 mg/L, the NO_2^--N removal rate decreased rapidly from 71.2% to 49.8%. It was found that non-biodegradable organic matter in the livestock wastewater had potential inhibition ability against ANAMMOX activity under high concentrations and long-term exposure conditions. The reason might be that the non-biodegradable organic matter, including cellulose and hemicelluloses in livestock wastewater, could disrupt the structure of the biofacies and lead to changes in extracellular polymeric substances (EPS) contents and granulation of ANAMMOX biomass [27].

The long-term performance of the treatment efficiency of the livestock wastewater after partial nitrification by using ANAMMOX technology were investigated at the different non-biodegradable COD concentration gradients. The effluents of IASBR with NH_4Cl adjustment (containing 1240 ± 45 mg/L of COD, 26 ± 3 mg/L of BOD_5, 420 ± 7 mg/L of NH_4^+-N and 552 ± 28 mg/L of NO_2^--N) were fed into the UASB after diluting 10 times from Day 1 to Day 17 (118 ± 13 mg COD/L), diluting 7.5-times from Day 18 to Day 35 (168 ± 9 mg COD/L) and diluting 5 times from Day 36 to Day 50 (242 ± 17 mg COD/L), respectively (Figure 9).

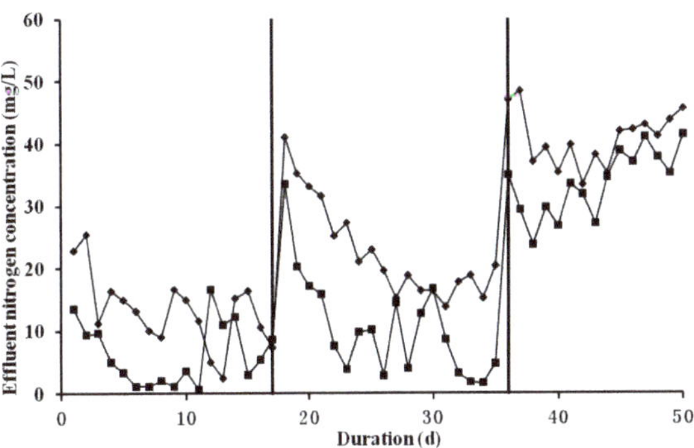

Figure 9. Profiles of ANAMMOX reactor treating the diluted effluents of IASBR (■: NH_4^+-N concentration; ♦: NO_2^--N concentration).

In the condition of 10-times dilution, 80.1% of NO_2^--N was removed immediately after 3 days. It shows that the treatment efficiency of cultured ANAMMOX bacteria was stable under the low concentration of non-biodegradable organic matter condition. In the condition of 7.5-times dilution, with the increases in influent COD and NO_2^--N concentrations, the performance of effluent NO_2^--N in UASB became worse from Day 18. However, the resumption of ANAMMOX activity was confirmed after 4 days and an average NO_2^--N removal rate of 87.4% was achieved in this period. At this COD loading rate, non-biodegradable organic matter did not severely affect the activity of ANAMMOX. In the third stage, effluent NO_2^--N and NH_4^+-N gradually increased and the NO_2^--N removal rate decreased to 60.5%. Potential inhibition of ANAMMOX biomass appeared with the increase in COD concentrations. One of the reasons was that ANAMMOX activity became worse under high concentrations of non-biodegradable organic matter according to Figure 8; another reason might be that heterotrophic denitrifiers existed in the effluents of IASBR which could compete with ANAMMOX bacteria at a high loading rate [28]. The long-term experimental results demonstrate that when the concentration of non-biodegradable organic matter was lower than 168 $\pm$ 9 mg COD/L, the application of ANAMMOX technology on nitrogen removal from the effluents of partial nitrification treating livestock wastewater was feasible.

4. Conclusions

The performance of nitrogen removal through partial nitrification coupled with ANAMMOX technology treating livestock wastewater was studied to evaluate the influences of non-biodegradable organic matter and microbial communities in this research. The obtained results are as follows:

Partial nitrification was successfully achieved for treating the livestock wastewater. At loading rates of 3.08 kg COD/(m^3·d) and 0.93 kg NH_4^+-N/(m^3·d), COD removal efficiency was 87.6 $\pm$ 1.7% and 89.7 $\pm$ 3.4%, and average nitrite accumulation efficiencies of 78.4 $\pm$ 1.3% and 64.7 $\pm$ 3.2% were obtained in IASBR and CASBR, respectively.

The optimal alternating non-aeration duration was 20–30 min for partial nitrification in IASBR, and the intermittent aeration strategy was discovered to induce a lag reaction for NO_2^--N oxidation in a short time, but had no further effects on the inhibition of NOB activity.

Nitrosomonas was the key bacteria for NH_4^+-N oxidation in the partial nitrification process. Under the intermittent aeration strategy, the quantity of *Nitrosomonas* increased from 9.3% to 11.8%, but only a 0.4% increase appeared in CASBR. Heterotrophic *Marichromatium* was the main NO_2^--N oxidation bacteria under the high-concentration non-biodegradable organic matter condition.

Feeding the effluents of partial nitrification into the ANAMMOX reactor, non-biodegradable organic matter showed potential inhibition effects on ANAMMOX activity with the extension of exposure reaction time and the increase in COD concentrations. When the concentration of non-biodegradable organic matter was lower than 168 $\pm$ 9 mg COD/L, an average NO_2^--N removal rate of 87.4% was achieved.

Author Contributions: Conceptualization, M.Z. and X.C.; methodology, M.Z. and X.X.; validation, X.X.; formal analysis, M.Z. and X.C.; investigation, M.Z. and X.C.; data curation, M.Z. and Z.F.; writing—original draft preparation, M.Z. and Z.F.; writing—review and editing, X.Z.; project administration and funding acquisition, M.Z. and X.X. All authors have read and agreed to the published version of the manuscript.

Funding: This research was funded by the National Natural Science Foundation of China (Ref. 51408104).

Institutional Review Board Statement: Not applicable.

Informed Consent Statement: Not applicable.

Data Availability Statement: Data are contained within the article.

Conflicts of Interest: The authors declare no conflict of interest.

References

1. Pandey, B.; Chen, L. Technologies to recover nitrogen from livestock manure—A review. *Sci. Total Environ.* **2021**, *784*, 147098. [CrossRef]
2. Meade, G.; Lalor, S.T.J.; Mc Cabe, T. An evaluation of the combined usage of separated liquid pig manure and inorganic fertiliser in nutrient programmes for winter wheat production. *Eur. J. Agron.* **2011**, *34*, 62–70. [CrossRef]
3. Xie, S.; Lawlor, P.G.; Frost, J.P.; Hu, Z.; Zhan, X. Effect of pig manure to grass silage ratio on methane production in batch anaerobic co-digestion of concentrated pig manure and grass silage. *Bioresour. Technol.* **2011**, *102*, 5728–5733. [CrossRef] [PubMed]
4. Karakashev, D.; Schmidt, J.E.; Angelidaki, I. Innovative process scheme for removal of organic matter, phosphorus and nitrogen from pig manure. *Water Res.* **2008**, *42*, 4083–4090. [CrossRef]
5. Ge, S.; Wang, S.; Yang, X.; Qiu, S.; Li, B.; Peng, Y. Detection of nitrifiers and evaluation of partial nitrification for wastewater treatment: A review. *Chemosphere* **2015**, *140*, 85–98. [CrossRef]
6. Wang, Z.; Zhang, L.; Zhang, F.; Jiang, H.; Ren, S.; Wang, W.; Peng, Y. A continuous-flow combined process based on partial nitrification-Anammox and partial denitrification-Anammox (PN/A + PD/A) for enhanced nitrogen removal from mature landfill leachate. *Bioresour. Technol.* **2020**, *297*, 122483. [CrossRef] [PubMed]
7. Regmi, P.; Miller, M.W.; Holgate, B.; Bunce, R.; Park, H.; Chandran, K.; Wett, B.; Murthy, S.; Bott, C.B. Control of aeration, aerobic SRT and COD input for mainstream nitritation/denitritation. *Water Res.* **2014**, *57*, 162–171. [CrossRef] [PubMed]
8. Rongsayamanont, C.; Limpiyakorn, T.; Khan, E. Effects of inoculum type and bulk dissolved oxygen concentration on achieving partial nitrification by entrapped-cell-based reactors. *Bioresour. Technol.* **2014**, *164*, 254–263. [CrossRef] [PubMed]
9. Wang, J.; Kang, J. The characteristics of anaerobic ammonium oxidation (ANAMMOX) by granular sludge from an EGSB reactor. *Process Biochem.* **2005**, *40*, 1973–1978. [CrossRef]
10. Li, J.; Li, J.; Peng, Y.; Wang, S.; Zhang, L.; Yang, S.; Li, S. Insight into the impacts of organics on anammox and their potential linking to system performance of sewage partial nitrification-anammox (PN/A): A critical review. *Bioresour. Technol.* **2020**, *300*, 122655. [CrossRef] [PubMed]
11. Liang, Z.; Liu, J. Landfill leachate treatment with a novel process: Anaerobic ammonium oxidation (Anammox) combined with soil infiltration system. *J. Hazard. Mater.* **2008**, *151*, 202–212. [CrossRef] [PubMed]
12. Lopez, H.; Puig, S.; Ganigue, R.; Ruscalleda, M.; Balaguer, M.D.; Colprim, J. Start-up and enrichment of a granular anammox SBR to treat high nitrogen load Wastewaters. *J. Chem. Technol. Biotechnol.* **2008**, *83*, 233–241. [CrossRef]
13. American Public Health Association. *Standard Methods for the Examination of Water and Wastewater*, 22nd ed.; American Public Health Association: Washington, DC, USA, 2012.
14. Yu, Z.; Morrison, M. Comparisons of different hypervariable regions of rrs genes for use in fingerprinting of microbial communities by PCR-denaturing gradient gel electrophoresis. *Appl. Environ. Microb.* **2004**, *70*, 4800–4806. [CrossRef] [PubMed]
15. McCabe, M.S.; Cormican, P.; Keogh, K.; O'Connor, A.; O'Hara, E.; Palladino, R.A.; Kenny, D.A.; Waters, S.M. Illumina miSeq phylogenetic amplicon sequencing shows a large reduction of an uncharacterised *Succinivibrionaceae* and an increase of the methanobrevibacter gottschalkii clade in feed restricted cattle. *PLoS ONE* **2015**, *10*, e0133234. [CrossRef]
16. Jiang, Y.; Dennehy, C.; Lawlor, P.G.; Hu, Z.; McCabe, M.; Cormican, P.; Zhan, X.; Gardiner, G.E. Exploring the roles of and interactions among microbes in dry co-digestion of food waste and pig manure using high-throughput 16S rRNA gene amplicon sequencing. *Biotechnol. Biofuels* **2019**, *12*, 5. [CrossRef]
17. Ciudad, G.; Gonzalez, R.; Bornhardt, C.; Antileo, C. Modes of operation and pH control as enhancement factors for partial nitrification with oxygen transport limitation. *Water Res.* **2007**, *41*, 4621–4629. [CrossRef]
18. Chen, J.; Zheng, P.; Yu, Y.; Mahmood, Q.; Tang, C. Enrichment of high activity nitrifiers to enhance partial nitrification process. *Bioresour. Technol.* **2010**, *101*, 7293–7298. [CrossRef]
19. Wei, D.; Xue, X.; Yan, L.; Sun, M.; Zhang, G.; Shi, L.; Du, B. Effect of influent ammonium concentration on the shift of full nitritation to partial nitrification in a sequencing batch reactor at ambient temperature. *Chem. Eng. J.* **2014**, *235*, 19–26. [CrossRef]
20. Ferrero, E.M.; de Godos, I.; Rodríguez, E.M.; García Encina, P.A.; Muñoz, R.; Bécares, E. Molecular characterization of bacterial communities in algal–bacterial photobioreactors treating piggery wastewaters. *Ecol. Eng.* **2012**, *40*, 121–130. [CrossRef]
21. Mao, Y.; Zhang, X.; Yan, X.; Liu, B.; Zhao, L. Development of group-specific PCR-DGGE fingerprinting for monitoring structural changes of *Thauera* spp. in an industrial wastewater treatment plant responding to operational perturbations. *J. Microbiol. Methods* **2008**, *75*, 231–236. [CrossRef]
22. Ordaz-Cortés, A.; Thalasso, F.; Salgado-Manjarrez, E.; Garibay-Orijel, C. Treatment of wastewater containing high concentrations of terephthalic acid by *Comamonas* sp. and *Rhodococcus* sp.: Kinetic and stoichiometric characterization. *Water Environ. J.* **2014**, *28*, 393–400. [CrossRef]
23. Wang, S.; Zhao, D.; Bai, X.; Zhang, W.; Lu, X. Identification and characterization of a large protein essential for degradation of the crystalline region of cellulose by *Cytophaga hutchinsonii*. *App. Environ. Microb.* **2017**, *83*, 2270. [CrossRef] [PubMed]
24. Qu, Z.; Bakken, L.R.; Molstad, L.; Frostegard, A.; Bergaust, L.L. Transcriptional and metabolic regulation of denitrification in *Paracoccus denitrificans* allows low but significant activity of nitrous oxide reductase under oxic conditions. *Environ. Microbiol.* **2016**, *18*, 2951–2963. [CrossRef] [PubMed]
25. Hong, X.; Chen, Z.; Zhao, C.; Yang, S. Nitrogen transformation under different dissolved oxygen levels by the anoxygenic phototrophic bacterium *Marichromatium gracile*. *World J. Microb. Biotechnol.* **2017**, *33*, 113. [CrossRef]

26. Brockmann, D.; Morgenroth, E. Evaluating operating conditions for outcompeting nitrite oxidizers and maintaining partial nitrification in biofilm systems using biofilm modelling and Monte Carlo filtering. *Water Res.* **2010**, *44*, 1995–2009. [CrossRef]
27. Li, Y.; Huang, Z.; Ruan, W.; Ren, H.; Zhao, M. ANAMMOX performance, granulation and microbial response under COD disturbance. *J. Chem. Technol. Biotechnol.* **2015**, *90*, 139–148. [CrossRef]
28. Molinuevo, B.; Garcia, M.C.; Karakashev, D.; Angelidaki, I. ANAMMOX for ammonia removal from pig manure effluents: Effect of organic matter content on process performance. *Bioresour. Technol.* **2009**, *100*, 2171–2175. [CrossRef]

Article

SEM-RCNN: A Squeeze-and-Excitation-Based Mask Region Convolutional Neural Network for Multi-Class Environmental Microorganism Detection

Jiawei Zhang [1], Pingli Ma [1], Tao Jiang [2,3,*], Xin Zhao [4], Wenjun Tan [5], Jinghua Zhang [1,6], Shuojia Zou [1], Xinyu Huang [6], Marcin Grzegorzek [6] and Chen Li [1,*]

1 Microscopic Image and Medical Image Analysis Group, College of Medicine and Biological Information Engineering, Northeastern University, Shenyang 110819, China
2 School of Intelligent Medicine, Chengdu University of Traditional Chinese Medicine, Chengdu 610075, China
3 International Joint Institute of Robotics and Intelligent Systems, Chengdu University of Information Technology, Chengdu 610225, China
4 School of Resources and Civil Engineering, Northeastern University, Shenyang 110819, China
5 School of Computer Science and Engineering, Northeastern University, Shenyang 110169, China
6 Institute of Medical Informatics, University of Luebeck, 23562 Luebeck, Germany
* Correspondence: jiang@cuit.edu.cn (T.J.); lichen201096@hotmail.com (C.L.)

Abstract: This paper proposes a novel Squeeze-and-excitation-based Mask Region Convolutional Neural Network (SEM-RCNN) for Environmental Microorganisms (EM) detection tasks. Mask RCNN, one of the most applied object detection models, uses ResNet for feature extraction. However, ResNet cannot combine the features of different image channels. To further optimize the feature extraction ability of the network, SEM-RCNN is proposed to combine the different features extracted by SENet and ResNet. The addition of SENet can allocate weight information when extracting features and increase the proportion of useful information. SEM-RCNN achieves a mean average precision (mAP) of 0.511 on EMDS-6. We further apply SEM-RCNN for blood-cell detection tasks on an open source database (more than 17,000 microscopic images of blood cells) to verify the robustness and transferability of the proposed model. By comparing with other detectors based on deep learning, we demonstrate the superiority of SEM-RCNN in EM detection tasks. All experimental results show that the proposed SEM-RCNN exhibits excellent performances in EM detection.

Keywords: environmental microorganisms; object detection; deep learning

Citation: Zhang, J.; Ma, P.; Jiang, T.; Zhao, X.; Tan, W.; Zhang, J.; Zou, S.; Huang, X.; Grzegorzek, M.; Li, C. SEM-RCNN: A Squeeze-and-Excitation-Based Mask Region Convolutional Neural Network for Multi-Class Environmental Microorganisms Detection. *Appl. Sci.* **2022**, *12*, 9902. https://doi.org/10.3390/app12199902

Academic Editors: Dino Musmarra and Paola Grenni

Received: 7 May 2022
Accepted: 25 September 2022
Published: 1 October 2022

1. Introduction

Environmental Microorganisms (EMs) collectively refer to all microorganisms that have an impact on the environment, including microorganisms living in the natural environment (such as oceans and deserts) and artificial environments (such as fisheries and wheat fields) [1]. There are about $10^{11} \sim 10^{12}$ types of EMs on Earth [2]. All of them play a positive or negative role in the task of environmental governance. For example, plant rhizosphere-promoting bacteria can help promote plants' healthy growth. It can also inhibit pathogenic microorganisms that harm plants. However, harmful rhizosphere bacteria can inhibit the normal growth of plants by producing phytotoxins [3]; the emergence of cyanobacteria will accelerate the process of eutrophication of water bodies and damage water quality, which will eventually lead to the death of a large number of aquatic organisms; aspidisca has a strong sensitivity to the chemical substances contained in the water body. Therefore, aspidisca is widely applied for evaluating the quality of the aquaculture water body in the water aquaculture industry. To better play the role of EMs in environmental governance, research on EM detection is essential. The methods of EM detection can be mainly grouped into manual microscope observation methods and computer-aided detection methods.

Manual microscope observation methods refer to the observation and record of EMs in the field of view by an experimenter with certain professional knowledge using a microscope. However, there exist some limitations and disadvantages with respect to manual microscope observation methods. First, the experimenter cannot make quick judgments and must consult many reference materials when facing a wide variety of EMs. Second, all experimenters have to spend a substantial amount of time when learning the basics of EMs and the operation of the microscope. Finally, the detection results obtained by different operators might be different, and the objectivity of the detection results is insufficient [4]. Therefore, manual microscope observation methods have great limitations for EM detection tasks.

Compared to manual microscope observation methods, computer-aided detection methods are more objective, accurate, and convenient. With rapid developments in computer vision and deep learning technologies, computer-assisted image analysis is broadly applied in many research fields, including fire emergency [5], histopathological image analysis [6–9], cytopathological image analysis [10–12], object detection [13–17], microorganism classification [18–23], microorganism segmentation [24–27], and microorganism counting [28,29]. In addition, with the advancement of computer hardware and the rapid development of computer-aided detection methods, the results obtained by computer-aided detection methods in EM detection are improving. Currently, the most popular computer-aided detection method is the EM detection method based on deep learning [30]. However, there is no relevant research on the detection of multi-class EMs. Therefore, we choose some classical detectors based on deep learning for multi-class EM detection and propose a novel detector called squeeze-and-excitation-based mask region convolutional neural network (SEM-RCNN). The flowchart of SEM-RCNN is shown in Figure 1.

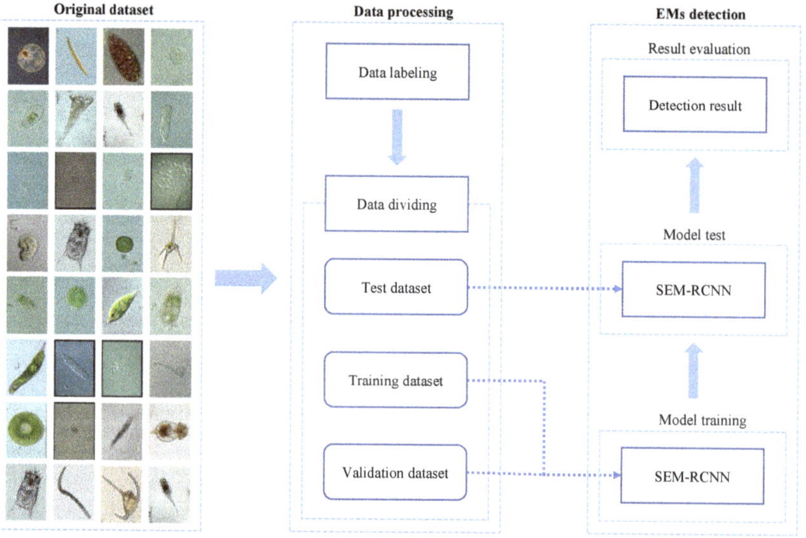

Figure 1. The flowchart of SEM-RCNN.

In Figure 1, three main parts are contained. Part one is the original dataset part, which includes enough images of EMs for model training and testing. We will introduce the specific information about the original dataset in detail in the experimental section. Part two is the data-processing part. In this part, the original dataset is firstly labeled in the format of the object detection dataset. Then, all data are grouped into the training set, validation set, and test set according to a certain proportion. Part three is the EM detection part. Firstly, the original SEM-RCNN is pre-trained on Microsoft Common Objects in Context (MS-COCO) dataset. Then the proposed model is finetuned and trained on the training and validation sets of EMDS-6. After that, the detection performance of the trained model is verified on the test set. Finally, we evaluate the detection results of SEM-RCNN by employing appropriate evaluation indicators.

The main contributions of this paper are listed as follows:

- A novel detector based on convolutional neural network (CNN): SEM-RCNN is proposed for multi-class EM detection;
- The block of SENet is designed to combine with ResNet as the backbone of the proposed SEM-RCNN, which can extract features with a self-attention mechanism;
- The proposed SEM-RCNN achieves the optimal detection performance both for small (EMDS-6) and large (blood cell) datasets.

To illustrate the proposed method clearer, the structure of this paper is designed as follows: In Section 2, the related research about computer-aided EM detection is summarized; In Section 3, detailed information about SEM-RCNN is introduced; In Section 4, the detailed operation of the experiments is introduced, including experimental data, experimental settings, evaluation criteria, detection results, and extensive experiment; In Section 5, the paper is summarized comprehensively.

2. Related Work

In this section, we group all computer-aided EM detection methods into classical image-processing-based methods, traditional machine-learning-based methods, and deep-learning-based methods. The detection methods are introduced based on relevant research studies.

2.1. Classical Image Processing Based Methods

Classical image-processing-based methods are the earliest computer-aided methods for EM detection. Classical image-processing-based methods contain two subcategories of detection methods, segmentation-based methods, and classification-based methods.

Thresholding-based methods are the most used technologies for image segmentation, such as in [31–48]. Thresholding methods are the most commonly used methods in image segmentation. In addition, thresholding methods can select an appropriate threshold for detection according to different EMs, which gives these methods strong generalization abilities. In [32], an area threshold was applied for *Chlamydomonas* and *Chlamydomonas bicuspidata* detection. In [37], the multiple thresholds method was employed for motile microorganisms. Multiple thresholds were firstly applied to binarize input images. Then, all white regions are regarded as EMs. In [40], the color threshold was applied for tubercle bacillus detection. In [43], adaptive threshold and global threshold were applied for nematode detection. An adaptive threshold was employed for binarizing the original image. Then, a global threshold was applied for extracting reference labels. By combining these two processed images, the nematode can be detected. Among all these works, the Otsu threshold is the most used one. In [33,42,44–46], the Otsu threshold is applied for different EMsdetection. The main idea of Otsu was to select an optimal threshold automatically from a gray-level histogram by a discriminant criterion [49]. The Otsu threshold can provide good results with simple calculations. Even if the gray value of the object to be segmented is similar to the gray value of the background, the Otsu threshold can achieve good segmentation results. However, due to the limitation of the calculation method, when the difference between the foreground and the background area is too large, the Otsu threshold cannot achieve a good segmentation result [50].

Classification-based methods apply shape features, geometric features, color features, texture features, and statistical features for EM detection, such as in [51–65]. In [51,52,55], shape features were selected as vital information for EM detection, including contour features, area features, squareness, angular, roundness, etc. In [56], a contour feature was used for detecting *Methanospirillum hungatei* and *Methanosarcina mazei*. In [52], a *C. elegans* nematode worm detection method based on angular features was designed. In [55], roundness was selected as the criterion for judging whether the detected objects are Rotavirus-A particles. In [57,59–63,66], geometric features were selected for EM detection. In [60], the area was regarded as an important indicator for judging the presence of bacteriophage. In [61], the most suitable combination of some kinds of geometric features was selected and applied for bacilli detection. In [62,63], an automatic detection method based on area-to-length ratio was proposed for six different airborne fungi spores. In [57,61], color features were employed for initial screening regions containing EMs. In [64], a *Anabaena* and *Oscillatoria* detection method based on texture features was proposed. From all these classification based methods, we can find that shape features are the most suitable feature for EM detection. In addition, detection methods by combining different features can achieve improved detection performances than a single feature.

2.2. Traditional Machine-Learning-Based Methods

Since 2006, traditional machine-learning-based methods have been gradually applied in the field of EM detection, such as in [67–79]. The main idea of this method is to determine the EMs category according to the acquired feature information and the corresponding network structure. In [67], a back propagation neural network was employed for bacteria detection. After several preprocessing steps such as threshold-based segmentation and denoising, morphological features of bacteria are extracted and then sent to a back propagation neural network for detection. In [68], a genetic algorithm-neural network method was presented for tubercle bacillus detection. By applying a color filter, moving k-means clustering, and region growing, a suitable segmented image was obtained, which is then sent to the color filter, moving k-means clustering and region growing for the final detection. In [69], a probabilistic neural network is applied for pathogens detection. First, the original image is processed by background correction and object isolation. Then, regions that may contain pathogens were selected. At last, a probabilistic neural network is built for pathogen detection.

Based on the research on traditional machine learning methods in this field, we find that the most widely used classification model is the support vector machine (SVM) classifier, mentioned in [70–79]. SVM can construct an optimal separation hyperplane in the feature space of the data to maximize the gap between positive and negative samples in the training set [80], which makes SVM an efficient classifier for binary classification tasks. Furthermore, SVM can efficiently use smaller training samples. This enables SVM in achieving higher classification accuracies on a smaller training set. Therefore, after the development of related technologies of SVM, it has been gradually applied to the detection of EMs. In [71], an SVM classifier is proposed for P. minimum species detection. In addition, to improve the accuracy of the detection results, the SVM classifier is combined with a random forest classifier. In [74], a multi-class SVM was proposed for EM detection. First, the Sobel edge detector is applied for image segmentation. After that, shape features, Fourier descriptors, and some other features were extracted from processed images and then sent to a multi-class SVM for detecting EMs. In [78], an SVM classifier is applied for planktonic organisms detection. The preprocessing step includes threshold segmentation, robust refocusing criterion, and re-segmentation. After that, the processed image is detected by an SVM classifier.

2.3. Deep-Learning-Based Methods

Compared with methods based on traditional machine learning, deep-learning-based methods have the advantages of the wide range of applications and high applicability. In the feature extraction step of detection processing, traditional machine-learning-based methods use manual feature engineering methods, which are labor-intensive and time-consuming. Deep-learning-based methods can achieve automatic feature learning through advanced network structures and complex features compared to simple ones. Therefore, with the development of deep learning technologies, increasing research about EM detection using deep learning methods is presented, such as in [81–90]. In [81–83], CNN was employed for EM detection. In [81], a tubercle bacillus detector was designed based on CNN. In [83], a CNN-based method was proposed for actinobacterial species detection. In [88], a region convolutional neural network (R-CNN)-based detector was proposed for diatom detection. In addition, a you only look once (YOLO)-based detector is prepared for comparisons. The result indicates that YOLO performs better than R-CNN in diatom detection. In [84–87], Faster R-CNN-based methods were employed for EM detection. In [85], a Faster R-CNN-based detector was proposed for parasite egg detection. In [87], Faster R-CNN was applied for algal detection. About 1859 samples were prepared for the test.

After consulting all these related research studies, we found that classical image-processing-based methods were mainly used as preprocessing methods in current EM detection studies. The most widely used methods in EM detection are traditional machine-learning-based methods. Although there are a few studies about deep-learning-based methods, deep-learning-based methods show great potential in EM detection. Therefore, we designed a deep-learning-based detector for EM detection called SEM-RCNN.

3. SEM-RCNN-Based EM Detection Method

The structure of the proposed SEM-RCNN is shown in Figure 2, which mainly includes the input step, feature extraction step, region proposal step, a mapping step between candidate boxes and feature maps, and the output step.

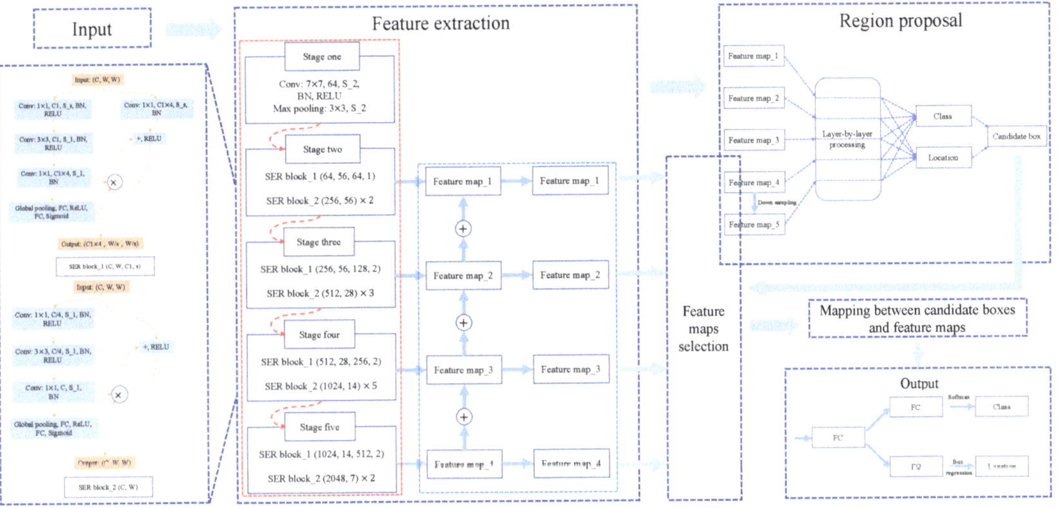

Figure 2. The structure of SEM-RCNN.

Figure 2: (a) Input: The dataset contains images of 21 types of EMs and their corresponding labeled images. There are 840 images in total, and each type has 40 original images (Sections 4.1 and 4.2.1 for details). (b) Feature extraction: A combination network based on SENet and feature pyramid network (FPN) is proposed for fuller and deeper feature extraction (Section 3.1 for details). (c) Region proposal: A region proposal network (RPN) was applied to obtain multi-candidate boxes of the object (Section 3.2 for details). (d) Mapping between candidate boxes and feature maps: The method based on the region of interesting align (ROI align) is applied for accurate mapping between candidate boxes and the feature map, as well as the mapping between the feature map and the fixed size feature map (Section 3.3 for details). (e) Output: A multi-branched structure is applied for feature maps regression, and the combined approach of the fully connected layer, bounding box regression, and Softmax is applied for object detection (Section 3.4 for details).

3.1. Feature Extraction Step

The feature extraction step is the basis for deep learning to perform all tasks. Therefore, whether a suitable feature extraction network is selected or not directly affects final detection results. After analyzing and comparing the existing networks, we finally chose the deep residual network (ResNet) [91] combined with a squeeze-and-excitation network (SENet) [92] as the basic backbone, together with the feature pyramid network(FPN) [93]. ResNet can solve the degradation problem occurring in the training process of CNN. SENet can effectively enhance the needed feature information while suppressing less useful feature information. FPN can achieve the accurate detection of multi-scale objects by making good use of shallow feature information and deep feature information.

3.1.1. ResNet

The main contribution of ResNet concerns an inevitable problem in the training of CNN, called the degradation problem. In general, as the number of layers of the network model increases, the overall detection effectiveness of the CNN model improves. However, when the number of layers deepens to a certain level, the effectiveness of the CNN model decreases, which is the degradation problem that occurs when CNN is trained. The idea of residual learning is introduced with conventional CNNs in ResNet to solve the degradation problem encountered in the deep training of CNN. The structure of a residual block is shown in Figure 3, where x denotes features learned by the shallow network; $F(x)$ is the residual function. The residual block allows the deep network to learn new features relative to the shallow network continuously. From Figure 3, the actual features learned by the network after the residual block are $F(x) + x$, which means that a deep network can obtain deeper and more complex features from the features extracted by the shallow network based on the introduction of the residual block.

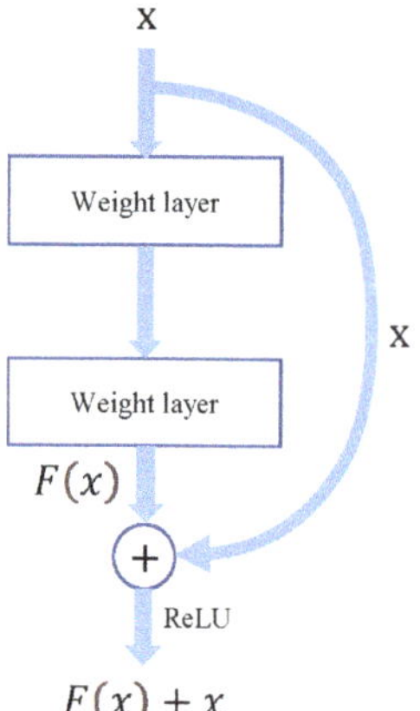

Figure 3. The structure of residual block.

3.1.2. SENet

SENet is a network structure that focuses on enhancing data channel information and enhancing desired feature information while suppressing less useful feature information based on the self-attention mechanism. SE block is the basic module of SENet and can be well integrated with many existing models. The corresponding experimental classification and detection results show that the combination with an SE block can increase the feature representation capability of the model and, thus, improve the classification or detection effects of the model. The basic structure of the SE block is shown in Figure 4. Among them, F_{tr} is the traditional convolution operation. X and U represent the input and output of this convolution operation, respectively; C, W, H, C', W', and H' represent the scales of data;

function F_{sq} and F_{ex} represent the two core processes of the SE block, squeeze processing and excitation processing; $\widetilde{X}$ represents the final output of the SE block.

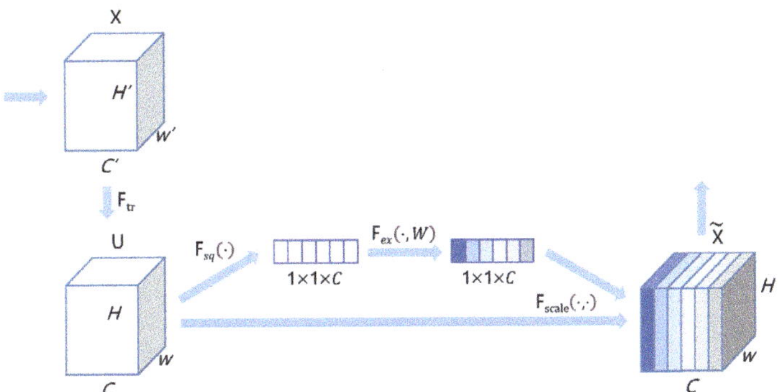

Figure 4. The structure of the SE block.

In the SE process, a squeeze operation is first performed on the convolution output. To make better use of the interconnected information between input data channels, the SE block first uses the idea of averaging to convert the information of all pixels involved in a plane into a specific value. The specific calculation procedure for averaging is shown in Equation (1).

$$z_c = F_{sq}(u_c) = \frac{1}{H \times W} \sum_{i=1}^{H} \sum_{j=1}^{W} u_c(i,j) \tag{1}$$

In Equation (1), z_c represents the squeeze output of the c channel data of input data; u_c represents the c channel data of input data. It can be seen from the formula that input data eventually become a column vector in the squeeze process. The length of this vector is the same as the number of channels, and each data value in this vector is closely related to the corresponding channel data.

After that, to further exploit the interlinked information between channels, SE designs an excitation process. The equation for this process is shown in Equation (2).

$$s = F_{ex}(z, W) = \sigma(g(z, W)) = \sigma(W_2 \sigma(W_1 z)) \tag{2}$$

In Equation (2), σ is the rectified linear unit (ReLU) activation function; $W_1 \in \mathbb{R}^{\frac{C}{r} \times C}$; $W_2 \in \mathbb{R}^{C \times \frac{C}{r}}$; W_1 is the dimensionality reduction layer with a dimensionality reduction ratio of r. W_2 is the proportionally identical data-dimensionality increase layer. After the excitation process, the complexity of the entire model is controlled. Moreover, vital features are enhanced, and weak features are limited based on the self-attention mechanism. Moreover, the generalization ability of the model is enhanced. The sigmoid function processes the final output of the excitation process to a value between zero and one.

The final output of the SE block multiplies the value obtained after compression and activation processing with the data of all the channels that U has. Based on such processing, the SE block can enhance features that have a greater impact on the experimental results while weakening features that have a smaller impact on experimental results.

3.1.3. FPN

FPN is a network component that assists CNN in detecting objects with different scales. From the comparison of the information contained in shallow and deep features, shallow features have richer location information and are more suitable for predicting the location coordinates of the object; deep features have richer category information and are more suitable for predicting the category of the object. Therefore, a suitable combination of shallow and deep features can achieve the accurate detection of EMs, which is the main idea of FPN. FPN consists of three main network structures: down–up processing, up–down processing, and horizontal linkage of feature layers, as shown in Figure 5.

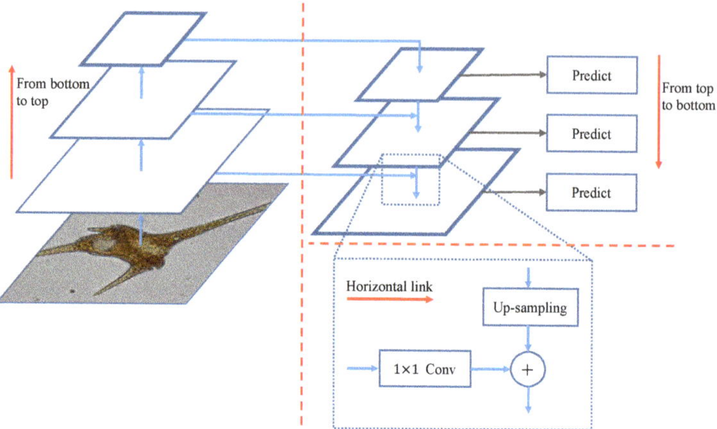

Figure 5. The structure of FPN.

Down–up processing is the feedforward process of CNN. By non-stop convolutional operations, multiple feature maps of different scales are obtained in this step. After that, the feature maps output by the deepest level network are up-sampled by the operation to keep the same size as the feature maps output by the previous level network. Then, upsampled feature maps are fused with the feature maps outputted by the previous layer network. Feature maps obtained based on FPN are richer in feature information than traditional feature maps. In addition, to better fuse the information of each map, FPN processes the fusion between feature maps by using the horizontal linkage operation. The horizontal linkage is performed by convolutional processing using a 1×1-sized convolutional kernel. In general, FPN can combine the rich location information of shallow feature maps and the rich category information of deep feature maps to provide accurate categories and locations of objects at different scales without increasing computational efforts.

3.1.4. Backbone of Feature Extraction Step

By fusing ResNet, SENet, and FPN, we finally design the backbone of SEM-RCNN, as shown in Figure 6. There are mainly five steps in the red dashed box in Figure 6. In stage one, 64 convolution kernels with a kernel size of 7×7 and a stride of 2, followed by batch normalization (BN) and ReLU operations, were applied for feature extraction. Then, max-pooling with a size of 3×3 and stride of 2 was applied to reduce the size of feature maps. After that, the following four steps are based on the combination approach of SER block_1 and SER block_2. The SER block_1 and SER block_2 are important parts of the backbone, and its structure are shown in Figure 7. From Figure 7, we can see that the SER block_1 is mainly used to deal with the case where the input and output dimensions are different; the SER block_2 is mainly used to deal with the case where input and output dimensions are the same. For the input parameters of SER block_1, the channel of the input image is denoted as C, the width (same as height) of the input image is denoted as W, the output channel of the feature map is denoted as C1, and the stride is denoted as s. For example, in stage two of Figure 6, input images are processed at the size of 54×54 with 64 channels. Hence, in the first step of SER block_1, 64 convolution kernels with the kernel size of 1×1 and stride of 1 (followed by BN and ReLU operations) were applied first. Then, the convolution kernels with the kernel size of 3×3 and 1×1 were applied sequentially to extract the feature and to adjust channels. At the same time, the input image is processed by 64 × 4 convolution kernels with the kernel size of 1 × 1 and stride of 1 (right part of SER block_1). Finally, the feature maps after global pooling and sigmoid (the left part of SER block_1) are residually connected with the right part of SER block_1. The output size of SER block_1 is 56 × 56 with channels of 64 × 4. The structure of SER block_2 is similar to SER block_1, but the output size is the same as the input size. The combination method of SER block_1 and SER block_2 repeats four times in SEM-RCNN for feature extraction.

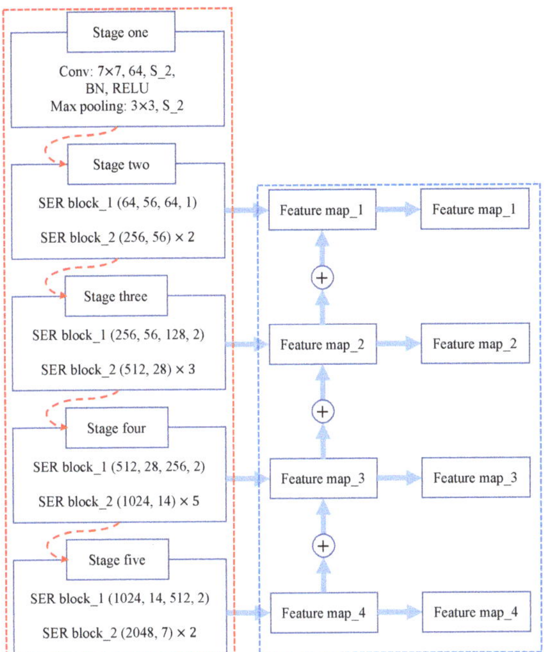

Figure 6. The backbone structure of SEM-RCNN. Red dashed box is SENet and ResNet part; blue dashed box is FPN part.

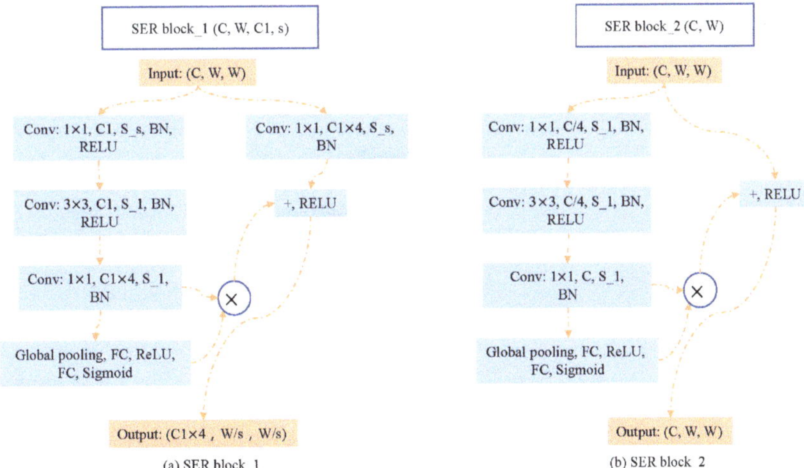

Figure 7. The structure of SER block_1 and SER block_2.

Based on our sufficient research foundation [15,20–22,24,29], it can be found that only a few studies employed deep learning methods to perform the detection task in microorganism image analysis. Since the detection task in microorganism image analysis has strong application background, almost all studies directly utilize existing deep learning models, such as RCNN [88] and Faster R-CNN [84,85]. Different from these studies, we proposed a novel self-attention-based two-stage detection framework, which is inspired by ResNet, SENet, and FPN. This framework achieves state-of-the-art performances on the detection task, which significantly promotes the development of the detection technology in the application of microorganism image analyses.

3.2. Region Proposal Step

Generating candidate boxes for objects is an important processing step of an object detector. In this step, a suitable target candidate frame needs to be generated based on the input feature map. Here, we choose the region proposal network (RPN) to accomplish the task of candidate boxes proposal for SEM-RCNN. RPN can generate prediction boxes for objects with different scales in a short period. RPN mainly includes three processing steps: generating anchor boxes, judging the category of generated anchor boxes, and adjusting the position of anchor boxes. The main flow of RPN is shown in Figure 8. First, RPN generates a certain amount of anchor boxes based on the input feature map; after that, the generated Anchor boxes are convolved with a convolution kernel of 3×3; then, the Softmax function and the border regression algorithm are used to distinguish prospect and background of boxes and obtain position coordinates of the predicted boxes respectively; finally, the candidate boxes are determined based on the obtained category scores and position coordinates.

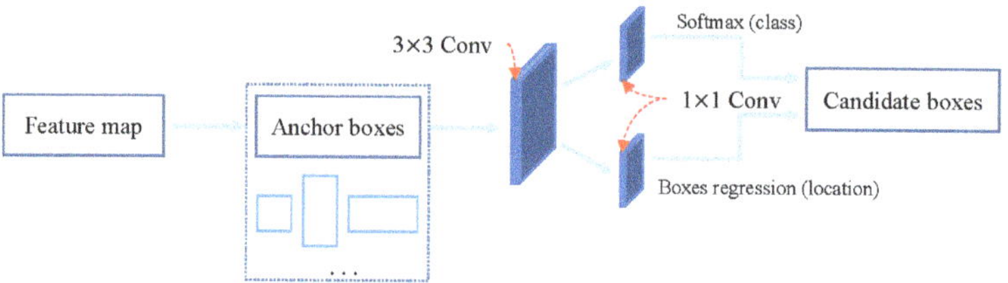

Figure 8. The structure of RPN.

3.3. RoI Align

After obtaining suitable boxes, the detector needs to associate the obtained boxes with feature maps. Here, RoI Align is employed to this end. The main idea of RoI Align is to use bilinear interpolation to obtain the value of floating-point coordinates. Therefore, RoI Align chooses to keep the floating-point coordinates in determining the corresponding area in the feature map based on the position coordinates of the candidate box. When dividing the area corresponding to the candidate box on the feature map equally into multiple small fixed-size feature maps, RoI Align chooses to maintain the segmented boundaries instead of performing quantization operations. Eventually, bilinear interpolation allows the RoI Align to obtain the feature values corresponding to the four coordinate positions of each small feature map.

Bilinear interpolation is the calculation of the value of a pixel point (floating point) that does not exist in the location image from the value of a known pixel point. The bilinear interpolation method can be computed by performing two horizontal interpolation operations and then one vertical interpolation or by performing two vertical interpolation operations and then one horizontal interpolation. Here, the specific calculation process of the bilinear interpolation method with lateral interpolation followed by vertical interpolation is introduced. The specific coordinates of Q_{11}, Q_{21}, Q_{12}, Q_{22}, R_1, R_2 and P involved in the formula are shown in Figure 9.

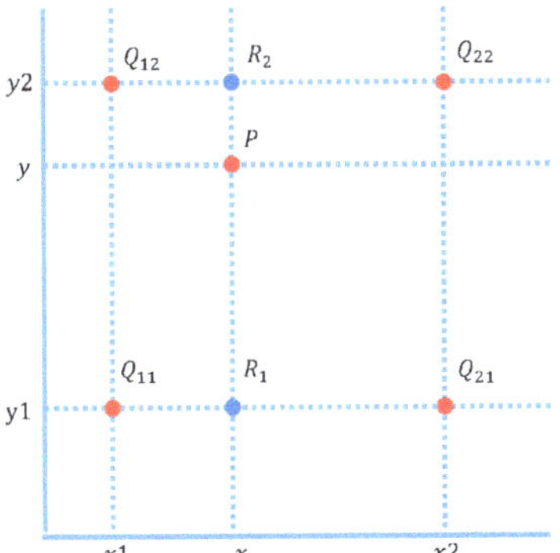

Figure 9. Schematic diagram of the bilinear interpolation method.

First is the first horizontal interpolation calculation, as shown in Equation (3).

$$f(R_1) \approx \frac{x2 - x}{x2 - x1} f(Q_{11}) + \frac{x - x1}{x2 - x1} f(Q_{21}) \tag{3}$$

This is followed by a second horizontal interpolation calculation, as shown in Equation (4).

$$f(R_2) \approx \frac{x2 - x}{x2 - x1} f(Q_{12}) + \frac{x - x1}{x2 - x1} f(Q_{22}) \tag{4}$$

The last step is calculation of the value of P point based on the two transverse interpolation results ($f(R_1)$ and $f(R_2)$) obtained from above calculation, as shown in Equation (5).

$$f(P) \approx \frac{y2 - y}{y2 - y1} f(R_1) + \frac{y - y1}{y2 - y1} f(R_2) \tag{5}$$

3.4. Output

The final goal of SEM-RCNN is to obtain the bounding box and class of object. Therefore, in the output part of SEM-RCNN, the feature map after RoI Align processing is first classified using a fully connected (FC) layer. Then, the output of the coordinate information of the object is realized by the FC layer and border regression; the output of the class information of the object is realized by the fully connected layer and Softmax function. Finally, the task of target detection is accomplished by combining border coordinate information and class information, as shown in Figure 10.

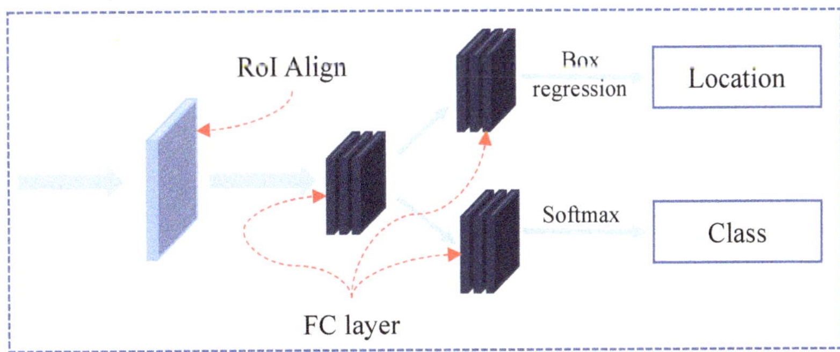

Figure 10. The output part of SEM-RCNN.

4. Experiment Results and Analysis

4.1. Dataset

We use the Environmental Microorganism Dataset Sixth Version (EMDS-6) [94]. The original EMs images with their corresponding ground truth (GT) images of 21 types of EMDS-6 are shown in Figure 11.

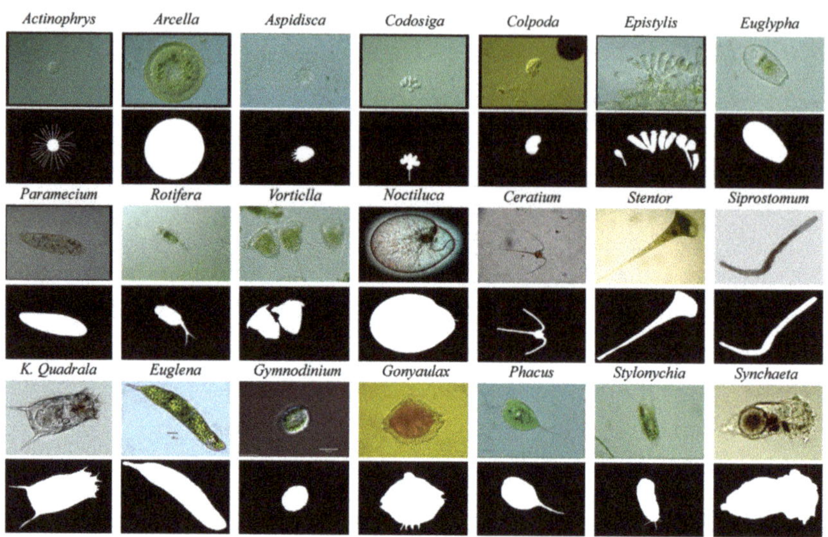

Figure 11. The EMs images in EMDS-6. The color images above are original images and the binary images below are the corresponding GT images.

EMDS-6 includes 21 types of EMs that have 840 original images and 840 ground truth images, respectively, as shown in Figure 11. There are 21 categories with 40 images in each category. In the training process, the dataset is evenly distributed to each category for balanced training. In our work, we perform object annotation using Lambelme software based on original and ground truth images provided by EMDS-6. Most of the images in EMDS-6 only contain one type of EM.

4.2. Experimental Settings

4.2.1. Data Settings

For EMDS-6, each type of EM is randomly grouped into training, validation, and test dataset with a ratio of 4:1, which means 32 images are applied for training (with 5-fold cross validation), and the last 8 images are applied for testing. Though the dataset is small for training a complex model, it can still achieve excellent detection performance by using transfer learning [94].

4.2.2. Hyper-Parameter Settings

In the process of training, the proposed SEM-RCNN model is pre-trained on the MS-COCO dataset, firstly. Then, the training step of SEM-RCNN contains two steps. The head is frozen and trained with an epoch of 150 and a learning rate of 0.0001. After that, the whole network is trained with an epoch of 150 and a learning rate of 0.001. The final model parameters have the lowest loss function obtained on the validation set. There are 98,793,356 trainable parameters of SEM-RCNN in total. The intersection over union (IoU) is the ratio of intersection and concatenation of the prediction box and true box. A result can be considered a correct detection only if the IoU value of the prediction box is greater than the set threshold, so the IoU threshold is a critical hyperparameter that may highly affect the performance of the proposed model. To systematically choose the IoU threshold, we test the detection performance of different IoU settings using SEM-RCNN based on SE-ResNet-101. The IoU threshold is set from 0.1 to 0.9 with the stride of 0.1, and the average detection indices based on 5-fold cross validation are shown in Table 1.

Table 1. The average detection indices (with 5-fold cross validation) of SEM-RCNN based on different IoU threshold settings.

IoU Threshold	Mean IoU	mAP	Precision	Recall	F1-Score
0.1	0.633	0.428	0.422	0.461	0.441
0.2	0.724	0.500	0.505	0.538	0.524
0.3	0.732	0.511	0.509	0.550	0.526
0.4	0.745	0.494	0.505	0.545	0.519
0.5	0.709	0.479	0.488	0.508	0.498
0.6	0.728	0.500	0.497	0.526	0.511
0.7	0.584	0.404	0.404	0.431	0.417
0.8	0.923	0.485	0.401	0.515	0.451
0.9	0.999	0.464	0.257	0.497	0.339

By reviewing Table 1, we find that the set of IoU thresholds is critical for the detection performance of the proposed model. The highest mAP, Precision, Recall, and F1-score can be obtained when the IoU threshold is set as 0.3. Though the mean IoU is higher when the threshold is set as 0.4, 0.8, and 0.9, the other indices (such as Precision) are not satisfactory. Therefore, by considering the aggregate detection performance, the IoU threshold is set as 0.3, which performs accurately and balanced.

4.3. Evaluation Criteria

We use mean average precision (mAP) as the evaluation metric in our experiments. mAP, as the best evaluation metric for the target detection task, combines the accuracy of detection category and location. The calculation of mAP is shown in Equation (6).

$$mAP = \frac{1}{n}\sum_{i=1}^{n} AP_i \tag{6}$$

In Equation (6), n presents the number of classes of EMs; AP is determined by the area under the accuracy-precision curve. The calculations of accuracy and precision are related to true positive (TP), true negative (TN), false positive (FP), false negative (FN), and intersection over union (IoU). The description of TP, TN, FP, and FN is shown in Table 2. The calculations of accuracy and precision are shown in Equations (7) and (8), respectively.

$$\text{Accuracy} = \frac{\text{TP} + \text{TN}}{\text{TP} + \text{TN} + \text{FP} + \text{FN}} \tag{7}$$

$$\text{Precision} = \frac{\text{TP}}{\text{TP} + \text{FP}} \tag{8}$$

Table 2. Description of TP, TN, FP, and FN.

		True Label	
		Positive	**Negative**
Predict label	Positive	TP	FP
	Negative	FN	TN

4.4. Detection Results and Analysis

The detection results of SEM-RCNN on EMDS-6 is shown in Figure 12.

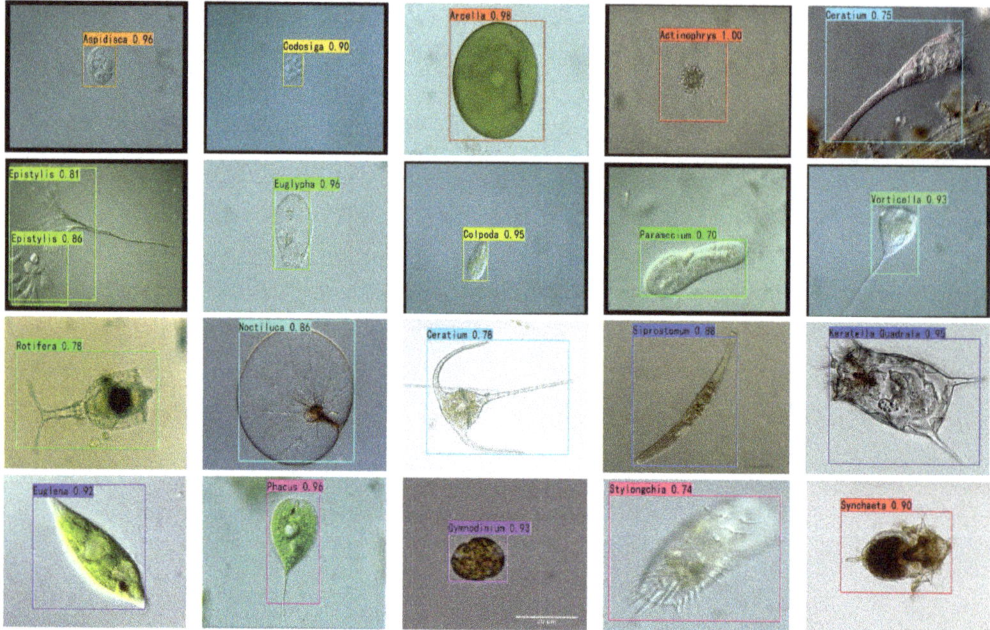

Figure 12. The detection results of SEM-RCNN on EMDS-6. The color boxes indicate the positions predicted by SEM-RCNN, for which its captions indicate the predicted categories and the probabilities.

The loss curves while training and validation are shown in Figure 13. It shows that the bounding box loss and classification loss curves of the proposed SEM-RCNN can converge steadily and quickly, which indicates that the proposed SEM-RCNN model has satisfactory performance on bounding box regression and object classification tasks.

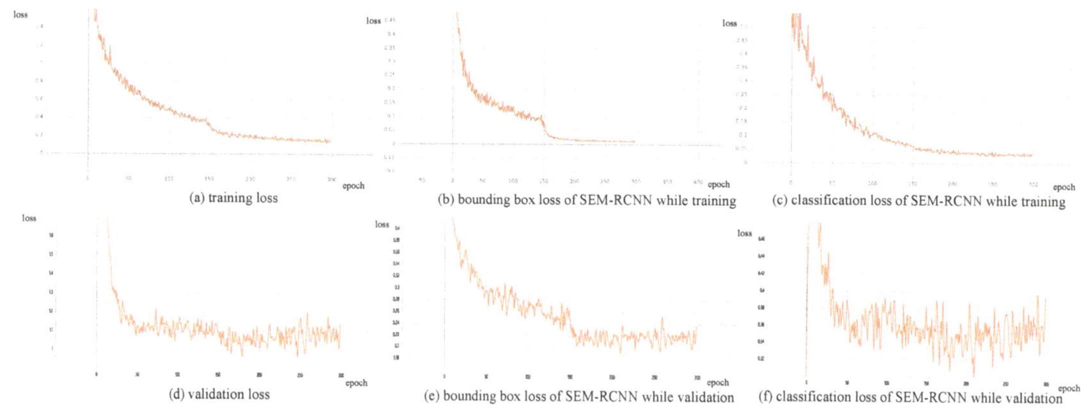

Figure 13. The loss curves of SEM-RCNN while training and validation on EMDS-6.

In order to show more intuitively how the SEM-RCNN compares with other detectors, we compare them in the form of a table in the following. In Table 3, we compared the average detection indices (with 5-fold cross validation) of SEM-RCNN and Mask RCNN on EMDS-6 when combined with ResNet-50 and ResNet-101.

Table 3. The average detection indices (with 5-fold cross validation) of SEM-RCNN and Mask RCNN.

Model	Mask RCNN		SEM-RCNN	
Backbone	ResNet-50	ResNet-101	SE-ResNet-50	SE-ResNet-101
mAP	0.440	0.488	0.450	0.511
precision	0.434	0.485	0.425	0.509
recall	0.458	0.511	0.451	0.550
F1-score	0.446	0.498	0.455	0.526

From Table 3, we can find that SEM-RCNN proposed in this paper can achieve better detection performance than Mask RCNN both in single-object and multi-object for EM detection. Overall, the models based on the backbone of ResNet-101 and SE-ResNet-101 perform better than those based on ResNet-50 and SE-ResNet. Compared with the original ResNet-based Mask RCNN, the performance increment of SEM-RCNN based on deeper SE-ResNet is much better than the Mask RCNN based on deeper ResNet both for mAP, precision, recall, and F1-score. Moreover, the mAP of the proposed SEM-RCNN achieves 0.511, which is much better than the result of Mask RCNN. The confusion matrix of the proposed SEM-RCNN is shown in Figure 14, showing the classification performance of the proposed model.

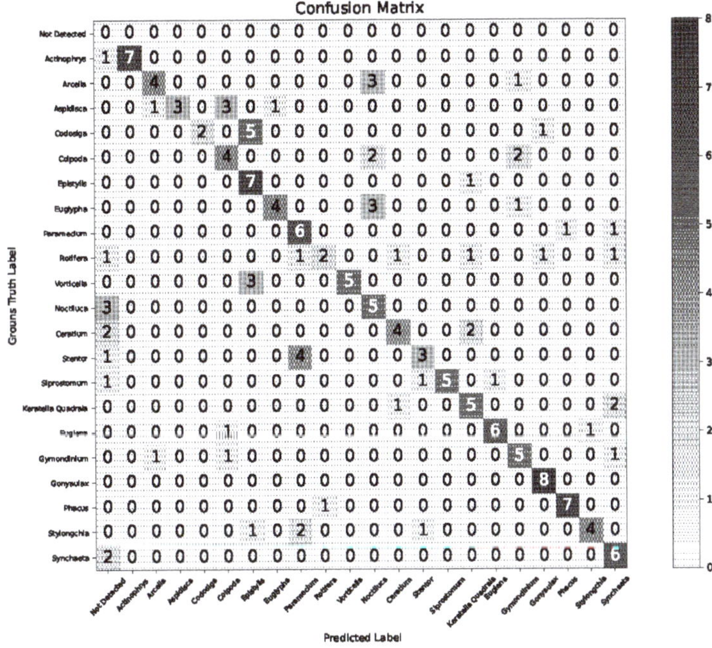

Figure 14. The confusion matrix of the detection results of SEM-RCNN on EMDS-6 (with 5-fold cross validation). Deeper color indicates the higher classification probability.

From Figure 14, most of the test images can be classified correctly. However, there still exists some mis-detected images. For example, 5 images of *Codosiga* are wrongly classified as *Epistylis*, showing the large error rate of the proposed model. By referring to the images of the two EMs in Figure 15, we find that the EMs have similar morphological features, and both are clustered microorganisms.

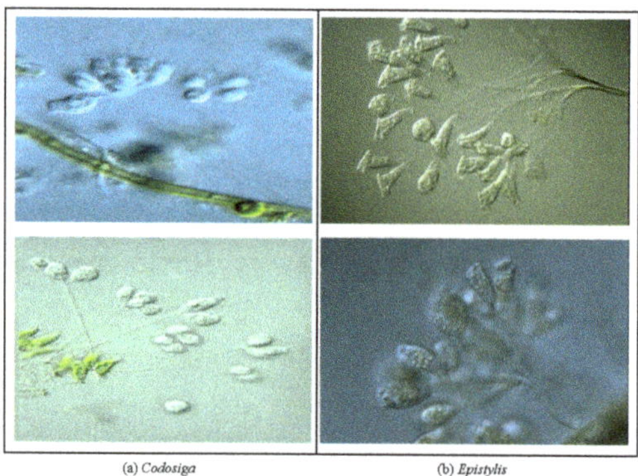

Figure 15. The example images of *Codosiga* and *Epistylis* in EMDS-6.

Moreover, by reviewing Figure 14, we find that *Stentor* is easily classified as *Paramecium*, an example is shown in Figure 16. Most EMs are colorless and transparent, so the morphological feature determines the classification result. However, though the EMs are in variance shapes, they may perform similar shapes at different growth phases. On the other hand, the quantity and quality of the dataset will have a greater impact on the model's performance. However, the satisfactory EM dataset is relatively difficult to obtain due to some objective reasons, such as the impurities in the acquisition environment, uneven natural light, and other adverse factors. So the model still cannot classify similar EMs accurately due to the small dataset.

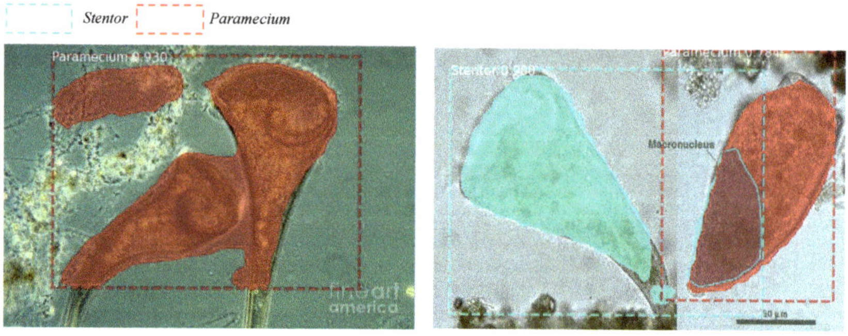

Figure 16. The example images of *Stentor* which are classified as *Paramecium* in EMDS-6.The color boxes indicate the positions predicted by SEM-RCNN, for which its captions indicate the predicted categories and the probabilities.

4.5. Extensive Experiment

To compare the detection performance of SEM-RCNN with existing deep learning-based detectors, we chose several classical detectors for our experiments. Table 4 introduces the detection results and population variances (with 5-fold cross validation) of some classical deep-learning-based detectors.

Table 4. Detection result of some classical deep-learning-based detectors (with 5-fold cross validation).

Model	Ours	SSD	Faster R-CNN	RetinaNet	YOLOv3	YOLOv4
mAP	0.511	0.421	0.377	0.401	0.425	0.436
Varp	1.46×10^{-5}	6.64×10^{-6}	1.38×10^{-5}	8.46×10^{-5}	3.78×10^{-5}	5.65×10^{-5}

From the final detection results, SEM-RCNN achieves better detection performance than SSD, Faster R-CNN, RetinaNet, YOLOv3, and YOLOv4. The results prove that it is feasible to improve the detection effectiveness of a detector by increasing its feature extraction capability.

To further prove the object detection performance of the proposed SEM-RCNN model, another dataset of EMs should be applied for the extensive experiment. However, by reviewing our previous works about EMs image analysis and EMs dataset [15,18–22,24–26,28,29,95], there are few proper open-access EMs dataset, which is caused by several objective reasons, including the uneven natural lighting and too many impurities while imaging. Hence, a dataset containing 8 types of blood cells is applied for the extensive experiment, including *erythrocytes, basophils, eosinophils, lymphocytes, monocytes, neutrophils, platelets, and immunoglobulins*. The reasons why we choose the blood cell images consisting four aspects: firstly, the images of EMs and blood cells are both microscopic images, which have strong morphological and shape similarities; secondly, both of them are non-directional images, where all objects in these images have no fixed positive direction; thirdly, both of them can be applied for multi-class detection tasks; finally, both of them have lots of noise and redundant impurities. Besides, the application of the blood cell image dataset can prove the strong generalization ability of the proposed SEM-RCNN. There are 17,090 labeled blood cell images in total, and the proposed SEM-RCNN is trained for 150 epochs with finetuning based on the pre-trained model for MS-COCO dataset. The result is shown in Table 5. By reviewing Table 5, we find that the proposed SEM-RCNN can achieve excellent blood cell detection performance. Most of the evaluation metrics are more than 0.9, which performs better than Mask RCNN significantly. The detection result is shown in Figure 17, which shows satisfactory detection performance of the proposed SEM-RCNN.

Table 5. Detection result of SEM-RCNN for blood cell detection.

Evaluation Metrics	IoU	mAP	Precision	Recall	F1-Score
SEM-RCNN	0.905	0.907	0.898	0.910	0.904
Mask RCNN	0.875	0.850	0.843	0.853	0.848

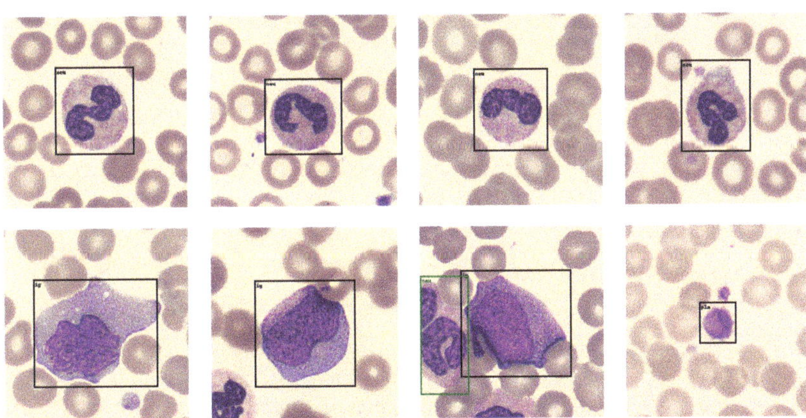

Figure 17. The results of SEM-RCNN for cell dataset detection.The color boxes indicate the positions predicted by SEM-RCNN, for which its captions indicate the predicted categories.

5. Conclusions and Future Work

The analysis and research of EMs are essential. Therefore, a suitable method for EM detection needs to be explored. After summarizing and analyzing the work related to EM detection, we designed the SEM-RCNN for the detection of EMs. In terms of applications, to fully demonstrate the feasibility of SEM-RCNN for EM detection, model training and testing were conducted in a small dataset of EMs and a large dataset of blood cells, respectively. The final detection results demonstrate the feasibility of SEM-RCNN for detecting EMs. In terms of technology, an improved method combining Mask RCNN with SENet is proposed in this paper. To verify the feasibility of the improved method, the detection results of SEM-RCNN and the original Mask RCNN are compared on the EMDS-6 dataset and blood cell dataset, respectively. The comparison results showed that the detection results of SEM-RCNN improved two to three points in mAP than that of the original Mask RCNN. Finally, SEM-RCNN achieved a 0.511 mAP on EMDS-6 and 0.907 mAP on the blood cell dataset.

This paper fills the gap in computer-aided multi-class EM detection research. However, considering the continuous innovation of related technologies and the challenges that need to be faced in practical applications, there is still more research potential and room for improvement in several aspects of this study. Regarding current research results, further research content with respect to our work will mainly be considered from EMs data. From the detection results on dataset EMDS-6 and blood cell dataset, it can be seen that sufficient training data can significantly improve the detection effect of the model. In contrast, insufficient training data can lead to poor detection effects from the model. Therefore, in the follow-up study, we will focus part of our efforts on expanding the existing microbial dataset to build an EMs dataset with more sufficient data and to better meet the training of detection models.

Author Contributions: J.Z. (Jiawei Zhang), Methodology, validation, data curation, writing—original draft preparation, review and editing, and visualization; P.M., methodology, software, validation, formal analysis, data curation, writing—original draft preparation, and visualization; T.J., investigation; funding acquisition; X.Z., resources; W.T., resources; J.Z. (Jinghua Zhang), validation; S.Z., validation; X.H., validation; M.G., investigation; C.L., conceptualization, methodology, data curation, writing—original draft preparation, writing—review and editing, supervision, and project administration. All authors have read and agreed to the published version of the manuscript.

Funding: This work is supported by the "Natural Science Foundation of China" (No. 61806047 and 61971118) and "Sichuan Science and Technology Program" (No. 2021YFH0069, 2021YFQ0057, and1614 2022YFS0565).

Data Availability Statement: EMDS-6 is at: https://figshare.com/articles/dataset/EMDS-6/17125025/1 (accessed on 15 February 2022).

Acknowledgments: We thank Zixian Li and Guoxian Li for their important discussions. We also thank M.E. Pingli Ma, due to his contribution is considered as important as the first author in this paper.

Conflicts of Interest: There is no conflict of interest in this paper.

Abbreviations

The following abbreviations are used in this manuscript:

EMs	environmental microorganisms;
CNN	convolutional neural network;
MS-COCO	microsoft common objects in context;
SEM-RCNN	squeeze-and-excitation-based mask region convolutional neural network;
RCNN	region convolutional neural network;
FC	fully connected;
ReLU	rectified linear unit;
BN	batch normalization;
SVM	support vector machine;
YOLO	you only look once;
ResNet	deep residual network;
SENet	squeeze-and-excitation network;

FPN	feature pyramid network;
RPN	region proposal network;
EMDS-6	the Environmental Microorganism Dataset Sixth Version;
mAP	mean average precision;
TP	true positive;
TN	true negative;
FP	false positive;
FN	false negative;
IoU	intersection over union;
GT	ground truth.

Reference

1. Pepper, I.L.; Gerba, C.P.; Gentry, T.J.; Maier, R.M. *Environmental Microbiology*; Academic Press: Cambridge, MA, USA, 2011.
2. Locey, K.J.; Lennon, J.T. Scaling laws predict global microbial diversity. *Proc. Natl. Acad. Sci. USA* **2016**, *113*, 5970–5975. [CrossRef] [PubMed]
3. Nehl, D.B.; Allen, S.J.; Brown, J.F. Deleterious rhizosphere bacteria: An integrating perspective. *Appl. Soil Ecol.* **1997**, *5*, 1–20. [CrossRef]
4. Van Deun, A.; Salim, A.H.; Cooreman, E.; Hossain, M.A.; Rema, A.; Chambugonj, N.; Hye, M.; Kawria, A.; Declercq, E. Optimal tuberculosis case detection by direct sputum smear microscopy: How much better is more? *Int. J. Tuberc. Lung Dis.* **2002**, *6*, 222–230. [PubMed]
5. Sharma, J.; Granmo, O.C.; Goodwin, M. Emergency Analysis: Multitask Learning with Deep Convolutional Neural Networks for Fire Emergency Scene Parsing. In Proceedings of the International Conference on Industrial, Engineering and Other Applications of Applied Intelligent Systems, Kuala Lumpur, Malaysia, 26–29 July 2021; Springer: Cham, Switzerland, 2021; pp. 101–112.
6. Li, X.; Li, C.; Rahaman, M.M.; Sun, H.; Li, X.; Wu, J.; Yao, Y.; Grzegorzek, M. A comprehensive review of computer-aided whole-slide image analysis: From datasets to feature extraction, segmentation, classification and detection approaches. *Artif. Intell. Rev.* **2022**, *55*, 4809–4878. [CrossRef]
7. Li, Y.; Li, C.; Li, X.; Wang, K.; Rahaman, M.M.; Sun, C.; Chen, H.; Wu, X.; Zhang, H.; Wang, Q. A Comprehensive Review of Markov Random Field and Conditional Random Field Approaches in Pathology Image Analysis. *Arch. Comput. Methods Eng.* **2021**, *29*, 609–639. [CrossRef]
8. Zhou, X.; Li, C.; Rahaman, M.M.; Yao, Y.; Ai, S.; Sun, C.; Wang, Q.; Zhang, Y.; Li, M.; Li, X.; et al. A comprehensive review for breast histopathology image analysis using classical and deep neural networks. *IEEE Access* **2020**, *8*, 90931–90956. [CrossRef]
9. Li, C.; Chen, H.; Li, X.; Xu, N.; Hu, Z.; Xue, D.; Qi, S.; Ma, H.; Zhang, L.; Sun, H. A review for cervical histopathology image analysis using machine vision approaches. *Artif. Intell. Rev.* **2020**, *53*, 4821–4862. [CrossRef]
10. Liu, W.; Li, C.; Rahaman, M.M.; Jiang, T.; Sun, H.; Wu, X.; Hu, W.; Chen, H.; Sun, C.; Yao, Y.; et al. Is the aspect ratio of cells important in deep learning? A robust comparison of deep learning methods for multi-scale cytopathology cell image classification: From convolutional neural networks to visual transformers. *Comput. Biol. Med.* **2022**, *141*, 105026. [CrossRef]
11. Rahaman, M.M.; Li, C.; Yao, Y.; Kulwa, F.; Wu, X.; Li, X.; Wang, Q. DeepCervix: A deep learning-based framework for the classification of cervical cells using hybrid deep feature fusion techniques. *Comput. Biol. Med.* **2021**, *136*, 104649. [CrossRef]
12. Rahaman, M.M.; Li, C.; Wu, X.; Yao, Y.; Hu, Z.; Jiang, T.; Li, X.; Qi, S. A survey for cervical cytopathology image analysis using deep learning. *IEEE Access* **2020**, *8*, 61687–61710. [CrossRef]
13. Zou, S.; Li, C.; Sun, H.; Xu, P.; Zhang, J.; Ma, P.; Yao, Y.; Huang, X.; Grzegorzek, M. TOD-CNN: An effective convolutional neural network for tiny object detection in sperm videos. *Comput. Biol. Med.* **2022**, *146*, 105543. [CrossRef] [PubMed]
14. Chen, A.; Li, C.; Zou, S.; Rahaman, M.M.; Yao, Y.; Chen, H.; Yang, H.; Zhao, P.; Hu, W.; Liu, W.; et al. SVIA dataset: A new dataset of microscopic videos and images for computer-aided sperm analysis. *Biocybern. Biomed. Eng.* **2022**, *42*, 204–214. [CrossRef]
15. Ma, P.; Li, C.; Rahaman, M.M.; Yao, Y.; Zhang, J.; Zou, S.; Zhao, X.; Grzegorzek, M. A state-of-the-art survey of object detection techniques in microorganism image analysis: From classical methods to deep learning approaches. *Artif. Intell. Rev.* **2022**, 1–72. [CrossRef] [PubMed]
16. Jung, H.K.; Choi, G.S. Improved YOLOv5: Efficient Object Detection Using Drone Images under Various Conditions. *Appl. Sci.* **2022**, *12*, 7255. [CrossRef]
17. Li, X.; Wang, C.; Ju, H.; Li, Z. Surface Defect Detection Model for Aero-Engine Components Based on Improved YOLOv5. *Appl. Sci.* **2022**, *12*, 7235. [CrossRef]
18. Zhao, P.; Li, C.; Rahaman, M.; Xu, H.; Yang, H.; Sun, H.; Jiang, T.; Grzegorzek, M. A Comparative Study of Deep Learning Classification Methods on a Small Environmental Microorganism Image Dataset (EMDS-6): From Convolutional Neural Networks to Visual Transformers. *Front. Microbiol.* **2022**, *13*, 792166. [CrossRef]
19. Kulwa, F.; Li, C.; Zhang, J.; Shirahama, K.; Kosov, S.; Zhao, X.; Jiang, T.; Grzegorzek, M. A new pairwise deep learning feature for environmental microorganism image analysis. *Environ. Sci. Pollut. Res.* **2022**, *29*, 51909–51926. [CrossRef]
20. Kosov, S.; Shirahama, K.; Li, C.; Grzegorzek, M. Environmental microorganism classification using conditional random fields and deep convolutional neural networks. *Pattern Recognit.* **2018**, *77*, 248–261. [CrossRef]

21. Li, C.; Shirahama, K.; Grzegorzek, M. Environmental microbiology aided by content-based image analysis. *Pattern Anal. Appl.* **2016**, *19*, 531–547. [CrossRef]
22. Li, C.; Shirahama, K.; Grzegorzek, M. Application of content-based image analysis to environmental microorganism classification. *Biocybern. Biomed. Eng.* **2015**, *35*, 10–21. [CrossRef]
23. Rahaman, M.M.; Li, C.; Yao, Y.; Kulwa, F.; Rahman, M.A.; Wang, Q.; Qi, S.; Kong, F.; Zhu, X.; Zhao, X. Identification of COVID-19 samples from chest X-Ray images using deep learning: A comparison of transfer learning approaches. *J. X-ray Sci. Technol.* **2020**, *28*, 821–839. [CrossRef] [PubMed]
24. Zhang, J.; Li, C.; Yin, Y.; Zhang, J.; Grzegorzek, M. Applications of artificial neural networks in microorganism image analysis: A comprehensive review from conventional multilayer perceptron to popular convolutional neural network and potential visual transformer. *Artif. Intell. Rev.* **2022**, 1–58. [CrossRef] [PubMed]
25. Kulwa, F.; Li, C.; Zhao, X.; Cai, B.; Xu, N.; Qi, S.; Chen, S.; Teng, Y. A state-of-the-art survey for microorganism image segmentation methods and future potential. *IEEE Access* **2019**, *7*, 100243–100269. [CrossRef]
26. Zhao, P.; Li, C.; Rahaman, M.M.; Xu, H.; Ma, P.; Yang, H.; Sun, H.; Jiang, T.; Xu, N.; Grzegorzek, M. EMDS-6: Environmental Microorganism Image Dataset Sixth Version for Image Denoising, Segmentation, Feature Extraction, Classification, and Detection Method Evaluation. *Front. Microbiol.* **2022**, 1334. [CrossRef] [PubMed]
27. Li, C.; Zhang, J.; Kulwa, F.; Qi, S.; Qi, Z. A SARS-CoV-2 Microscopic Image Dataset with Ground Truth Images and Visual Features. In Proceedings of the Chinese Conference on Pattern Recognition and Computer Vision (PRCV), Nanjing, China, 16–18 October 2020; Springer: Cham, Switzerland, 2020; pp. 244–255.
28. Zhang, J.; Xu, N.; Li, C.; Rahaman, M.M.; Yao, Y.D.; Lin, Y.H.; Zhang, J.; Jiang, T.; Qin, W.; Grzegorzek, M. An application of Pixel Interval Down-sampling (PID) for dense tiny microorganism counting on environmental microorganism images. *arXiv* **2022**, arXiv:2204.01341.
29. Zhang, J.; Li, C.; Rahaman, M.M.; Yao, Y.; Ma, P.; Zhang, J.; Zhao, X.; Jiang, T.; Grzegorzek, M. A Comprehensive Review of Image Analysis Methods for Microorganism Counting: From Classical Image Processing to Deep Learning Approaches. *Artif. Intell. Rev.* **2022**, *55*, 2875–2944. [CrossRef]
30. Prada, P.; Brunel, B.; Reffuveille, F.; Gangloff, S.C. Technique Evolutions for Microorganism Detection in Complex Samples: A Review. *Appl. Sci.* **2022**, *12*, 5892. [CrossRef]
31. Bloem, J.; Veninga, M.; Shepherd, J. Fully automatic determination of soil bacterium numbers, cell volumes, and frequencies of dividing cells by confocal laser scanning microscopy and image analysis. *Appl. Environ. Microbiol.* **1995**, *61*, 926–936. [CrossRef]
32. Qing, S.; Wu, Y.; Juan, J.; Zhao, X.; Que, X. Application of Microscopic Color Image Processing in Algae Recognition and Statistics. *Agric. Mech. Res.* **2006**, *6*, 199–203.
33. Zhang, C.; Chen, W.; Liu, W.; Chen, C. An automated bacterial colony counting system. In Proceedings of the 2008 IEEE International Conference on Sensor Networks, Ubiquitous, and Trustworthy Computing (SUTC 2008), Taichung, Taiwan, 11–13 June 2008; pp. 233–240.
34. Rizvandi, N.B.; Pizurica, A.; Philips, W.; Ochoa, D. Edge linking based method to detect and separate individual C. Elegans worms in culture. In Proceedings of the 2008 Digital Image Computing: Techniques and Applications, Canberra, ACT, Australia, 1–3 December 2008; pp. 65–70.
35. Rizvandi, N.B.; Pizurica, A.; Rooms, F.; Philips, W. Skeleton analysis of population images for detection of isolated and overlapped nematode C. elegans. In Proceedings of the 2008 16th European Signal Processing Conference, Lausanne, Switzerland, 25–29 August 2008; pp. 1–5.
36. Zhou, B.T.; Baek, J.H. Using Machine Vision to Detect Distinctive Behavioral Phenotypes of Thread-shape Microscopic Organism. In *Applications of Computational Intelligence in Biology*; Springer: Berlin/Heidelberg, Germany, 2008; pp. 161–182.
37. Wang, P.; Wen, C.; Li, W.; Chen, Y. Motile microorganism tracking system using micro-visual servo control. In Proceedings of the 2008 3rd IEEE International Conference on Nano/Micro Engineered and Molecular Systems, Sanya, China, 6–9 January 2008; pp. 178–182.
38. Fernandez, H.; Hintea, S.; Csipkes, G.; Pellow, A.; Smith, H. Machine vision application to the detection of micro-organism in drinking water. In Proceedings of the International Conference on Knowledge-Based and Intelligent Information and Engineering Systems, Zagreb, Croatia, 3–5 September 2008; Springer: Cham, Switzerland, 2008; pp. 302–309.
39. Zhai, Y.; Liu, Y.; Zhou, D.; Liu, S. Automatic identification of mycobacterium tuberculosis from ZN-stained sputum smear: Algorithm and system design. In Proceedings of the 2010 IEEE International Conference on Robotics and Biomimetics, Tianjin, China, 14–18 December 2010; pp. 41–46.
40. Raof, R.A.A.; Mashor, M.Y.; Ahmad, R.B.; Noor, S.S.M. Image segmentation of Ziehl-Neelsen sputum slide images for tubercle bacilli detection. *Image Segm.* **2011**, *2011*, 365–378.
41. Shi, H.; Shi, Y.; Yin, Y. Food bacteria auto identification method based on image treatment. *J. Jilin Univ. (Eng. Technol. Ed.)* **2012**, *42*, 1049–1053.
42. Badsha, S.; Mokhtar, N.; Arof, H.; Lim, Y.A.L.; Mubin, M.; Ibrahim, Z. Automatic Cryptosporidium and Giardia viability detection in treated water. *EURASIP J. Image Video Process.* **2013**, *2013*, 56. [CrossRef]
43. Kowalski, M.; Kaczmarek, P.; Kabaciński, R.; Matuszczak, M.; Tranbowicz, K.; Sobkowiak, R. A simultaneous localization and tracking method for a worm tracking system. *Int. J. Appl. Math. Comput. Sci.* **2014**, *24*, 599–609. [CrossRef]

44. Rachna, H.B.; Swamy, M.S.M. Detection of Tuberculosis bacilli using image processing techniques. *Int. J. Soft Comput. Eng.* **2013**, *3*, 47–51.
45. Kurtulmuş, F.; Ulu, T.C. Detection of dead entomopathogenic nematodes in microscope images using computer vision. *Biosyst. Eng.* **2014**, *118*, 29–38. [CrossRef]
46. Goyal, A.; Roy, M.; Gupta, P.; Dutta, M.K.; Singh, S.; Garg, V. Automatic detection of mycobacterium tuberculosis in stained sputum and urine smear images. *Arch. Clin. Microbiol.* **2015**, *6*, 1.
47. Javidi, B.; Moon, I.; Yeom, S.; Carapezza, E. Three-dimensional imaging and recognition of microorganism using single-exposure on-line (SEOL) digital holography. *Opt. Express* **2005**, *13*, 4492–4506. [CrossRef]
48. Fernandez-Canque, H.; Beggs, B.; Smith, E.; Boutaleb, T.; Smith, H.; Hintea, S. Micro-organisms detection in drinking water using image processing. *Cell* **2006**, *15*, 4-2.
49. Otsu, N. A threshold selection method from gray-level histograms. *IEEE Trans. Syst. Man Cybern.* **1979**, *9*, 62–66. [CrossRef]
50. Lee, H.; Park, R. Comments on" An optimal multiple threshold scheme for image segmentation. *IEEE Trans. Syst. Man Cybern.* **1990**, *20*, 741–742. [CrossRef]
51. Javidi, B.; Yeom, S.; Moon, I.; Daneshpanah, M. Real-time automated 3D sensing, detection, and recognition of dynamic biological micro-organic events. *Opt. Express* **2006**, *14*, 3806–3829. [CrossRef] [PubMed]
52. Huang, K.M.; Cosman, P.; Schafer, W.R. Automated detection and analysis of foraging behavior in Caenorhabditis elegans. *J. Neurosci. Methods* **2008**, *171*, 153–164. [CrossRef] [PubMed]
53. Moon, I.; Yi, F.; Javidi, B. Automated three-dimensional microbial sensing and recognition using digital holography and statistical sampling. *Sensors* **2010**, *10*, 8437–8451. [CrossRef] [PubMed]
54. Javidi, B.; Moon, I.; Daneshpanah, M. Detection, identification and tracking of biological micro/nano organisms by computational 3D optical imaging. In Proceedings of the Biosensing III. International Society for Optics and Photonics, Nanjing, China, 13 July 2010; Volume 7759, p. 77590R. [CrossRef]
55. Hiremath, P.S.; Bannigidad, P.; Hiremath, M. Segmentation and identification of rotavirus—A in digital microscopic images using active contour model. In *Thinkquest~2010*; Springer: Berlin/Heidelberg, Germany, 2011; pp. 177–181.
56. Dubuisson, M.; Jain, A.K.; Jain, M.K. Segmentation and classification of bacterial culture images. *J. Microbiol. Methods* **1994**, *19*, 279–295. [CrossRef]
57. Fang, S.P.; Hsu, H.J.; Hung, L.L.; Wu, Y.S. *Automatic Identification of Mycobacterium Tuberculosis in Acid-Fast Stain Sputum Smears with Image Processing and Neural Networks*; Department of Electronic Engineering: Tainan, Taiwan, 2008.
58. Ogawa, M.; Tani, K.; Ochiai, A.; Yamaguchi, N.; Nasu, M. Multicolour digital image analysis system for identification of bacteria and concurrent assessment of their respiratory activity. *J. Appl. Microbiol.* **2005**, *98*, 1101–1106. [CrossRef] [PubMed]
59. Liu, P.Y.; Chin, L.K.; Ser, W.; Ayi, T.C.; Yap, P.H.; Bourouina, T.; Leprince-Wang, Y. Virus infectivity detection by effective refractive index using optofluidic imaging. In Proceedings of the 18th International Conference on Miniaturized Systems for Chemistry and Life Sciences, MicroTAS, San Antonio, TX, USA, 2014, 26–30 October 2014.
60. Yu, J.Q.; Huang, W.; Chin, L.K.; Lei, L.; Lin, Z.P.; Ser, W.; Chen, H.; Ayi, T.C.; Yap, P.H.; Chen, C.H.; et al. Droplet optofluidic imaging for λ-bacteriophage detection via co-culture with host cell Escherichia coli. *Lab Chip* **2014**, *14*, 3519–3524. [CrossRef] [PubMed]
61. Forero, M.; Cristobal, G.; Alvarez-Borreg, J. Automatic identification techniques of tuberculosis bacteria. In *Applications of Digital Image Processing XXVI*; International Society for Optics and Photonics: Bellingham, WA, USA, 2003; Volume 5203, pp. 71–81.
62. Perner, P.; Perner, H.; Janichen, S.; Buhring, A. Recognition of airborne fungi spores in digital microscopic images. In Proceedings of the 17th International Conference on Pattern Recognition, ICPR 2004, Cambridge, UK, 26 August 2004, Volume 3, pp. 566–569.
63. Sklarczyk, C.; Perner, H.; Rieder, H.; Arnold, W.; Perner, P. Image acquisition and analysis of hazardous biological material in air. In Proceedings of the International Conference on Mass Data Analysis of Images and Signals in Medicine, Biotechnology, and Chemistry, Leipzig, Germany, 18 July 2007; Springer: Berlin/Heidelberg, Germany, 2007; pp. 1–14.
64. Thiel, S.; Wiltshire, R.J. The automated detection of cyanobacteria using ddigital image processing techniques. *Environ. Int.* **1995**, *21*, 233–236. [CrossRef]
65. Jan, Z.; Rafiq, M.; Muhammad, H.; Zada, N. Detection of tuberculosis bacteria in sputum slide image using morphological features. In Proceedings of the International Conference: Beyond Databases, Architectures and Structures, Ustroń, Poland, 26–29 May 2015; Springer: Cham, Switzerland, 2015; pp. 408–414.
66. Liu, P.Y.; Chin, L.K.; Ser, W.; Ayi, T.C.; Yap, P.H.; Bourouina, T.; Leprince-Wang, Y. An optofluidic imaging system to measure the biophysical signature of single waterborne bacteria. *Lab Chip* **2014**, *14*, 4237–4243. [CrossRef]
67. Yin, Y.; Ding, Y. Rapid method for enumeration of total viable bacteria in vegetables based on computer vision. *Trans. CSAE* **2009**, *25*, 249–254.
68. Osman, M.K.; Ahmad, F.; Saad, Z.; Mashor, M.Y.; Jaafar, H. A genetic algorithm-neural network approach for Mycobacterium tuberculosis detection in Ziehl-Neelsen stained tissue slide images. In Proceedings of the 2010 10th International Conference on Intelligent Systems Design and Applications, Cairo, Egypt, 29 November–1 December 2010; pp. 1229–1234.
69. Kumar, S.; Mittal, G.S. Rapid detection of microorganisms using image processing parameters and neural network. *Food Bioprocess Technol.* **2010**, *3*, 741–751. [CrossRef]

70. White, A.G.; Cipriani, P.G.; Kao, H.; Lees, B.; Geiger, D.; Sontag, E.; Gunsalus, K.C.; Piano, F. Rapid and accurate developmental stage recognition of C. elegans from high-throughput image data. In Proceedings of the 2010 IEEE Computer Society Conference on Computer Vision and Pattern Recognition, San Francisco, CA, USA, 13–18 June 2010; pp. 3089–3096.
71. Verikas, A.; Gelzinis, A.; Bacauskiene, M.; Olenina, I.; Olenin, S.j.; Vaiciukynas, E. Phase congruency-based detection of circular objects applied to analysis of phytoplankton images. *Pattern Recognit.* **2012**, *45*, 1659–1670. [CrossRef]
72. Khutlang, R.; Krishnan, S.; Whitelaw, A.; Douglas, T.S. Automated detection of tuberculosis in Ziehl-Neelsen-stained sputum smears using two one-class classifiers. *J. Microsc.* **2010**, *237*, 96–102. [CrossRef] [PubMed]
73. Chang, J.; Arbeláez, P.; Switz, N.; Reber, C.; Tapley, A.; Davis, J.L.; Cattamanchi, A.; Fletcher, D.; Malik, J. Automated tuberculosis diagnosis using fluorescence images from a mobile microscope. In Proceedings of the International Conference on Medical Image Computing and Computer-Assisted Intervention, Nice, France, 1–5 October 2012; Springer: Berlin/Heidelberg, Germany, 2012; pp. 345–352.
74. Li, C.; Shirahama, K.; Czajkowsk, J.; Grzegorzek, M.; Ma, F.; Zhou, B. A multi-stage approach for automatic classification of environmental microorganisms. In Proceedings of the International Conference on Image Processing, Computer Vision, and Pattern Recognition (IPCV), Las Vegas, NV, USA, 15 June 2013; The Steering Committee of the World Congress in Computer Science, Computer Engineering and Applied Computing (WorldComp): Las Vegas, NV, USA, 2013; p. 1.
75. Santiago-Mozos, R.; Pérez-Cruz, F.; Madden, M.G.; Artés-Rodríguez, A. An automated screening system for tuberculosis. *IEEE J. Biomed. Health Inform.* **2013**, *18*, 855–862. [CrossRef] [PubMed]
76. Li, C.; Shirahama, K.; Grzegorzek, M. Environmental microorganism classification using sparse coding and weakly supervised learning. In Proceedings of the 2nd International Workshop on Environmental Multimedia Retrieval, Shanghai China, 23–26 June 2015; pp. 9–14.
77. Verikas, A.; Gelzinis, A.; Bacauskiene, M.; Olenina, I.; Vaiciukynas, E. An integrated approach to analysis of phytoplankton images. *IEEE J. Ocean. Eng.* **2014**, *40*, 315–326. [CrossRef]
78. Zetsche, E.; Mallahi, A.E.; Dubois, F.; Yourassowsky, C.; Kromkamp, J.C.; Meysman, F.J.R. Imaging-in-Flow: Digital holographic microscopy as a novel tool to detect and classify nanoplanktonic organisms. *Limnol. Oceanogr. Methods* **2014**, *12*, 757–775. [CrossRef]
79. Shan-e-Ahmed Razaa, M.Q.; Marjanb, M.A.; Farhana Buttc, F.S.; Rajpoota, N.M. Anisotropic Tubular Filtering for Automatic Detection of Acid-Fast Bacilli in Digitized Microscopic Images of Ziehl-Neelsen Stained Sputum Smear Samples. In *Progress in Biomedical Optics and Imaging-Proceedings of SPIE*; SPIE: Bellingham, WA, USA, 2015.
80. Boser, B.E.; Guyon, I.M.; Vapnik, V.N. A training algorithm for optimal margin classifiers. In Proceedings of the Fifth Annual Workshop on Computational Learning Theory, Pittsburgh, PA, USA, 27–29 July 1992; pp. 144–152.
81. Panicker, R.O.; Kalmady, K.S.; Rajan, J.; Sabu, M.K. Automatic detection of tuberculosis bacilli from microscopic sputum smear images using deep learning methods. *Biocybern. Biomed. Eng.* **2018**, *38*, 691–699. [CrossRef]
82. Tahir, M.W.; Zaidi, N.A.; Rao, A.A.; Blank, R.; Vellekoop, M.J.; Lang, W. A fungus spores dataset and a convolutional neural network based approach for fungus detection. *IEEE Trans. Nanobiosci.* **2018**, *17*, 281–290. [CrossRef]
83. Sajedi, H.; Mohammadipanah, F.; Rahimi, S.A.H. Actinobacterial strains recognition by Machine learning methods. *Multimed. Tools Appl.* **2019**, *78*, 20285–20307. [CrossRef]
84. Hung, J.; Carpenter, A. Applying faster R-CNN for object detection on malaria images. In Proceedings of the IEEE Conference on Computer Vision and Pattern Recognition (CVPR) Workshops, Honolulu, HI, USA, 21–26 July 2017; pp. 56–61.
85. Viet, N.Q.; ThanhTuyen, D.T.; Hoang, T.H. Parasite worm egg automatic detection in microscopy stool image based on Faster R-CNN. In Proceedings of the 3rd International Conference on Machine Learning and Soft Computing, Da Lat Viet Nam, Vietnam, 25–28 January 2019; pp. 197–202.
86. Baek, S.; Pyo, J.; Pachepsky, Y.; Park, Y.; Ligaray, M.; Ahn, C.; Kim, Y.; Chun, J.A.; Cho, K.H. Identification and enumeration of cyanobacteria species using a deep neural network. *Ecol. Indic.* **2020**, *115*, 106395. [CrossRef]
87. Qian, P.; Zhao, Z.; Liu, H.; Wang, Y.; Peng, Y.; Hu, S.; Zhang, J.; Deng, Y.; Zeng, Z. Multi-Target Deep Learning for Algal Detection and Classification. In Proceedings of the 2020 42nd Annual International Conference of the IEEE Engineering in Medicine & Biology Society (EMBC), Montreal, QC, Canada, 20–24 July 2020; pp. 1954–1957.
88. Pedraza, A.; Bueno, G.; Deniz, O.; Ruiz-Santaquiteria, J.; Sanchez, C.; Blanco, S.; Borrego-Ramos, M.; Olenici, A.; Cristobal, G. Lights and pitfalls of convolutional neural networks for diatom identification. In *Optics, Photonics, and Digital Technologies for Imaging Applications V*; International Society for Optics and Photonics: Bellingham, WA, USA, 2018; Volume 10679, p. 106790G.
89. Salido, J.; Sánchez, C.; Ruiz-Santaquiteria, J.; Cristóbal, G.; Blanco, S.; Bueno, G. A Low-Cost Automated Digital Microscopy Platform for Automatic Identification of Diatoms. *Appl. Sci.* **2020**, *10*, 6033. [CrossRef]
90. Ruiz-Santaquiteria, J.; Bueno, G.; Deniz, O.; Vallez, N.; Cristobal, G. Semantic versus instance segmentation in microscopic algae detection. *Eng. Appl. Artif. Intell.* **2020**, *87*, 103271. [CrossRef]
91. He, K.; Zhang, X.; Ren, S.; Sun, J. Deep residual learning for image recognition. In Proceedings of the IEEE Conference on Computer Vision and Pattern Recognition, Las Vegas, NV, USA, 27–30 June 2016; pp. 770–778.
92. Hu, J.; Shen, L.; Sun, G. Squeeze-and-excitation networks. In Proceedings of the IEEE Conference on Computer Vision and Pattern Recognition, Salt Lake City, UT, USA, 18–22 June 2018; pp. 7132–7141.
93. Lin, T.; Dollár, P.; Girshick, R.; He, K.; Hariharan, B.; Belongie, S. Feature pyramid networks for object detection. In Proceedings of the IEEE Conference on Computer Vision and Pattern Recognition, Honolulu, HI, USA, 21–26 July 2017; pp. 2117–2125.

94. Zhuang, F.; Qi, Z.; Duan, K.; Xi, D.; Zhu, Y.; Zhu, H.; Xiong, H.; He, Q. A comprehensive survey on transfer learning. *Proc. IEEE* **2020**, *109*, 43–76. [CrossRef]
95. Xu, H.; Li, C.; Rahaman, M.M.; Yao, Y.; Li, Z.; Zhang, J.; Kulwa, F.; Zhao, X.; Qi, S.; Teng, Y. An enhanced framework of generative adversarial networks (EF-GANs) for environmental microorganism image augmentation with limited rotation-invariant training data. *IEEE Access* **2020**, *8*, 187455–187469. [CrossRef]

Article

Comparative Study for Patch-Level and Pixel-Level Segmentation of Deep Learning Methods on Transparent Images of Environmental Microorganisms: From Convolutional Neural Networks to Visual Transformers

Hechen Yang [1], Xin Zhao [2], Tao Jiang [3,4,*], Jinghua Zhang [1,5], Peng Zhao [1], Ao Chen [1], Marcin Grzegorzek [5], Shouliang Qi [1], Yueyang Teng [1] and Chen Li [1,*]

1 Microscopic Image and Medical Image Analysis Group, College of Medicine and Biological Information Engineering, Northeastern University, Shenyang 110169, China
2 School of Resources and Civil Engineering, Northeastern University, Shenyang 110819, China
3 School of Intelligent Medicine, Chengdu University of Traditional Chinese Medicine, Chengdu 610075, China
4 International Joint Institute of Robotics and Intelligent Systems, Chengdu University of Information Technology, Chengdu 610225, China
5 Institute of Medical Informatics, University of Luebeck, 23562 Luebeck, Germany
* Correspondence: jiang@cuit.edu.cn (T.J.); lichen@bmie.neu.edu.cn (C.L.)

Abstract: Currently, the field of transparent image analysis has gradually become a hot topic. However, traditional analysis methods are accompanied by large amounts of carbon emissions, and consumes significant manpower and material resources. The continuous development of computer vision enables the use of computers to analyze images. However, the low contrast between the foreground and background of transparent images makes their segmentation difficult for computers. To address this problem, we first analyzed them with pixel patches, and then classified the patches as foreground and background. Finally, the segmentation of the transparent images was completed through the reconstruction of pixel patches. To understand the performance of different deep learning networks in transparent image segmentation, we conducted a series of comparative experiments using patch-level and pixel-level methods. In two sets of experiments, we compared the segmentation performance of four *convolutional neural network* (CNN) models and a *visual transformer* (ViT) model on the transparent environmental microorganism dataset fifth version. The results demonstrated that U-Net++ had the highest accuracy rate of 95.32% in the pixel-level segmentation experiment followed by ViT with an accuracy rate of 95.31%. However, ResNet50 had the highest accuracy rate of 90.00% and ViT had the lowest accuracy of 89.25% in the patch-level segmentation experiments. Hence, we concluded that ViT performed the lowest in patch-level segmentation experiments, but outperformed most CNNs in pixel-level segmentation. Further, we combined patch-level and pixel-level segmentation results to reduce the loss of segmentation details in the EM images. This conclusion was also verified by the environmental microorganism dataset sixth version dataset (EMDS-6).

Keywords: patch-level; pixel-level; image classification; image segmentation; transparent images; deep learning; convolutional neural network; visual transformer; environmental microorganism

Citation: Yang, H.; Zhao, X.; Jiang, T.; Zhang, J.; Zhao, P.; Chen, A.; Grzegorzek, M.; Qi, S.; Teng, Y.; Li, C. Comparative Study for Patch-Level and Pixel-Level Segmentation of Deep Learning Methods on Transparent Images of Environmental Microorganisms: From Convolutional Neural Networks to Visual Transformers. *Appl. Sci.* **2022**, *12*, 9320. https://doi.org/10.3390/app12189321

Academic Editors: Bernhard Baumann and Jan Egger

Received: 31 May 2022
Accepted: 13 September 2022
Published: 17 September 2022

Publisher's Note: MDPI stays neutral with regard to jurisdictional claims in published maps and institutional affiliations.

1. Introduction

With the advent of science and technology, the application of transparent images has increasingly been used in various fields around humans, such as the segmentation of renal transparent cancer cell nuclei in medicine [1]. The shape and location information of the cell nucleus are of great significance in the segmentation and diagnosis of benign and malignant tumors [2,3]. Another example is the identification of the number of transparent microorganisms in an environment to determine the degree of environmental pollution [4]. In recent years, the segmentation of transparent objects in images has become a hot topic in vision research [5,6]. It is not easy to detect whether there are transparent or translucent objects in the images because the transparent object area to be observed is generally small or thin, and the colors and contrasts of the foreground and background are similar [7]. Only the residual edge can lead to a low resolution of the foreground or background, which largely depends on the background and lighting conditions. Therefore, there is an urgent need for effective methods for identifying transparent or translucent images.

In recent years, deep learning has achieved a good performance in the field of computer vision. We consider the excellent performance of computer vision in image analysis [8], such as its high speed, high accuracy, low consumption, high degree of quantification, and strong objectivity [9]. Additionally, compared with traditional manual methods, computer image analysis can reduce manual effort from time-consuming tasks. Consequently, computer vision-based approaches can overcome the limitations of traditional methods and demonstrate great potential for transparent image analyses [10]. Particularly, when the object is transparent or has low contrast in the image, more foreground information is required; therefore, more visual details are found to recover the lost information from patches or pixels. As shown in Figure 1, the foreground and background of microorganisms are similar. There is only a small amount of information on the edges; which makes it difficult for traditional convolutional neural network (CNN) algorithms to distinguish transparent objects in images globally [11]. To address this problem, it is necessary to analyze transparent images from patches. We cropped the image into fixed-size patches and created a deep learning network to learn the features of the visual information of foreground and background patches. The network trained in this manner is sensitive to the foreground and background, which helps to distinguish transparent objects and achieve segmentation.

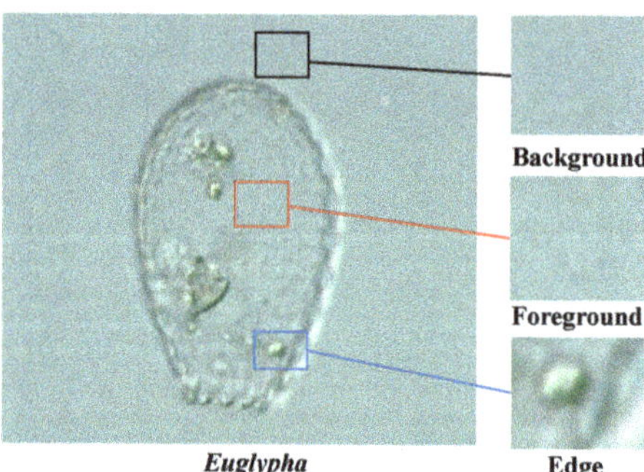

Figure 1. An example of transparent images (a low contrast environmental microorganism image).

In recent years, machine vision has been widely used for image processing [12,13]. Deep learning is a more effective method in the field of machine vision, such as the popular *Convolutional Neural Network* (CNN) Xception, VGG-16, Resnet50, Inception-V3, U-Net, and novel *Visual Transformers* (ViT) [14]. CNNs gradually expand the receptive field by increasing the size of the convolution kernel until it covers the entire image; thus, CNNs complete the image extraction from local to global information. Contrarily, transformers can obtain global information from the beginning, making learning more challenging, but their ability to retain long-term dependence is more potent than that of CNN [15]. Therefore, CNNs and transformers have both advantages and disadvantages when handling visual information. Therefore, this study compares the patch-level and pixel-level segmentation performance of transparent images with different CNN and visual transformer methods. This study aims to determine the adaptability of various deep learning models in this research domain.

The main contributions of this paper are as follows:

(1) A comparative study on patch-level transparent image segmentation was conducted to help analyze transparent images.

(2) The segmentation performances of multiple CNN and ViT deep learning networks under patch-level and pixel-level images were compared, which is convenient in performing further ensemble learning.

2. Related Work

This section briefly introduces related research on transparent images in practical analysis tasks and classical deep learning models.

2.1. Introduction to Transparent Image Analysis

Object analysis is one of the essential branches in robot vision; particularly, the analysis of transparent images of objects (transparent images) is challenging [16]. In the traditional machine learning method, the multiclass fusion algorithm can only extract the shallow features of a transparent image, and the obtained feature layer is incomplete. Practically, it is difficult for multiclass fusion algorithms to detect transparent objects. For example, home robots cannot see objects at all when they detect transparent glassware. The ClearGrasp machine learning algorithm performs well in analyzing transparent objects [17]. It can estimate the high-precision data of transparent objects from RGB-D transparent images, thereby improving the accuracy of transparent object detection.

Photoelectric sensors are widely used in industrial automation, mechanization, and intelligence as a necessary technical means for analyzing objects. In [18], it uses the properties of light to detect the position and changes of an object. Sensors and smart systems are used to separate recyclable materials (transparent materials and metals) into different bins, without using manpower.

There are many transparent objects in the industrial field, such as transparent plastics, colloids, and liquid drops. These transparent objects bring considerable uncertainty to products. For factories to have high-quality products, it is sometimes essential to analyze these transparent objects and control their shapes. However, it is difficult to segment the shapes of transparent objects using morphological methods. For instance, Hata et al. used a genetic algorithm to segment a transparent paste-drop-shape in the industry and obtained good performance [19].

Segmentation of transparent objects is significantly useful for computer vision applications. However, the foreground of a transparent image is usually similar to its background environment, which leads to commonly used image segmentation methods handling transparent images generally. The light-field image segmentation method can accurately and automatically segment transparent images with a small depth-of-field difference and improve the accuracy of the segmentation with less calculations [20]. Hence, they are widely used for the segmentation of transparent images.

The correct segmentation of zebrafish in biology has extensively promoted the development of the life sciences. However, zebrafish transparency makes the edges blurred during segmentation. The mean-shift algorithm can enhance the color representation in the image and improve the discrimination of the specimen against the background [21]. This method improves the efficiency and accuracy of zebrafish specimen segmentation.

Visual object analysis is vital for robotics and computer vision applications. Commonly used statistical analysis methods, such as the bag-of-features, are often applied to image segmentation [22]. The principle is to extract local features of the image for segmentation. However, the foreground transparent objects in transparent images do not have complete features; therefore, it is difficult to accurately segment transparent images. The more popular method is the light field distortion feature [23], which can describe transparent objects without knowing the texture of the scene, thus improving the segmentation accuracy of transparent images.

2.2. Introduction Classic of Deep Learning Network Models

Simonyan et al. proposed a VGG series of deep learning network models (VGG-Net), of which VGG-16 is the most representative [24]. VGG-Net can imitate a larger receptive field by using multiple 3 × 3 filters, enhancing nonlinear mapping, reducing parameters, and improving the network to better its analytics. Meanwhile, VGG-16 continues to deepen the previous VGG-Net with 13 convolutional layers and three fully connected layers. With a continuous increase in the convolution kernel and convolution layer, the nonlinear ability of the model becomes stronger. VGG-16 can better learn the features in images and achieve good performance in image classification, segmentation, and detection. Simonyan proved that as the depth of the network increases, the accuracy of the image analysis increases [24]. Nevertheless, this increase in depth is limited. An excessively increase in the depth of the network may lead to network degradation problems. Therefore, the optimal network depth of the VGG-Net is set to 16–19 layers. Moreover, VGG-16 has three fully connected layers, which causes more memory to be occupied, a long training time, and difficulty in tuning the parameters.

He et al. proposed a ResNet series of networks and added a residual structure in the networks to solve the problem of network degradation [25]. The ResNet model introduces the jumpy connection method (shortcut connection). This connection method allows the residual structure to skip levels that are not fully trained in the feature extraction process and increases the model's utilization of feature information during the training process. As the most classical model in the ResNet series, ResNet50 has a 50-layer network structure. This model adopts a highway network structure, which allows the network to have strong expression capabilities and acquire more advanced features. Therefore, it is widely used in image analysis. However, the network model is significantly deep and complicated; therefore, determining the layers in the deep network that are not thoroughly trained and then optimizing the network is a complex problem.

Szegedy et al. proposed a GoogLeNet network model, which has the advantage of reducing the complexity of networks based on ResNet. They first proposed Inception-V1, whose network is 22 layers deep and comprises multiple inception structure cascades as basic modules; each inception module consists of a 1 × 1, 3 × 3, 5 × 5 convolution kernel and a 3 × 3 maximum pooling, which is similar to the idea of multiscale, and increases the adaptability of the network to different scales [26]. With the continuous improvement of the inception module, the Inception-V2 network uses two 3 × 3 convolutions instead of 5 × 5 convolutions. This improves the *Batch Normalization* (BN) method, which rescales the data distribution to accelerate the model convergence [27]. Inception-V3 network introduces the idea of decomposing convolution, splitting a larger two-dimensional convolution into two smaller one-dimensional convolutions, further reducing the number of calculations [28]. Concurrently, Inception-V3 optimizes the inception module, embeds the branch in the branch, and improves the model's accuracy.

Xception is another improvement after Inception-V3 [29]. It mainly uses depthwise separable convolution to replace the convolution operation in Inception-V3. The Xception model uses deep separable convolution to increase the width of the network, which improves the accuracy of the classification and ability to learn subtle features. Xception adds a residual mechanism similar to ResNet to significantly improve the speed of convergence during training and accuracy of the model. However, Xception is relatively fragmented during the calculation process and results in a slower iteration speed during training.

U-Net is a CNN that was initially used to perform medical image segmentation. The U-Net architecture is symmetrical. It comprises a contracting and an expansive paths [30]. U-Net makes two significant contributions. The first is the extensive use of data augmentation to solve the problem of insufficient training data. The second is its end-to-end structure, which can help the network retrieve information from the shallow layers. Owing to its outstanding performance, U-Net has been widely used in semantic segmentation.

The transformer is a deep neural network based on the self-attention mechanism, enabling the model to be trained in parallel and obtain global information from the training data. Owing to its computational efficiency and scalability, it is widely used in *natural language processing* (NLP). Recently, Dosovitskiy et al. proposed a *vision transformer* (ViT) model that performs significantly well in image classification tasks [31]. In the first training step, the ViT model divides pictures into fixed-size image patches and uses their linear sequence as the input of the transformer model. In the second step, position embeddings are added to the embedding patches to retain the position information, and the image features are extracted through the multihead attention mechanism. Finally, the classification model is trained. ViT overcomes the limitation that the CNN model cannot be calculated in parallel, and self-attention can produce a more interpretable model. ViT is suitable for solving image-processing tasks, but experiments have proven that large data samples are required to improve the training effect.

Currently, deep learning methods are used to solve practical application problems in various fields. For example, in [32], a deep learning model was developed to detect and track sperm, which can effectively assist doctors in determining male reproductive health. In [33–36], a deep learning network is used to identify areas of cervical cancer that helps doctors to analyze cervical histopathological images. Owing to the spread of coronavirus disease 2019 (COVID-19), medical resources have continuously been depleted. In [37], the detection performance of 15 different deep learning models for COVID-19 X-ray image identification are compared, which can help reduce the workload of doctors. In [38], a multiple network model was proposed for the analysis of intracranial pressure (ICP) and heart rate (HR) after severe traumatic brain injury in pediatric patients. In [39], to help pathologists detect cancer subtypes and genetic mutations, a deep learning model was developed. In [40], to predict the response to immune checkpoint inhibitors in advanced melanoma and effectively assist doctors in diagnosis, a deep learning model was trained on clinical data. In [41], machine-learning methods are used to investigate, predict, and discriminate COVID-19.

3. Comparative Experiment

This section introduces patch-level and pixel-level segmentation experiments and the segmentation results of transparent images under several deep learning networks. The patch-level and pixel-level image segmentation workflows are shown in Figure 2.

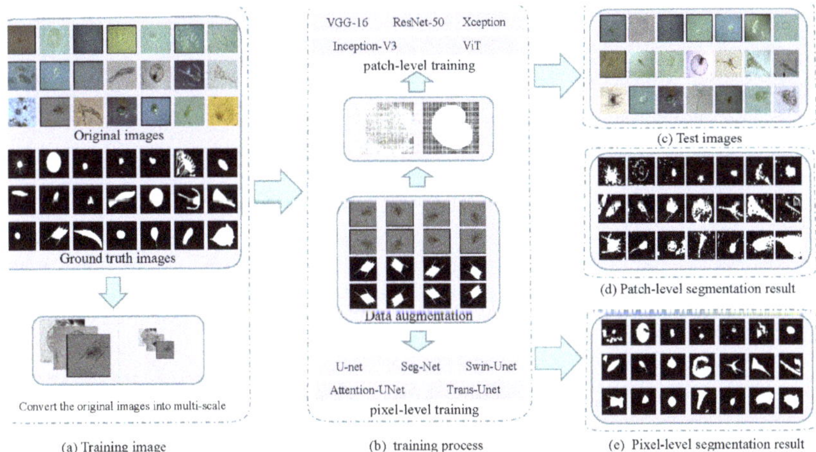

Figure 2. Workflow of patch-level and pixel-level segmentation in transparent images (using environmental microorganism EMDS-5 images as examples) ((**a**) is the image of the training set and the grayscale of the original image. (**b**) is the patch-level and pixel-level training process. In (**c**) is the test set image. (**d**,**e**) are patch-level and pixel-level segmentation results, respectively).

3.1. Experiment Setting

3.1.1. Data Settings

In this study, we used the environmental microorganism dataset fifth version (EMDS-5) as transparent images for the analysis [4]. The effectiveness and robustness of deep learning methods based on the small dataset, EMDS-5, are provided in detail in [42]. Table 1 shows the data distribution of EMDS-5 in the experiment. It is a newly released version of the EMDS series, which includes 21 types of EMs, each of which contains 20 original microscopic images and their corresponding ground-truth (GT) images (examples are shown in Figure 3). We randomly divided each category of the EMDS-5 into training, validation, and test datasets in a ratio of 1:1:2. Thus, we obtained 105 original images and their corresponding GT images for training and validation, respectively, and 210 original images for testing.

Table 1. EMDS-5 experimental data setting.

	Training Set	Validation Set	Test Set
Actinophrys	5	5	10
Arcella	5	5	10
Aspidisca	5	5	10
Codosiga	5	5	10
Colpoda	5	5	10
Epistylis	5	5	10
Euglypha	5	5	10
Paramecium	5	5	10
Rotifera	5	5	10
Vorticlla	5	5	10
Noctiluca	5	5	10
Ceratium	5	5	10
Stentor	5	5	10
Siprostomum	5	5	10
K.Quadrala	5	5	10
Euglena	5	5	10
Gymnodinium	5	5	10
Gonyaulax	5	5	10
Phacus	5	5	10
Stylonychia	5	5	10
Synchaeta	5	5	10
total	105	105	210

3.1.2. Data Preprocessing

Patch-Level Data Preprocessing

In the first step, considering that the light source had a significantly large effect on the color of the microscope image and that the same sample show different colors under different light sources, the color information was less important for the sign extraction of the microscopic image [43]. Therefore, we converted the EM microscopic images into grayscale to reduce the computational workload of the network model. In the second step, we converted all image sizes into 256×256 pixels because the microscopic images were of various sizes. In the third step, the training and validation images, and their corresponding GT images were cropped into patches (8×8 pixels), and $105 \times 1024 =$ 107,520 patches were obtained. Based on the corresponding GT image small patches, we divided these small patches into two categories: foreground and background. The partition criterion was based on whether the area of interest in the patch comprises half of the entire patch. If so, we assigned the foreground as the label of this patch; otherwise, it was annotated as the background. Finally, we found that the 8×8 pixel patches with foreground and background were 16,554 and 90,966, respectively. During the training process, we found that the model weights were heavily biased towards negative samples owing to the imbalance between positive and negative samples. To avoid data imbalance during training, we rotated the training set image small patches by 0°, 90°, 180°, and 270°, and mirrored them for data augmentation. We then further obtained 16,544 $\times 8 =$ 132,432 patches, from which 90,966 patches were randomly selected as the patches that were finally used in the training set. Details of the image patches are listed in Table 2.

Table 2. Patch-Level data preprocessing. FG (foreground) and BG (background).

Data Set	Training Set	Validation Set	Test Set
8×8 pixels FG	90,966	17,356	32,445
8×8 pixels BG	90,966	90,164	182,595
8×8 Total	181,932	107,520	215,040

Pixel-Level Data Preprocessing

The image was converted to grayscale and resized to 256×256 pixels for the pixel-level segmentation experiments.

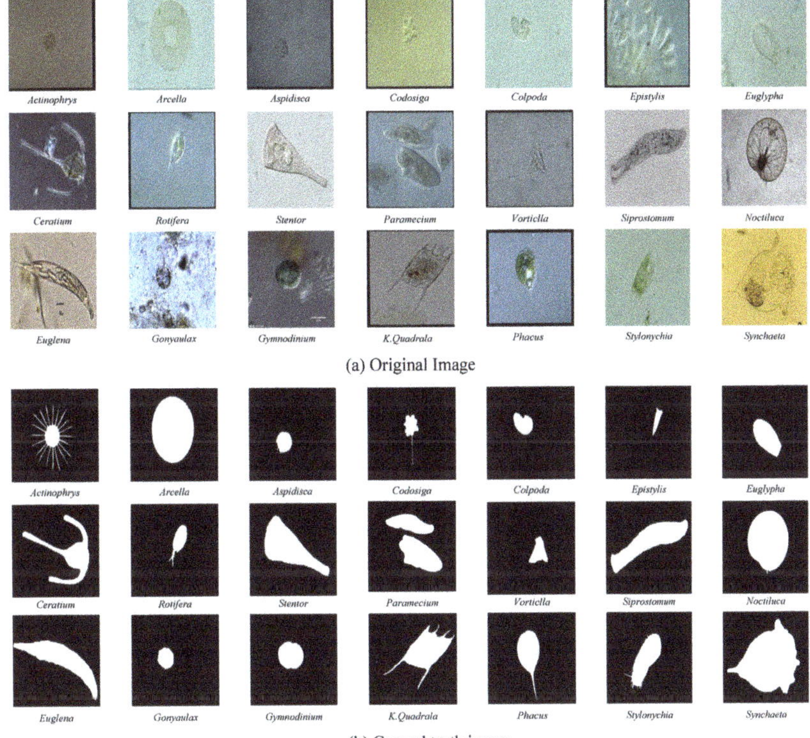

Figure 3. Examples of the environmental microorganism image in EMDS-5. (**a**) is the original images of EMDS-5, each image contains one or more EM objects of the same species, and one image is selected for each species as a representative. (**b**) correspond to the real segmentation images of microorganisms in each image in the (**a**). The pixel value of the background part in the microorganism image is set to 0, and the foreground part is set to 1.

3.1.3. Hyper Parameters

The patch-level experiment used the Adam optimizer with a learning rate of 0.0002, and the batch size was set to 32. During the training, we used the cross-entropy loss function to optimize the deep learning model [44]. Figure 4 shows the accuracy and loss curves of the different deep learning models used in this experiment. The epoch was determined based on the convergence of the loss curve. In our pretest, we tried to train

100 epochs and maintained the best training model weights, and found that the best model appeared between 40 and 50, where too much training caused overfitting and too little training was not able to train the optimal model. Therefore, considering the computational performance of the workstation, we finally set 50 epochs for training. Deep neural networks have a strong expressive ability compared to traditional models that require more data to avoid overfitting. Because our experiment was conducted on a small dataset, we employed a transfer learning approach to avoid the overfitting problem [45]. Meanwhile, because of the outstanding classification ability of CNN in ImageNet and the significant performance of transfer learning with a limited training data set [24], we used the limited EM training data to fine-tune the CNN model pretrained by ImageNet [46,47]. It has been proven that using CNN pretrained on ImageNet is useful for classification tasks through the concept of transfer learning and fine-tuning [48]. Before fine-tuning the pretrained CNN, we froze the parameters of the pretrained model. Subsequently, we used patch-level data to fine-tune the dense layers of the CNN. We retained the backbone network of the CNN classification network to extract image features and replace the last fully connected layer of the CNN model with Global Average Pooling2D + dense + dense + SoftMax. Global average Pooling2D simplifies many parameter operations. The purpose of the dense layer is to extract the correlation between these features through nonlinear changes in the dense layer and map them to the output space. Finally, the class probability result was outputted using SoftMax. We also compared the validation set accuracy of the ViT model with and without pre-trained weights. In both sets of experiments, we trained three times and then averaged the results. We found that ViT without pretrained weights and ViT with ImageNet pretrained weights had accuracies of 0.8923 and 0.8926 on the validation set, respectively. During training, ViT takes approximately 2G less memory for loading than the ImageNet pretrained weight model. To compare the performance of the two methods, we used ViT without pretraining as the optimization option. We set the network depth to six, heads to 16, mlp_dim to 3000, and dropout and emb_dropout to 0.1. The pixel-level experiment used the Adam optimizer with a learning rate of 0.001, and the batch size was set to 4. Figure 5 shows the loss curves of the different deep learning models in this experiment. The training curves began to converge after 90 epochs of iterations for the five models. To prevent overfitting, we set 100 epochs for training.

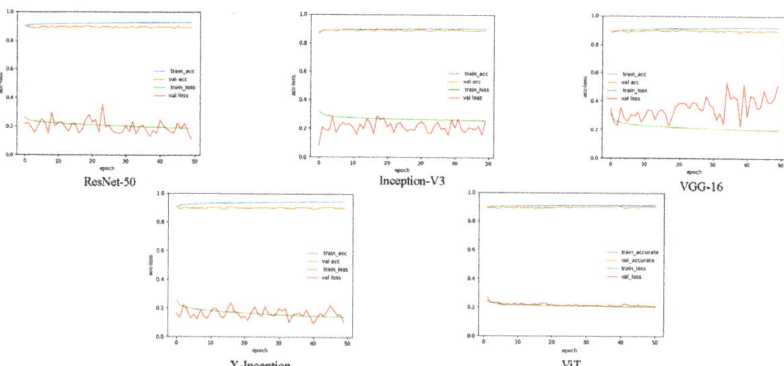

Figure 4. A comparison of the image segmentation results of the loss and accuracy curves of deep learning on 8 × 8 pixels training and validation sets. (Each legend has four curves, respectively, the accuracy and loss values of the training set, and the accuracy and loss values of the validation set).

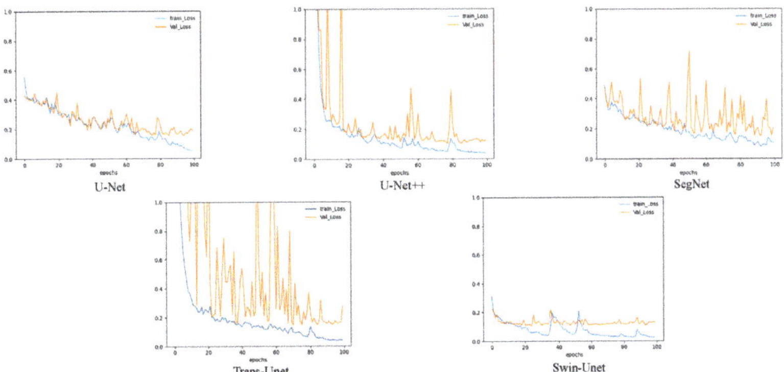

Figure 5. A comparison of the image segmentation results of the loss curves of deep learning on pixel-level training and validation sets.

3.2. Evaluation Metrics

To compare the classification foreground and background performances of different methods, we used the commonly used deep learning classification indexes—accuracy (Acc), precision (Pre), recall (Rec), specificity (Spe), and F1-Score (F1)—to evaluate the patch-level results [49]. Acc reflects the ratio of correct classification samples to total samples. Pre reflects the proportion of correctly predicted positive samples in the model classification of positive samples. Rec reflects the correct proportion of model classification for all positive samples. Spe reflects the proportion of the model that correctly classifies negative samples among the total negative samples. F1 is a calculation result that comprehensively considers the Pre and Rec of the model. In addition, we employed Dice, Jaccard, Pre, Acc, and Rec to evaluate the results of pixel-level segmentation [38]. V_{pred} represents the foreground predicted by the model. V_{gt} represents the foreground in the ground-truth image. From Table 3, we can determine that the higher the values of the first four metrics (Dice, Jaccard, recall, and accuracy), the better the segmentation results. True positive (TP), false negative (FN), false positive (FP), and true negative (TN) are concepts in the confusion matrix.

Table 3. Evaluation metrics.

Metrics	Formula	Metrics	Formula
Acc	$\frac{TP+TN}{TP+TN+FP+FN}$	Dice	$\frac{2\times\lvert V_{pred}\cap V_{gt}\rvert}{\lvert V_{pred}\rvert+\lvert V_{gt}\rvert}$
Pre (P)	$\frac{TP}{TP+FP}$	Jaccard	$\frac{\lvert V_{pred}\cap V_{gt}\rvert}{\lvert V_{pred}\cup V_{gt}\rvert}$
Rec (R)	$\frac{TP}{TP+FN}$	F1	$\frac{2\times P\times R}{P+R}$
Spe	$\frac{TN}{TN+FP}$		

3.3. Comparative Experiment

To ensure the reliability of the deep learning models, we performed five-fold cross-validation in all experiments in this study [50]. We took the average of the experimentally obtained model performance indicators as the data for the final evaluation model (precision, recall, F1-Score, accuracy, time, size, Dice, and Jaccard).

3.3.1. Comparative Experiment of Patch-Level Segmentation

Comparison on Training and Validation Sets

To compare the classification performance of the CNNs and ViT models, we calculated the precision, recall, F1-Score, and maximum accuracy to evaluate the models. The segmentation results of the 8 × 8 pixel patches in the validation set are presented in Table 4. Overall, the Pre of the deep learning network that classifies the transparent image background was higher than that of the foreground image. In addition, the Pre of the five models for classifying transparent image backgrounds was approximately 97%; the highest was the VGG-16 value of 97.6%, and the lowest was the Xception and ViT value of 96.7%. The Pre rate of foreground classification VGG-16 was the best, and the Pre rate was 63.1%. Inception-V3 showed the lowest value (53.3%). For transparent image foreground classification, the highest Rec rate was obtained with Xception (89.2%), and the lowest was ViT (84.1%). For transparent image background classification, the highest Rec rate was the ViT value of 90.3%, and the lowest was the Xception value of 85.0%. The Spe obtained by the five models in the classification background was opposite to the Rec rate obtained in the classification foreground. Among the five models, the highest Acc was ResNet50 with 92.87%, and the lowest was ViT with 89.26%.

Table 4. Classification performance of models of five-fold cross-validation experiment on validation set of 8 × 8 pixels patches. MAcc (Max Acc), FG (foreground) and BG (background) (In [%]).

Model	Class	Avg.Pre	Avg.Rec	Avg.Spe	Avg.F1	MAcc
ResNet50	FG	62.3	88.2	89.7	73.0	92.87
	BG	97.5	89.7	88.2	93.4	
Inception-V3	FG	61.8	88.6	89.5	72.8	90.24
	BG	97.6	89.5	88.6	93.4	
VGG-16	FG	63.1	88.6	90.0	73.7	92.09
	BG	97.6	90.0	88.6	93.6	
Xception	FG	53.3	89.2	85.0	66.7	91.10
	BG	96.7	85.0	89.2	90.9	
ViT	FG	62.4	84.1	90.3	71.6	89.26
	BG	96.7	90.3	84.1	93.4	

Comparison on Test Set

Table 5 summarizes the results of the five network predictions. We found that the Acc of ResNet50 was the highest (90.00%), and that of Xception was the lowest at 85.85%. Furthermore, the lowest predicted Acc of the transparent foreground was the Xception with 51.8%, and the highest was the ResNet50 (62.2%).

To express the classification results of the CNN and ViT models for transparent image patches more intuitively, we summarize the confusion matrices predicted by the five models, as presented in Figure 6. Inception-V3 correctly classified 2509 more foreground patches than the ViT model. However, Inception-V3 correctly classified only 260 more foreground patches than Xception. In classifying background patches, the Resnet50 model exhibited the best performance, correctly classifying 165,369 background patches. Resnet50 correctly classifies 10,183 background patches better than the Xception model, but only 625 more background patches than ViT. The classification details are shown in Figure 6. In addition, the number of correctly classified backgrounds in ResNet50 was 165,369, accounting for 90.57% of the total correct background patches, and the Pre of the classified background patches was 97.55%. Among the five models, ResNet50 exhibited the highest prediction accuracy rate of 90.06%. The classification accuracies of the Xception and Inception models were lower than those of the other five models at 85.85% and 86.30%, respectively. Moreover, we found that the Xception and Inception models exhibited poor background, but better foreground recognition performances. The Inception-V3 model correctly classified approximately 29,688 foreground patches, accounting for 91.50% of the total number of foreground patches. The Xception model misclassified a maximum of 27,409 background patches, accounting for 15.01% of the total background patches. Among the five models, the classification performance of the VGG model was relatively moderate. To better illustrate the classification results, we reconstructed the transparent image after dicing, as shown in Figure 7.

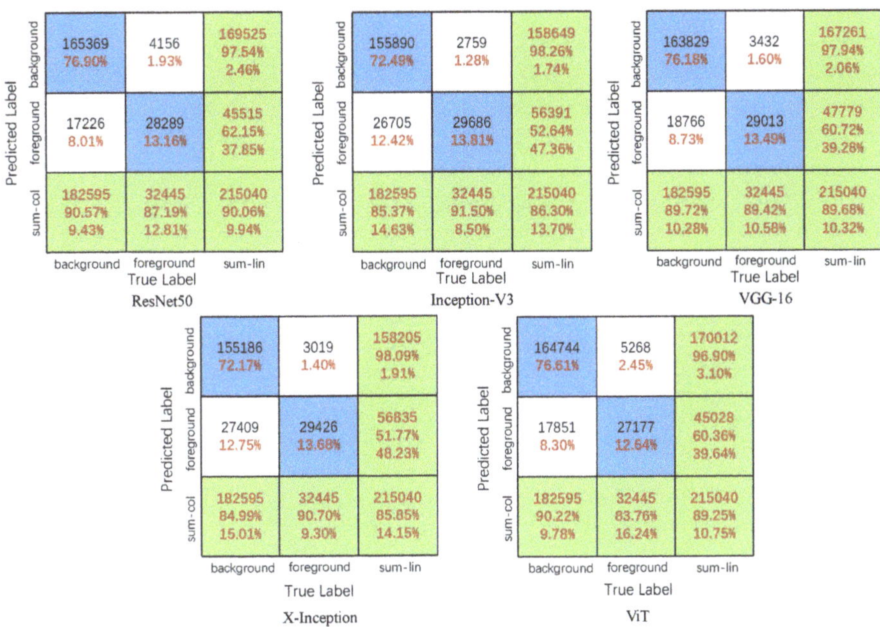

Figure 6. Predict the confusion matrix on test set of 8 × 8 pixels' patches.

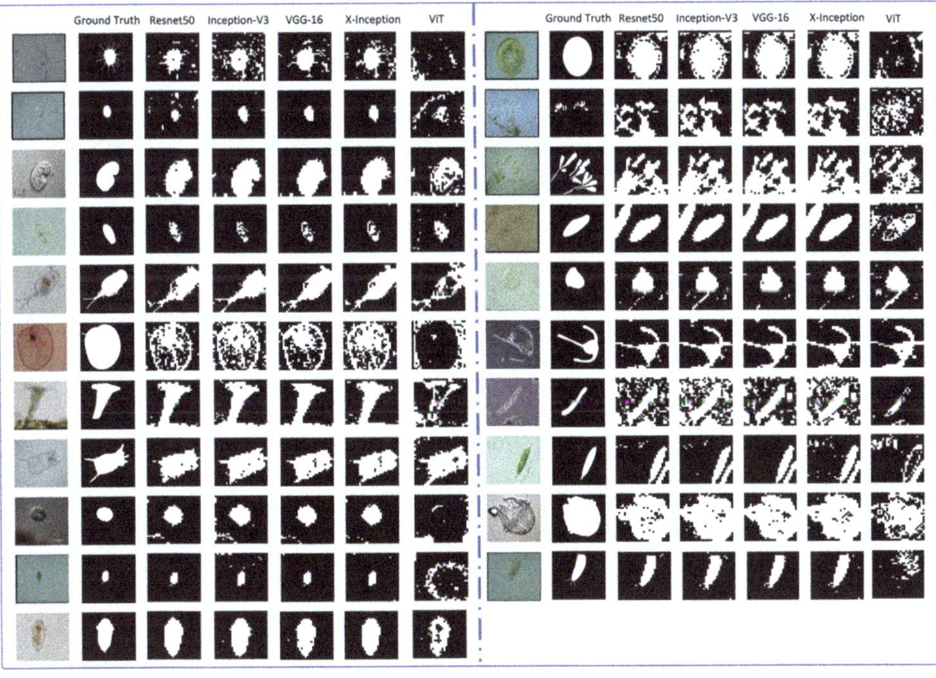

Figure 7. Reconstruct the 8 × 8 pixel patch transparent image segmentation results. (The figure contains the original image, ground truth image and Resnet50, Inception-V3, VGG-16, Xception, ViT network model predicted segmentation results).

Table 5. Classification performance of models of five-fold cross-validation experiment on test set of 8 × 8 pixels' patches. PAcc (prediction accuracy), FG (foreground) and BG (background) (In [%]).

Model	Class	Avg.Pre	Avg.Rec	Avg.Spe	Avg.F1	Avg.PAcc
ResNet50	FG	62.2	87.2	90.6	72.6	90.0
	BG	97.5	90.6	87.2	93.9	
Inception-V3	FG	52.6	91.5	85.4	66.8	86.29
	BG	98.3	85.4	91.5	91.4	
VGG-16	FG	60.7	89.4	89.7	72.6	89.6
	BG	97.9	89.7	89.4	93.6	
Xception	FG	51.8	90.7	85.0	65.9	85.85
	BG	98.1	85.0	90.7	91.1	
ViT	FG	60.4	83.8	90.2	70.2	89.25
	BG	96.9	90.2	83.8	93.4	

3.3.2. Comparison Experiment of Pixel-Level Segmentation

To compare the effect of path-level segmentation, we conducted extended experiments using pixel-level segmentation. We applied five networks for the comparative experiments: U-Net, U-Net++, SegNet, TransUnet, and Swin-UNet. We used these five networks to compare the performance of the CNN and ViT for pixel-level segmentation. U-Net, U-Net++, and SegNet stand for CNN network, Swin-UNet stands for transformer networks, and TransUnet stands for CNNs joining the transformer. Table 6 presents the outcome of the five model prediction metrics. We found that U-Net++ exhibited the highest segmentation performance overall, but also had the longest training time. U-Net exhibited the worst segmentation performance. The segmentation result of the vision transformer network (Swin-UNet) was second after that of U-Net++. The Jaccard and precision values were 71.26% and 85.00%, respectively, which were higher than those of the other network models. To compare the segmentation results more intuitively, we presented the pixel-level segmentation results in Figure 8. Clearly, the pixel-level segmentation results are better than those of the patch level. However, the patch-level segmentation effect was better for multiobject transparent microorganism images. Compared to the 8 × 8 patch-level segmentation, the network model with a transformer structure (Swin-UNet) at the pixel level performed well, and the ViT was higher than the accuracy of the CNN model. However, in the 8 × 8 patch-level experiment, the accuracy of the CNN was higher than that of ViT. It is shown in Figure 5 that the loss curve stability of Swin-UNet is significantly better than that of the other four models during training. The training loss stability of the ViT model was better than that of the CNN model. To reflect the training process of the model more intuitively, the *Intersection-over-Union* (IOU) curves of the five models on the training and validation sets in the pixel-level experiment is presented in Figure 9.

Table 6. Segmentation performance of models of five-fold cross-validation experiment on the test set.

Model	Avg.Dice	Avg.Jaccard	Avg.Precision	Avg.Recall	Avg.Acc
U-Net	71.82	59.23	68.98	76.06	91.93
U-Net++	82.51	73.51	83.42	85.98	95.32
SegNet	78.21	67.70	77.45	84.66	74.06
Trans-Unet	75.50	64.13	72.52	86.75	93.44
Swin-UNet	81.00	71.26	85.00	82.08	95.31

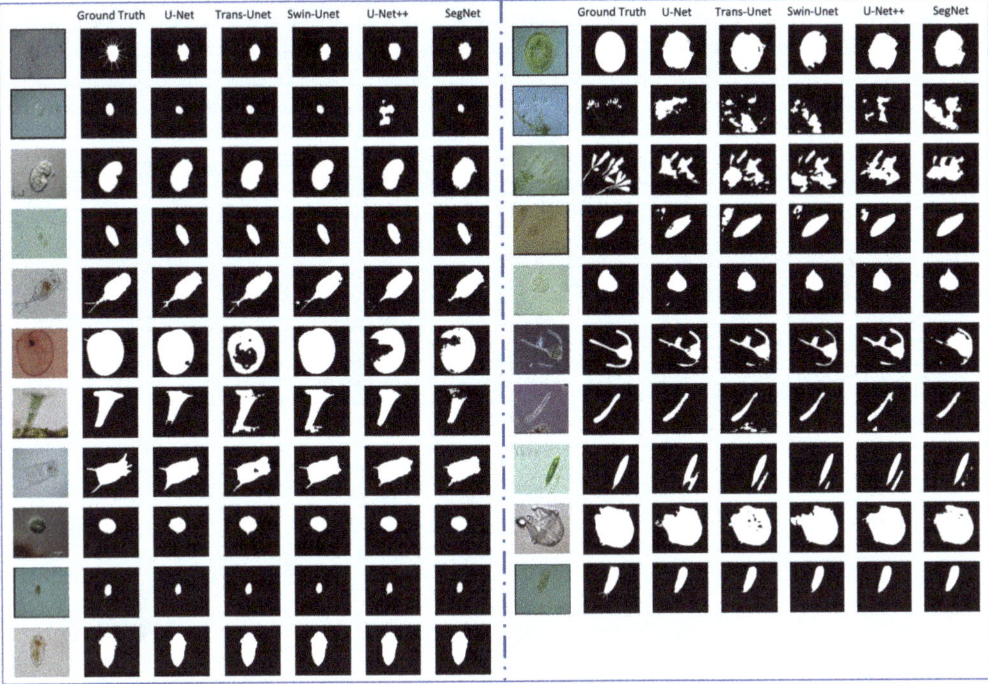

Figure 8. Reconstruction of pixel-level segmentation results on transparent images of the test set. (The figure contains the original image, ground truth image and U-net, Trans-Unet, Swin-Unet, U-Net++, and SegNet network model predicted segmentation results).

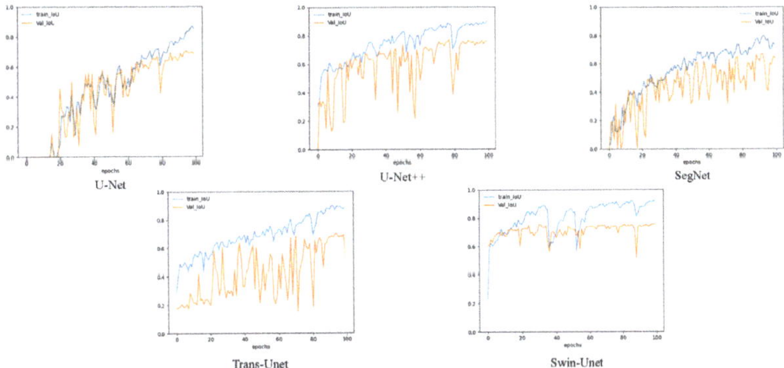

Figure 9. A comparison of the image segmentation results of the IOU curves of deep learning on pixel-level training and validation sets.

3.3.3. Additional Experiment Based on EMDS-6 Dataset

To demonstrate the applicability of the models in our comparative experiments, we compared five models in pixel-level experiments on the EMDS-6 dataset. A partial EM sample of EMDS-6 is shown in Figure 10. EMDS-6 contains 840 EM's microscopic images within 21 classes. We divided the dataset into training, validation, and test sets in a 1:1:2 ratio. We trained five models using the same parameters as those used in the EMDS-5 experiments. The performance metrics of the experimental results of the five models are presented in Table 7. We found that on EMDS-6, the pixel-level segmentation performance is consistent with the segmentation performance of EMDS-5. The segmentation performance of the U-Net, U-Net++, and Swin-Unet models was similar, and the segmentation accuracy was approximately 95%. SegNet exhibited the worst segmentation performance, with an accuracy of 91.21%. In addition, the number of images in EMDS-6 was twice that of EMDS-5; therefore, the model learns more EM information during training, leading to an overall improvement in the segmentation performance of the five models. The loss and IOU curves trained using the five models are shown in Figures 11 and 12, respectively. We found that the loss and IOU curves trained by the five models were similar to the results in EMDS-5; therefore, the five models are suitable for comparison experiments. The segmentation results are shown in Figure 13.

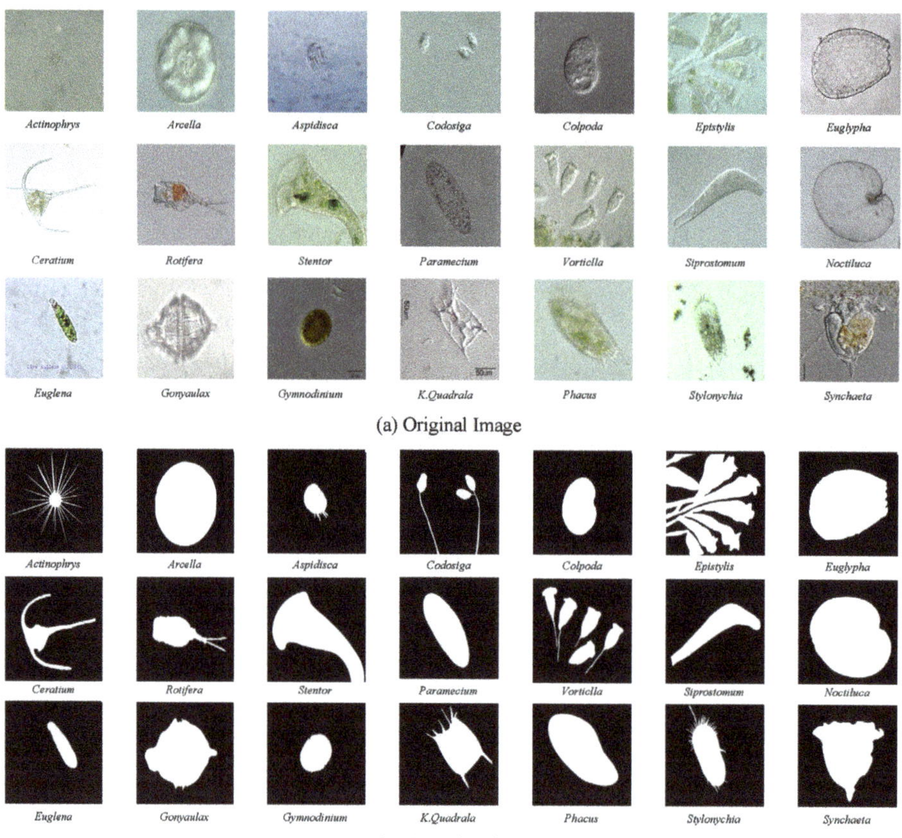

Figure 10. Examples of the environmental microorganism image in EMDS-6. (**a**) is the original images of EMDS-6, each image contains one or more EM objects of the same species, and one image is selected for each species as a representative. (**b**) correspond to the real segmentation images of microorganisms in each image in the (**a**). The pixel value of the background part in the microorganism image is set to 0, and the foreground part is set to 1.

Table 7. Segmentation performance of models of five-fold cross-validation experiment on the EMDS-6 test set.

Model	Dice	Jaccard	Precision	Recall	Acc
U-Net	84.81	76.24	88.83	83.53	95.43
U-Net++	86.48	78.25	89.02	87.08	95.80
SegNet	74.63	62.50	73.88	83.59	91.21
Trans-Unet	84.66	76.09	86.04	86.88	94.98
Swin-UNet	86.11	78.05	89.46	85.79	95.49

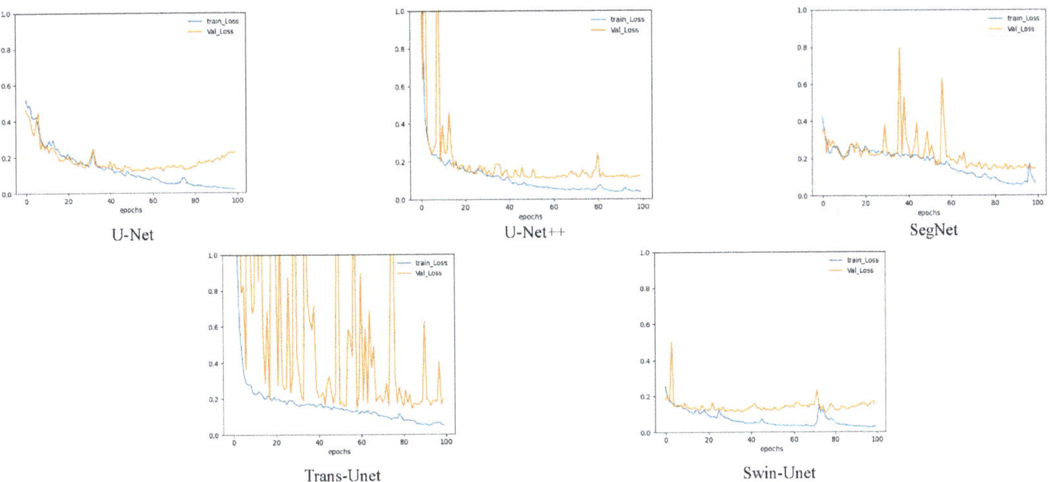

Figure 11. A comparison of the image segmentation results of the loss and accuracy curves of deep learning on pixel-level training and validation set of EMDS-6.

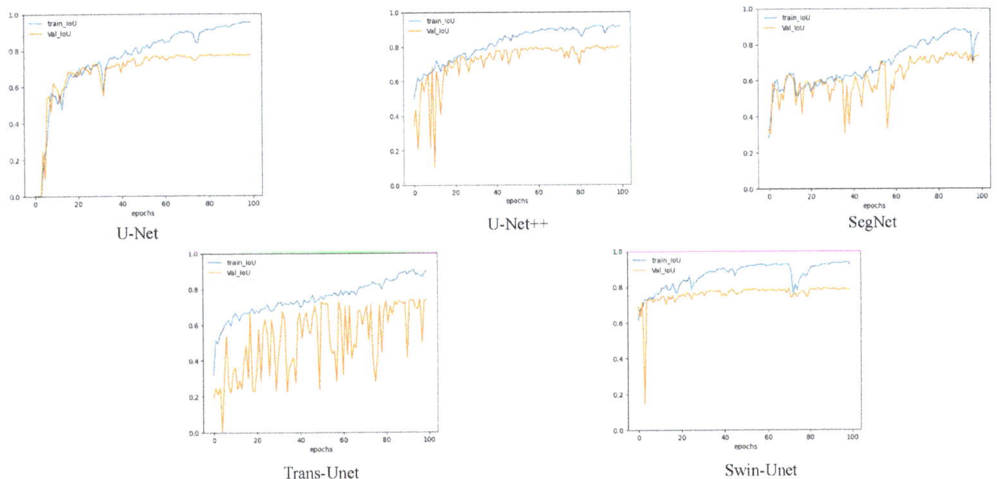

Figure 12. A comparison of the image segmentation results of the IOU curves of deep learning on pixel-level training and validation sets of EMDS-6.

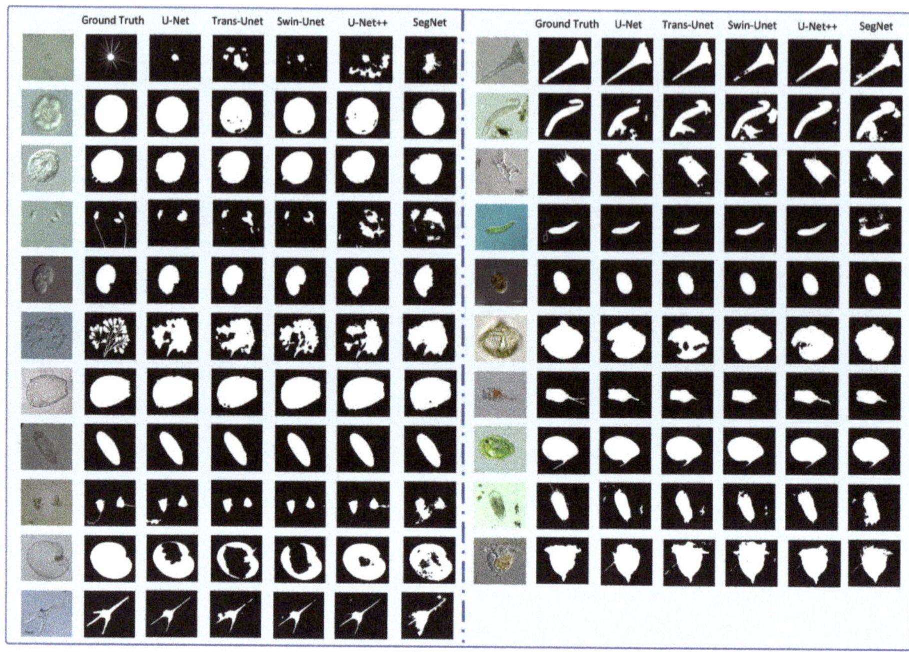

Figure 13. Reconstruction of pixel-level segmentation results on transparent images the EMDS-6 test set. (The figure contains the original image, ground truth image and U-net, Trans-Unet, Swin-Unet, U-Net++, SegNet network model predicted segmentation results).

3.3.4. Experimental Environment

A comparative experiment was conducted using a local computer, with a running memory of 16 GB. The computer used the Win10 Professional operating system equipped with an 8 GB NVIDIA Quadro RTX 4000 GPU. In the patch-level experiment, the four CNN network models were imported from Keras version 2.3.1, and used TensorFlow 2.0.0 as the background. The experimental frameworks for ViT and pixel level were Pytorch 1.7.1 and Torchvision 8.0.2. Table 8 presents the model training and prediction time, and the size of the model during the experiment. From the perspective of model training time, the ViT model was much lower than CNN models, where the ViT training time was 13,992 s, and the Xception training time was the longest, 46,383 s. From the perspective of the model size, the minimum size of the ViT model was 31.2 M, and the maximum size of the ResNet50 model was 114 M. We calculated the times required for the five prediction models. The fastest prediction time for Inception-V3 was 583 s, and the prediction time for a single picture was 0.0027 s. The slowest time for ViT was 1308 s, and the prediction time of a single image was 0.0061 s.

Table 8. A comparison of the classification results of five-fold cross-validation experiment on train and test sets of 8 × 8 pixels patches. Train (Average training time), Test (Average test times) and Avg.p (Single picture prediction time) (In [s]).

Model	Train	Test	Avg.p	Size (MB)
ResNet50	36,754	878	0.0041	114
Inception-V3	24,064	583	0.0027	107
VGG-16	34,736	781	0.0036	62.2
Xception	46,383	1014	0.0047	103
ViT	13,992	1308	0.0061	31.2

3.4. In-Depth Analysis

In the predicted 215,040 patches, we compared the performance of five types of network classification: foreground and background. From Figure 6, Inception-V3 has the largest number of correct foregrounds under 8 × 8 pixel patches, whereas ResNet50 has the largest number of correctly classified backgrounds. Furthermore, ViT network models misclassified foreground patches more than CNNs models. Consequently, the number of correctly classified foregrounds in the CNNs network was greater than that in the ViT network. Moreover, the ability of Swin-UNet to segment the foreground outperformed most models. Therefore, the ViT model was found as outstanding for low-transparency image recognition. Furthermore, in [32,36,49], on other datasets, the classification performance of the CNN model was better than that of the ViT network, and the training time of the ViT network model was less than that of the CNN model.

We found that the segmentation effect of the pixel level was higher than that of the patch level, but the segmentation result of the patch level can compensate for the loss of details of pixel-level segmentation. Therefore, we combined the patch-level and pixel-level segmentation results to obtain the optimal segmentation results. Figure 14 compares the GT, pixel-level, patch-level, and combined segmentation results. We set the segmentation region to a red mask and found that the combined results were significantly better than the single pixel-level or patch-level segmentation results.

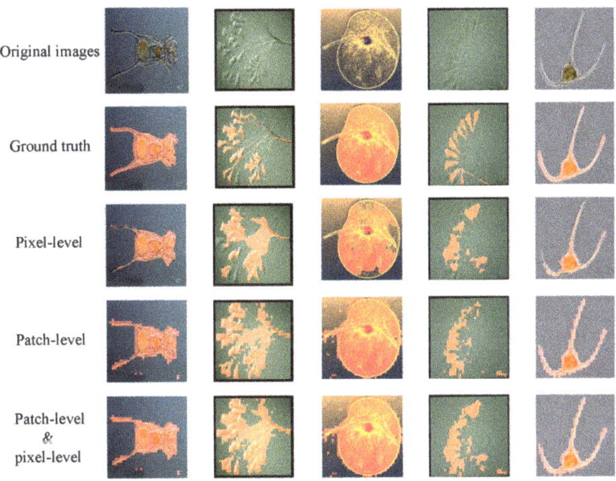

Figure 14. Valid examples in EMDS-5 that fuse pixel-level segmentation and patch-level segmentation. From top to bottom, the images in each original image of EM, the EM represent GT image, pixel-level segmentation result, patch-level segmentation result, and combined result, respectively. (The red part in the figure is the segmentation result).

4. Discussion

This study investigates the patch-level and pixel-level segmentation performance of five deep learning models on the EMDS-5 [4]. The comparison results based on the evaluation indicators are listed in Tables 4, 5, 6 and 8. In addition, to verify the generalization of the model, we used the same method to perform expansion experiments on the EMDS-6 dataset. The results are summarized in Table 4 and Figures 11–13. To improve the reliability of the conclusion, all experimental results in this study were repeated five times, and then averaged for the final result [50].

In the patch-level segmentation experiments, the performance indicators of the five models were compared [49]. We found that all five deep learning models were better at classifying transparent image backgrounds than foregrounds. We can speculate that the foreground features of the transparent image are similar to those of the background, and all models had a strong ability to classify the background. Further, we compare the CNN and ViT models among the five models [14]. We found that ResNet50 [25] model had the highest segmentation performance; however, the ViT [31] model was better than the CNN model in terms of training time and model convergence speed. The ViT network has evident advantages in the time of training the model, and the time consumption is much less than that in other models. We can speculate that the ViT model may further expand its advantages when trained using more training data.

In the pixel-level segmentation experiment, we also conducted a comparative experiment between the CNN and ViT models. The model with the highest segmentation performance was U-Net++ [51]. Similarly, the Swin-UNet model [52], represented by the ViT model, had a better convergence speed than the CNN model. In addition, the accuracy of the Swin-UNet segmentation was better than that of U-Net, SegNet [53], and Trans-Unet [54].

The magnitude and index system of patch-level and pixel-level segmentation are different. Therefore, we cannot numerically compare the segmentation performance of the two groups of experiments. However, we made a full comparison between the patch-level and pixel-level visual segmentation results and fused the patch-level and pixel-level segmentation results to obtain a more complete segmentation result.

In computer vision tasks, image segmentation for multi-size EMs [5,55] and for weakly visible EMs [56,57] are introduced in recent years, where multi-scale CNN and pair-wise CNN methods are developed. However, the transparent EM image has not been developed and studied, so this paper used patch-level and pixel-level methods to segment transparent images, and a total of eight CNN and two ViT models were used to test the performance of the model. This study provides an analysis table of differences between the patch-level and pixel-level models. Our research and conclusions significantly reduce the workload of the researcher's choice of experimental augmentation method. This reference is of great significance.

In this paper, we investigate deep learning methods for analyzing transparent EM images. In Section 2.1, we mentioned some existing techniques for analyzing transparent objects, and these techniques also have great potential in analyzing transparent EM images. For example, cleargrasp uses a deep convolutional network to infer surface normals, masks and occlusion boundaries on the surface of transparent objects. These outputs are then used to optimize the initial depth estimates of all transparent surfaces in the scene [17]. The leargrasp algorithm has great potential for transparent EM edge segmentation. In the industrial field, optical sensors are widely used to detect transparent objects by using the difference of light propagation rate through different media to determine transparent objects [19]. Optical sensors have potential value for transparent EM torso segmentation [20]. Meanwhile, the mean-shift algorithm and genetics algorithm also perform well in analyzing transparent objects, and we can also apply it to the segmentation of transparent EM blur and occluded locations [19,21]. The above methods of analyzing images have been proposed with the development of image processing technology, and these methods have the potential to contribute to the analysis of transparent EM images.

5. Conclusions and Future Work

In this study, we aimed to address the segmentation problems in transparent images by cropping the image into patches and classifying their foreground and background. We used CNNs and ViT deep learning methods to compare the patch-level and pixel-level performances of the transparent image segmentation. In segmenting transparent microorganism images, we found that the pixel-level generally outperforms the patch-level segmentation. However, the patch-level method works better in multiobject segmentation. Moreover, in the patch-level segmentation experiment, CNNs were better than the ViT models, but in the pixel-level experiment, the ViT model segmentation performed better than that of most CNNs. The smaller the patch pixel is, the more the regions perceived by the ViT model, and the stronger the ability to combine contextual information. Furthermore, the loss convergence and stability of the ViT model during training were better than those of the CNN model. In conclusion, the CNN and ViT models have more advantages in image classification. CNN is better at extracting the local features of images, whereas ViT is better at extracting the global features of images combined with contextual information. The ViT model has great potential for the future.

In the future, we plan to increase the amount of data to improve the stability of the comparisons. Meanwhile, images reconstructed by deep learning classification can be extended to the positioning, recognition, and detection of transparent images. We will further strengthen the applicability of these results.

Author Contributions: Conceptualization, C.L.; methodology, H.Y. and C.L.; software, H.Y.; validation, P.Z., A.C. and H.Y.; formal analysis, Y.T. and H.Y.; investigation, M.G. and T.J.; resources, X.Z.; data curation, C.L. and H.Y.; writing—original draft preparation, H.Y. and C.L.; writing—review and editing, C.L., J.Z., S.Q. and H.Y.; visualization, H.Y.; supervision, C.L.; project administration, C.L.; funding acquisition, C.L. and T.J. All authors have read and agreed to the published version of the manuscript.

Funding: National Natural Science Foundation of China (No.61806047) and Sichuan Science and Technology Plan (No. 2021YFH0069, 2021YFQ0057, 2022YFS056).

Institutional Review Board Statement: Not applicable.

Informed Consent Statement: Not applicable.

Data Availability Statement: Not applicable.

Acknowledgments: We thank Zixian Li and Guoxian Li for their important discussion.

Conflicts of Interest: The authors declare no conflict of interest.

References

1. Liao, S.Y.; Aurelio, O.N.; Jan, K.; Zavada, J.; Stanbridge, E.J. Identification of the mn/ca9 protein as a reliable diagnostic biomarker of clear cell carcinoma of the kidney. *Cancer Res.* **1997**, *57*, 2827–2831. [CrossRef] [PubMed]
2. Xue, D.; Zhou, X.; Li, C.; Yao, Y.; Rahaman, M.M.; Zhang, J.; Qi, S.; Sun, H. An application of transfer learning and ensemble learning techniques for cervical histopathology image classification. *IEEE Access* **2020**, *8*, 104603–104618. [CrossRef]
3. Zhou, X.; Li, C.; Rahaman, M.M.; Yao, Y.; Ai, S.; Sun, C.; Wang, Q.; Zhang, Y.; Li, M.; Li, X.; et al. A comprehensive review for breast histopathology image analysis using classical and deep neural networks. *IEEE Access* **2020**, *8*, 90931–90956. [CrossRef]
4. Li, Z.; Li, C.; Yao, Y.; Zhang, J.; Rahaman, M.M.; Xu, H.; Kulwa, F.; Lu, B.; Zhu, X.; Jiang, T. Emds-5: Environmental microorganism image dataset fifth version for multiple image analysis tasks. *PLoS ONE* **2021**, *16*, e0250631.
5. Zhang, J.; Li, C.; Kosov, S.; Grzegorzek, M.; Shirahama, K.; Jiang, T.; Sun, C.; Li, Z.; Li, H. Lcu-net: A novel low-cost u-net for environmental microorganism image segmentation. *Pattern Recognit.* **2021**, *115*, 107885. [CrossRef]
6. Kulwa, F.; Li, X.; Zhao, C.; Cai, B.; Xu, N.; Qi, S.; Chen, S.; Teng, Y. A state-of-the-art survey for microorganism image segmentation methods and future potential. *IEEE Access* **2019**, *7*, 100243–100269. [CrossRef]
7. Khaing, M.P.; Masayuki, M. Transparent object detection using convolutional neural network. In Proceedings of the International Conference on Big Data Analysis and Deep Learning Applications, Miyazaki, Japan, 14–15 May 2018; Springer: Berlin/Heidelberg, Germany, 2018; pp. 86–93.
8. Kosov, S.; Shirahama, K.; Li, C.; Grzegorzek, M. Environmental microorganism classification using conditional random fields and deep convolutional neural networks. *Pattern Recognit.* **2018**, *77*, 248–261. [CrossRef]
9. Yoshua, B.; Yann, L.; Geoffrey, H. Deep learning. *Nature* **2015**, *521*, 436–444.
10. Zhang, J.; Yang, K.; Constantinescu, A.; Peng, K.; Müller, K.; Stiefelhagen, R. Trans4trans: Efficient transformer for transparent object segmentation to help visually impaired people navigate in the real world. In Proceedings of the IEEE/CVF International Conference on Computer Vision, Montreal, QC, Canada, 10–17 October 2021; pp. 1760–1770.
11. Yan, Z.; Zhan, Y.; Zhang, S.; Metaxas, D.; Zhou, X.S. Multi-instance multi-stage deep learning for medical image recognition. In *Deep Learning for Medical Image Analysis*; Elsevier: Amsterdam, The Netherlands, 2017; pp. 83–104.
12. Ai, S.; Li, C.; Li, X.; Jiang, T.; Grzegorzek, M.; Sun, C.; Rahaman, M.M.; Zhang, J.; Yao, Y.; Li, H. A state-of-the-art review for gastric histopathology image analysis approaches and future development. *BioMed Res. Int.* **2021**, *2021*, 6671417. [CrossRef]
13. Chen, H.; Li, C.; Li, X.; Rahaman, M.M.; Hu, W.; Li, Y.; Liu, W.; Sun, C.; Sun, H.; Huang, X.; et al. Il-mcam: An interactive learning and multi-channel attention mechanism-based weakly supervised colorectal histopathology image classification approach. *Comput. Biol. Med.* **2022**, *143*, 105265. [CrossRef]
14. Dong, S.; Wang, P.; Abbas, K. A survey on deep learning and its applications. *Comput. Sci. Rev.* **2021**, *40*, 100379. [CrossRef]

15. Raghu, M.; Unterthiner, T.; Kornblith, S.; Zhang, C.; Dosovitskiy, A. Do vision transformers see like convolutional neural networks? *Adv. Neural Inf. Process. Syst.* **2021**, *34*, 12116–12128.
16. Zeng, A.; Yu, K.T.; Song, S.; Suo, D.; Walker, E.; Rodriguez, A.; Xiao, J. Multi-view self-supervised deep learning for 6d pose estimation in the amazon picking challenge. In Proceedings of the 2017 IEEE International Conference on Robotics and Automation (ICRA), Singapore, 29 May–3 June 2017; IEEE: Piscataway, NJ, USA, 2017; pp. 1383–1386.
17. Sajjan, S.; Moore, M.; Pan, M.; Nagaraja, G.; Lee, J.; Zeng, A.; Song, S. Clear grasp: 3d shape estimation of transparent objects for manipulation. In Proceedings of the 2020 IEEE International Conference on Robotics and Automation (ICRA), Paris, France, 31 May–31 August 2020; IEEE: Piscataway, NJ, USA, 2020; pp. 3634–3642.
18. Senturk, S.F.; Gulmez, H.K.; Gul, M.F.; Kirci, P. Detection and separation of transparent objects from recyclable materials with sensors. In Proceedings of the International Conference on Advanced Network Technologies and Intelligent Computing, Varanasi, India, 17–18 December 2021; Springer: Berlin/Heidelberg, Germany, 2021; pp. 73–81.
19. Hata, S.; Saitoh, Y.; Kumamura, S.; Kaida, K. Shape extraction of transparent object using genetic algorithm. In Proceedings of 13th International Conference on Pattern Recognition, Vienna, Austria, 25–29 August 1996; IEEE: Piscataway, NJ, USA, 1996; Volume 4, pp. 684–688.
20. Xu, Y.; Nagahara, H.; Shimada, A.; Taniguchi, R.I. Transcut: Transparent object segmentation from a light-field image. In Proceedings of the IEEE International Conference on Computer Vision, Santiago, Chile, 7–13 December 2015; pp. 3442–3450.
21. Guo, Y.; Xiong, Z.; Verbeek, F.J. An efficient and robust hybrid method for segmentation of zebrafish objects from bright-field microscope images. *Mach. Vis. Appl.* **2018**, *29*, 1211–1225. [CrossRef] [PubMed]
22. Nasirahmadi, A.; Ashtiani, S.-H.M. Bag-of-feature model for sweet and bitter almond classification. *Biosyst. Eng.* **2017**, *156*, 51–60. [CrossRef]
23. Xu, Y.; Maeno, K.; Nagahara, H.; Shimada, A.; Taniguchi, R.I. Light field distortion feature for transparent object classification. *Comput. Vis. Image Underst.* **2015**, *139*, 122–135. [CrossRef]
24. Simonyan, K.; Zisserman, A. Very deep convolutional networks for large-scale image recognition. *arXiv* **2014**, arXiv:1409.1556.
25. He, K.; Zhang, X.; Ren, S.; Sun, J. Deep residual learning for image recognition. In Proceedings of the IEEE Conference on Computer Vision and Pattern Recognition, Las Vegas, NV, USA, 27–30 June 2016; pp. 770–778.
26. Szegedy, C.; Liu, W.; Jia, Y.; Sermanet, P.; Reed, S.; Anguelov, D.; Erhan, D.; Vanhoucke, V.; Rabinovich, A. Going deeper with convolutions. In Proceedings of the IEEE Conference on Computer Vision and Pattern Recognition, Boston, MA, USA, 7–12 June 2015; pp. 1–9.
27. Ioffe, S.; Szegedy, C. Batch normalization: Accelerating deep network training by reducing internal covariate shift. In Proceedings of the International Conference on Machine Learning, Lille, France, 6–11 July 2015; PMLR: New York City, NY, USA, 2015; pp. 448–456.
28. Szegedy, C.; Vanhoucke, V.; Ioffe, S.; Shlens, J.; Wojna, Z. Rethinking the inception architecture for computer vision. In Proceedings of the IEEE Conference on Computer Vision and Pattern Recognition, Las Vegas, NV, USA, 27–30 June 2016; pp. 2818–2826.
29. Chollet, F. Xception: Deep learning with depthwise separable convolutions. In Proceedings of the IEEE Conference on Computer Vision and Pattern Recognition, Honolulu, HI, USA, 21–26 July 2017; pp. 1251–1258.
30. Ronneberger, O.; Fischer, P.; Brox, T. U-net: Convolutional networks for biomedical image segmentation. In Proceedings of the International Conference on Medical Image Computing and Computer-Assisted Intervention, Munich, Germany, 5–9 October 2015; Springer: Berlin/Heidelberg, Germany, 2015; pp. 234–241.
31. Dosovitskiy, A.; Beyer, L.; Kolesnikov, A.; Weissenborn, D.; Zhai, X.; Unterthiner, T.; Dehghani, M.; Minderer, M.; Heigold, G.; Gelly, S.; et al. An image is worth 16x16 words: Transformers for image recognition at scale. *arXiv* **2020**, arXiv:2010.11929.
32. Chen, A.; Li, C.; Zou, S.; Rahaman, M.M.; Yao, Y.; Chen, H.; Yang, H.; Zhao, P.; Hu, W.; Liu, W.; et al. Svia dataset: A new dataset of microscopic videos and images for computer-aided sperm analysis. *Biocybern. Biomed. Eng.* **2022**, *42*, 204–214. [CrossRef]
33. Li, C.; Chen, H.; Li, X.; Xu, N.; Hu, Z.; Xue, D.; Qi, S.; Ma, H.; Zhang, L.; Sun, H. A review for cervical histopathology image analysis using machine vision approaches. *Artif. Intell. Rev.* **2020**, *53*, 4821–4862. [CrossRef]
34. Rahaman, M.M.; Li, C.; Wu, X.; Yao, Y.; Hu, Z.; Jiang, T.; Li, X.; Qi, S. A survey for cervical cytopathology image analysis using deep learning. *IEEE Access* **2020**, *8*, 61687–61710. [CrossRef]
35. Rahaman, M.M.; Li, C.; Yao, Y.; Kulwa, F.; Wu, X.; Li, X.; Wang, Q. Deepcervix: A deep learning-based framework for the classification of cervical cells using hybrid deep feature fusion techniques. *Comput. Biol. Med.* **2021**, *136*, 104649. [CrossRef] [PubMed]
36. Liu, W.; Li, C.; Rahaman, M.M.; Jiang, T.; Sun, H.; Wu, X.; Hu, W.; Chen, H.; Sun, C.; Yao, Y.; et al. Is the aspect ratio of cells important in deep learning? a robust comparison of deep learning methods for multi-scale cytopathology cell image classification: From convolutional neural networks to visual transformers. *Comput. Biol. Med.* **2021**, *141*, 105026. [CrossRef] [PubMed]
37. Rahaman, M.M.; Li, C.; Yao, Y.; Kulwa, F.; Rahman, M.A.; Wang, Q.; Qi, S.; Kong, F.; Zhu, X.; Zhao, X. Identification of COVID-19 samples from chest x-ray images using deep learning: A comparison of transfer learning approaches. *J. X-ray Sci. Technol.* **2020**, *28*, 821–839. [CrossRef] [PubMed]
38. Taha, A.A.; Hanbury, A. Metrics for evaluating 3d medical image segmentation: Analysis, selection, and tool. *BMC Med. Imaging* **2015**, *15*, 1–28. [CrossRef]

39. Dimitri, G.M.; Agrawal, S.; Young, A.; Donnelly, J.; Liu, X.; Smielewski, P.; Hutchinson, P.; Czosnyka, M.; Lió, P.; Haubrich, C. A multiplex network approach for the analysis of intracranial pressure and heart rate data in traumatic brain injured patients. *Appl. Netw. Sci.* **2017**, *2*, 1–12. [CrossRef]
40. Cicaloni, V.; Spiga, O.; Dimitri, G.M.; Maiocchi, R.; Millucci, L.; Giustarini, D.; Bernardini, G.; Bernini, A.; Marzocchi, B.; Braconi, D.; et al. Interactive alkaptonuria database: Investigating clinical data to improve patient care in a rare disease. *FASEB J.* **2019**, *33*, 12696–12703. [CrossRef]
41. Kwekha-Rashid, A.S.; Abduljabbar, H.N.; Alhayani, B. Coronavirus disease (COVID-19) cases analysis using machine-learning applications. *Appl. Nanosci.* **2021**, 1–13. [CrossRef] [PubMed]
42. Zhao, P.; Li, C.; Rahaman, M.M.; Xu, H.; Yang, H.; Sun, H.; Jiang, T.; Grzegorzek, M. A comparative study of deep learning classification methods on a small environmental microorganism image dataset (emds-6): From convolutional neural networks to visual transformers. *Front. Microbiol.* **2022**, *13*, 792166. [CrossRef]
43. Li, C. *Content-Based Microscopic Image Analysis*; Logos Verlag Berlin GmbH: Berlin, Germany, 2016; Volume 39.
44. Wang, Y.; Ma, X.; Chen, Z.; Luo, Y.; Yi, J.; Bailey, J. Symmetric cross entropy for robust learning with noisy labels. In Proceedings of the IEEE/CVF International Conference on Computer Vision, Seoul, Korea, 27–28 October 2019; pp. 322–330.
45. Wang, Y.; Yao, Q.; Kwok, J.T.; Ni, L.M. Generalizing from a few examples: A survey on few-shot learning. *ACM Comput. Surv.* **2020**, *53*, 1–34. [CrossRef]
46. Deng, J.; Dong, W.; Socher, R.; Li, L.J.; Li, K.; Fei-Fei, L. Imagenet: A large-scale hierarchical image database. In Proceedings of the 2009 IEEE Conference on Computer Vision and Pattern Recognition, Miami, FL, USA, 20–25 June 2009; IEEE: Piscataway, NJ, USA, 2009; pp. 248–255.
47. Zhu, H.; Jiang, H.; Li, S.; Li, H.; Pei, Y. A novel multispace image reconstruction method for pathological image classification based on structural information. *BioMed Res. Int.* **2019**, *2019*, 3530903. [CrossRef]
48. Shin, H.C.; Roth, H.R.; Gao, M.; Lu, L.; Xu, Z.; Nogues, I.; Yao, J.; Mollura, D.; Summers, R.M. Deep convolutional neural networks for computer-aided detection: Cnn architectures, dataset characteristics and transfer learning. *IEEE Trans. Med. Imaging* **2016**, *35*, 1285–1298. [CrossRef]
49. Zhao, P.; Li, C.; Rahaman, M.M.; Xu, H.; Ma, P.; Yang, H.; Sun, H.; Jiang, T.; Xu, N.; Grzegorzek, M. Emds-6: Environmental microorganism image dataset sixth version for image denoising, segmentation, feature extraction, classification, and detection method evaluation. *Front. Microbiol.* **2022**, 1334. [CrossRef] [PubMed]
50. Wong, T.-T.; Yeh, P.-Y. Reliable accuracy estimates from k-fold cross validation. *IEEE Trans. Knowl. Data Eng.* **2019**, *32*, 1586–1594. [CrossRef]
51. Zhou, Z.; Siddiquee, M.M.R.; Tajbakhsh, N.; Liang, J. Unet++: Redesigning skip connections to exploit multiscale features in image segmentation. *IEEE Trans. Med. Imaging* **2019**, *39*, 1856–1867. [CrossRef] [PubMed]
52. Cao, H.; Wang, Y.; Chen, J.; Jiang, D.; Zhang, X.; Tian, Q.; Wang, M. Swin-unet: Unet-like pure transformer for medical image segmentation. *arXiv* **2021**, arXiv:2105.05537.
53. Badrinarayanan, V.; Kendall, A.; Cipolla, R. Segnet: A deep convolutional encoder-decoder architecture for image segmentation. *IEEE Trans. Pattern Anal. Mach. Intell.* **2017**, *39*, 2481–2495. [CrossRef]
54. Chen, J.; Lu, Y.; Yu, Q.; Luo, X.; Adeli, E.; Wang, Y.; Lu, L.; Yuille, A.L.; Zhou, Y. Transunet: Transformers make strong encoders for medical image segmentation. *arXiv* **2021**, arXiv:2102.04306.
55. Zhang, J.; Li, C.; Kulwa, F.; Zhao, X.; Sun, C.; Li, Z.; Jiang, T.; Li, H.; Qi, S. A multiscale cnn-crf framework for environmental microorganism image segmentation. *BioMed Res. Int.* **2020**, *2020*, 4621403. [CrossRef]
56. Kulwa, F.; Li, C.; Zhang, J.; Shirahama, K.; Kosov, S.; Zhao, X.; Jiang, T.; Grzegorzek, M. A new pairwise deep learning feature for environmental microorganism image analysis. *Environ. Sci. Pollut. Res.* **2022**, *29*, 51909–51926. [CrossRef]
57. Kulwa, F.; Li, C.; Grzegorzek, M.; Rahaman, M.M.; Shirahama, K.; Kosov, S. Segmentation of weakly visible environmental microorganism images using pair-wise deep learning features. *arXiv* **2022**, arXiv:2208.14957.

applied sciences

Article

An Application of Pixel Interval Down-Sampling (PID) for Dense Tiny Microorganism Counting on Environmental Microorganism Images

Jiawei Zhang [1], Xin Zhao [2], Tao Jiang [3,4,*], Md Mamunur Rahaman [1], Yudong Yao [5], Yu-Hao Lin [6], Jinghua Zhang [1], Ao Pan [7], Marcin Grzegorzek [8] and Chen Li [1,*]

1 Microscopic Image and Medical Image Analysis Group, College of Medicine and Biological Information Engineering, Northeastern University, Shenyang 110819, China; jiaweime111@163.com (J.Z.); md_mamunur.rahaman@unsw.edu.au (M.M.R.); zjh@nudt.edu.cn (J.Z.)
2 School of Resources and Civil Engineering, Northeastern University, Shenyang 110819, China; zhaoxin@mail.neu.edu.cn
3 School of Control Engineering, Chengdu University of Information Technology, Chengdu 610225, China
4 School of Intelligent Medicine, Chengdu University of TCM, Chengdu 611137, China
5 Stevens Institute of Technology, Hoboken, NJ 07030, USA; yyao@stevens.edu
6 Department of Environmental Engineering, National Chung Hsing University, 145 Xingda Rd., Taichung 402, Taiwan; 11844@yahoo.com.tw
7 School of Life Science and Technology, Huazhong University of Science and Technology, Wuhan 430074, China; u201912638@hust.edu.cn
8 Institute of Medical Informatics, University of Luebeck, 23562 Luebeck, Germany; marcin.grzegorzek@uni-luebeck.de
* Correspondence: jiang@cuit.edu.cn (T.J.); lichen201096@hotmail.com (C.L.)

Citation: Zhang, J.; Zhao, X.; Jiang, T.; Rahaman, M.M.; Yao, Y.; Lin, Y.-H.; Zhang, J.; Pan, A.; Grzegorzek, M.; Li, C. An Application of Pixel Interval Down-Sampling (PID) for Dense Tiny Microorganism Counting on Environmental Microorganism Images. *Appl. Sci.* **2022**, *12*, 7314. https://doi.org/10.3390/app12147314

Academic Editor: Bart Van der Bruggen

Received: 12 June 2022
Accepted: 16 July 2022
Published: 21 July 2022

Publisher's Note: MDPI stays neutral with regard to jurisdictional claims in published maps and institutional affiliations.

Abstract: This paper proposes a novel pixel interval down-sampling network (PID-Net) for dense tiny object (yeast cells) counting tasks with higher accuracy. The PID-Net is an end-to-end convolutional neural network (CNN) model with an encoder–decoder architecture. The pixel interval down-sampling operations are concatenated with max-pooling operations to combine the sparse and dense features. This addresses the limitation of contour conglutination of dense objects while counting. The evaluation was conducted using classical segmentation metrics (the Dice, Jaccard and Hausdorff distance) as well as counting metrics. The experimental results show that the proposed PID-Net had the best performance and potential for dense tiny object counting tasks, which achieved 96.97% counting accuracy on the dataset with 2448 yeast cell images. By comparing with the state-of-the-art approaches, such as Attention U-Net, Swin U-Net and Trans U-Net, the proposed PID-Net can segment dense tiny objects with clearer boundaries and fewer incorrect debris, which shows the great potential of PID-Net in the task of accurate counting.

Keywords: yeast counting; image segmentation; pixel interval down-sampling; tiny objects

1. Introduction

With the development of industrialization, environmental pollution has become a vital problem to be resolved urgently. Compared with the classical physical and chemical approaches, the novel biological methods are more efficient and transparent, causing no secondary pollution, which has become the preference for environmental pollution. The research of *Environmental Microorganisms* (EMs) is helpful to focus on the interrelationship among microorganisms, pollutants and the environment. It is essential to use microorganisms to degrade the increasingly severe and diverse environmental pollutants effectively.

Yeast is a kind of single-celled eukaryotic microorganism that is highly adaptable to the environment. It is widely applied to produce alcohol, glycerol and organic acids, which are closely linked to the life and production activity of humanity. Until now, yeast have also been used in the treatment of toxic industrial wastewater and solid waste, which plays an important role in treating environmental pollution [1,2].

In the research of yeast applied in industrial production and environmental pollution control, biomass is the basic evaluation method and can quantitatively consider the performance of yeast in various tasks [3]. At present, there are mainly two types of counting methods. The first is manual counting methods, such as plate counting and hemocytometry; another is semi-automatic counting methods, such as flow cytometry [4,5].

Manual counting is straightforward and stable to use with high accuracy when the number of cells is limited. However, when the number of cells becomes larger, it will be time-consuming, and the accuracy will be lowered due to the subjective influence of the operator. Semi-automatic counting is more accurate and can obtain ideal results in the case of large biomass; however, it is not portable and requires expensive equipment [6]. Therefore, these classical methods have non-negligible limitations in practice.

Due to the rapid developments of computer vision and deep learning technologies, computer-assisted image analysis is broadly applied in many research fields, including histopathological image analysis [7–10], cytopathological image analysis [11–13], object detection [14–16], microorganism classification [17–22], microorganism segmentation [23–26] and microorganism counting [27,28].

However, by reviewing the works of microorganism counting from the 1980s until now [28], we find that, in the process of image segmentation, all existing segmentation approaches use traditional technologies, such as thresholding [29], edge detection [30] and watershed [31]. Most of the deep-learning approaches are only applied for microorganism classification but not for microorganism segmentation in the task of microorganism counting [32]. Here, we propose a novel *Pixel Interval Down-sampling Network* (PID-Net) for the yeast counting task with higher accuracy.

The PID-Net is an improved Convolutional Neural Network (CNN) based on an encoder–decoder architecture, pixel interval down-sampling and concatenate operations. By comparing with the traditional SegNet [33] and U-Net [34]-based object counting algorithms, the accuracy of counting is improved. The workflow of the proposed PID-Net counting method is shown in Figure 1.

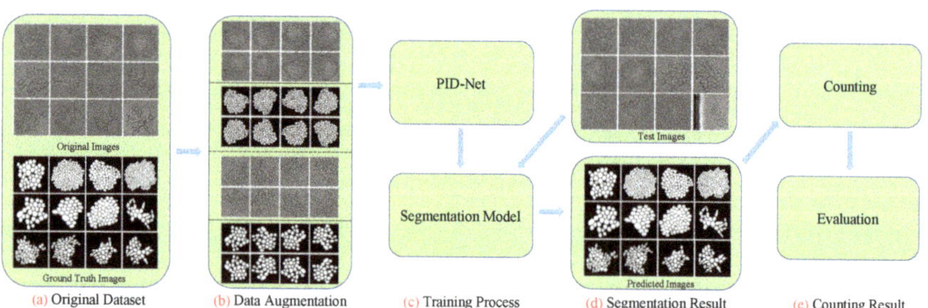

Figure 1. The workflow diagram of the proposed yeast image counting method using PID-Net.

In Figure 1, (a) Original Dataset: The dataset contains images of yeast cells and their ground truth (GT). The range is from 1 to 256 yeast cells in each image. (b) Data Augmentation: Mirror and rotation operations are applied to augment the original dataset. (c) Training Process: PID-Net is trained for image segmentation and the best model is generated. (d) Segmentation Result: Test images are processed using the trained PID-Net model and output the predicted segmentation results. (e) Counting Result: The number of yeast cells is counted by using connected domain detection.

The main contributions of this paper are as follows:

- We propose PID-Net for dense tiny object counting. MaxPooling and pixel interval down-sampling are concatenated as down-sampling to extract spatial local and global features.

- The operation of max-pooling may lose some local features of tiny objects while segmentation, and the edge lines may not be connected after max-pooling. However, the PID-Net can cover a more detailed region.
- The proposed PID-Net achieves better counting performance than other models on the EM (yeast) counting task.

The paper is organized as follows: Section 2 is the related work of existing image analysis-based microorganism counting methods. Section 3 describes the architecture of the proposed PID-Net in detail. Section 4 consists of the experimental setting, evaluation metrics and results. Section 5 is the conclusion of this paper.

2. Related Work

In this section, related approaches to image analysis-based microorganism counting methods are summarized in Table 1, which consist of classical counting methods and machine-learning-based counting methods. More detailed research can be referred to in our survey paper [28].

Table 1. Microorganism image counting methods.

Category	Subcategory	Related Work
Classical Methods	Thresholding-Based Methods	[35–37]
	Edge Detection-based Methods	[38–40]
	Watershed-Based Methods	[41–44]
Machine-Learning-Methods	Hough Transformation	[45–47]
	Classical Machine-Learning-Based Methods	[48–51]
	Deep-Learning-Based Methods	[52–55]

2.1. Classical Counting Methods

Image segmentation is the most significant part in microorganism-counting task. As shown in Table 1, the classical methods contain thresholding, edge detection and watershed methods [56]. For thresholding approaches, the selection of threshold determines the result of segmentation. The most used approaches are iterative thresholding and Otsu thresholding at present. Otsu thresholding can achieve satisfactory segmentation results for most images [57].

Edge detection approaches can extract all boundaries in an image, and then each close region can be separated [58]. Watershed is one kind of region-based segmentation approach, which can be calculated by iterative labeling [59]. The satisfactory segmentation result can be received though in an image with weak edges. Hough transformation was proposed for line or circle detection tasks in image with strong anti-noise capability and high accuracy, which can be applied for circular microorgansim counting tasks [60].

In [35–37], various thresholding methods were applied for microorganism counting. In [35], an adaptive thresholding was used for microorganism segmentation, and after that, the minima function was applied to locate the center of each colony for counting. The work [36] applied Otsu thresholding for bacteria segmentation, and the hypothesis testing was then applied for debris erasing. In [37], the contrast-limited adaptive histogram equalization was applied to enhance the plate contour first. Then, the Otsu threshold was applied to detect the plate region and binarize the images automatically. Finally, the colonies were separated and counted.

The works [38–40] used several edge detection methods for microorganism counting. In [38], five different combination methods applying for microorganism segmentation were compared, such as Gaussian Laplacian and Canny filters. Then, the concave surface between the connected colonies was detected for counting. The works [39,40] used Sobel and Laplacian filters for the edge detection of bacteria images.

The works [41–44] use watershed-based methods for microorganism counting. In [41], watershed was applied for separation of clustered colonies of bacteria images. After that, the circularity ratio was calculated for colony counting. In [42], marker-controlled watershed was applied for bacteria segmentation. Then, the number of colonies was estimated as the ratio of cluster area to an average colony area. In [43], Otsu and adaptive thresholding were applied for image binarization. Then, the combination method of distance transform and watershed was applied for bacteria counting. In [44], watershed was used for image segmentation. After that, the gray level co-occurrence matrix (GLCM) of the image was extracted and classified using SVM.

The works [45–47] used Hough-transformation-based methods for microorganism counting. In [45], the iterative local threshold was used for bacteria segmentation, and then a Hough circle transformation was applied to separate clustered colonies into a single colony. In [46], a median filter was applied for denoising. Then, the circular area was detected using the Hough transform to obtain only the inner area of the dish. Afterward, Gaussian adaptive thresholding was applied for bacteria segmentation. Finally, cross correlation-based granulometry was used to count the bacteria colonies. In [47], Otsu thresholding and a Laplacian filter were applied for edge detection. Then, a circular Hough transform was used to detect circular bacteria colonies.

2.2. Machine-Learning-Based Counting Methods

As shown in Table 1, the machine-learning-based microorganism counting approaches consist of machine-learning- and deep-learning-based methods. The classical machine-learning-based approaches contain Principal Component Analysis (PCA) [61] and Support Vector Machine (SVM) [62]. Deep-learning methods are usually based on CNN [63], Back Propagation Neural Network (BPNN) [64] and Artificial Neural Network (ANN) [65] algorithms.

The works [48–51] used classical machine learning for microorganism counting. In [48], PCA was applied for separation the bacteria with other debris. After that, the nearest neighbor searching algorithm was applied for clustered colony separation. In [49], the shape features were extracted for training, then the SVM was applied for microorganism classification and counting after Otsu thresholding.

In [50], the histogram local equalization was applied to enhance the contours of protozoa images. Then, morphological erosion and reconstruction were used to eliminate the flocs of the protozoa silhouette. Finally, PCA was applied to classify different protozoa, and the number of the various species of protozoa was counted. In [51], local auto-correlational masks were used for image enhancement, and PCA was applied for plankton counting.

The works [52–55] used deep learning for microorganism classification and counting. In [52], the Marr–Hildreth operator and thresholding were applied for edge detection and binarization of bacteria images. Then, ANN was designed for classification and counting. In [53], the contrast-limited adaptive histogram equalization was applied for bacteria image segmentation, then four convolutional and one fully connected was trained for classification and counting. In [54], contrast limited adaptive histogram equalization was used for image enhancement.

Then, CNN was applied for bacteria classification, and the watershed algorithm was applied for colony separation and counting. In [55], a classification-type convolutional neural network (cCNN) was proposed for automatic bacteria classification and counting. First, the original images were segmented with an adaptive binary thresholding method, and images with individual cells or cell clusters were cropped. Then, the images were classified using cCNN. The network can output the number of bacteria in given clusters, and the total count can be calculated.

By reviewing all related works of microorganism counting, we found that deep learning technologies are widely applied for microorganism classification; however, few machine-learning-based image segmentation methods are applied. Since the development of deep learning and computer vision technologies, CNN-based image segmentation approaches have be applied for accurate microorganism segmentation, such as SegNet [33], U-Net [34], Attention U-Net [66], Trans U-Net [67] and Swin U-Net [68]. Though the methods above have not been applied for microorganism-counting tasks, they have enough potential to extract the microorganism before counting, which can be inferred with better performance.

Since the yeast images in our dataset range from 1 to 256 yeast cells in each image, and the boundaries of the cells are not clear, it can be inferred that the classical segmentation methods may show poor performance in this counting task. Therefore, we propose an encoder–decoder model that concentrates on the dense and tiny object counting task.

3. PID-Net-Based Yeast Counting Method

Although the existing image segmentation models, such as SegNet and U-Net, have been widely applied in semantic segmentation and biomedical image segmentation, they still cannot meet the requirements of accurate segmentation in the microorganism-counting task. To this end, we propose PID-Net, a CNN-based on pixel interval down-sampling, MaxPooling and concatenate operations to obtain a better performance. The process of microorganism counting mainly contains two parts, the first is microorganism image segmentation, whose purpose is to classify the foreground and background at the pixel-level. The second part is microorganism counting, whose purpose is to count the number of segmented objects after post-processing.

3.1. Basic Knowledge of SegNet

SegNet is a CNN-based image segmentation network with the structure of an encoder and decoder. The innovation of SegNet is that the dense feature maps of high resolution images can be calculated by the encoder, and the up-sampling operation for low-resolution feature maps can be performed by the decoder network [33]. The structure of SegNet can be considered as an encoder network and a corresponding decoder network. The last part is a pixel-level classification layer.

The first 13 convolutional layers of VGG16 [69] is applied in encoder network of SegNet, which consist of convolutional layers, pooling layers and Batch Normalization layers. In the encoder network, two sequences, which consist of one 3×3 convolution operation, followed by a Batch Normalization and a ReLU operation, are applied in each step. After that, the feature maps are down-sampled by using a max-pooling operation with the size of 2×2 and stride of 2 pixels. After pooling, the size of the feature map is changed into half of the initial.

What is noteworthy is that the Pooling Indices are saved while pooling, which records the initial position of the maximum value in the input feature maps. In the decoder network, the up-sampling operation is applied for feature maps, and then the convolution operation is performed three times to fix the detail loss while pooling. The same operation is replicated five times to change the feature maps into the initial image size. The saved Pooling Indices are applied while up-sampling to set the feature points into correct positions. A Softmax layer is applied finally for feature map classification.

3.2. Basic Knowledge of U-Net

U-Net is an U-shape CNN model based on an encoder–decoder and skip connection. U-Net is first designed for the segmentation of biomedical images. The max-pooling with the size of 2×2 and stride of 2 pixels is applied for down-sampling. There are two 3×3 convolution operations (each followed by a ReLU) between two down-sampling operations. The down-sampling operation repeats four times, and the number of feature map channels is modified to 1024.

In the decoder network, the result after up-convolution operation (a 2×2 up-sampling and a 2×2 convolution operation) is concatenated with the corresponding feature maps of encoder, which can combine the high-level semantics with the low-level fine-grained information of the image. After that, two 3×3 convolution operations (each followed by a ReLU) are applied. The size of each feature map is changed into the size of input after four up-sampling operations. Finally, a Sigmoid function is applied for classification.

3.3. The Structure of PID-Net

Following the basic idea of SegNet and U-Net for image segmentation, the structure of the proposed PID-Net is shown in Figure 2, which is an end-to-end CNN structure based on the encoder and decoder. There are four blocks in the encoder network.

The first parts in each block are two convolution operations with a kernel size of 3×3 (each followed by a ReLU operation), and then the max-pooling with the size of 2×2 and stride of 2 pixels is applied to reduce the size of feature maps by half, followed by a convolution and ReLU operation. The channel of feature maps is denoted as C. Pixel interval down-sampling is applied for down-sampling, which is shown in Figure 2. Each pixel is sampled with the pixels apart, and the size of each feature map is replaced by half.

The classical down-sampling methods, such as max-pooling (with the kernel size of 2) will drop $\frac{3}{4}$ data of the original image. It can retain the main information but not be fit to the task of tiny object counting (the edge lines may be lost while max-pooling). Though there are several learnable pooling layers that have proposed, such as Fractional pooling [70], Stochastic pooling [71] and learned-norm pooling approaches [72]; however, they still cannot meet the requirement of accurate segmentation for dense tiny yeast cells.

Thus, a new down-sampling method, pixel interval down-sampling, is proposed here, which can reduce the size of feature maps without dropping data. Afterward, four pixel interval down-sampling feature maps and the features after max-pooling are concatenated to $5C$-dimensional features. Finally, a convolutional filter with C channels is applied to reduce $5C$-dimensional features to C-dimensional features. Hereto, the initial feature maps with size $H \times W$ and channel C are changed to feature maps with size $\frac{H}{2} \times \frac{W}{2}$ and channel C. The procedure is repeated four times with output resolutions of $\frac{H}{16} \times \frac{W}{16}$ and channel of $8C$.

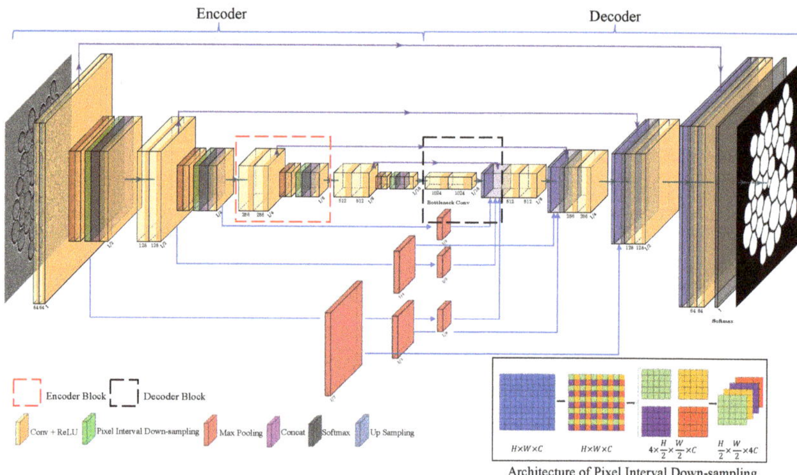

Figure 2. The structure of the proposed PID-Net.

In the decoder network, four blocks are applied for up-sampling. Two convolution operations with a kernel size of 3×3 (each followed by a ReLU operation) are applied first. Then, the transposed convolution operation with a kernel size of 3, a stride of 2 and padding of 1 is applied for up-sampling. The transposed convolution operation is widely applied in GANs to expand the size of images [73]. The count of channels after up-sampling of the bottleneck is 512, which is calculated by using the transposed convolutional filter with 512 channels [74].

The params in the transposed convolution filter can be learned while training. After that, the high resolution feature maps of encoder network are transformed to low-resolution feature maps using 2×, 4× and 8× max-pooling, which is shown in Figure 2. Then, the feature maps after up-sampling and max-pooling are concatenated with the feature maps generated by the corresponding layer from the encoder.

For instance, the 8× max-pooling features of the first block, 4× max-pooling of the second block and 2× max-pooling of the third block in the encoder are concatenated with the copied features of the fourth encoder block and the features after up-sampling (five parts of feature maps are concatenated in the first decoder block). In the same way, there are 4, 3 and 2 parts of features are concatenated in the second, third and fourth level of the decoder, respectively. After the concatenated operation, two convolutions and ReLU operations are applied to change the number of channels. The up-sampling operation is repeated four times with output resolutions of $H \times W$ and channel of C, which has the same size as the encoder's input features. Finally, a Softmax layer with two output channels is applied for feature map classification.

3.4. Counting Approach

A post-processing method is applied to eliminate the effect of noises after segmentation. First, a morphological filter is applied to remove useless debris, which can improve the performance of counting prominently. Then, the eight neighborhood search algorithm is applied to count the connected regions of segmented images after denoising [75]. The process of counting is shown in Figure 3. A binary matrix is traversed in line and a mark matrix is applied to mark the connected domain [76]. Finally, the number of connected domain in mark matrix is the number of yeast cells.

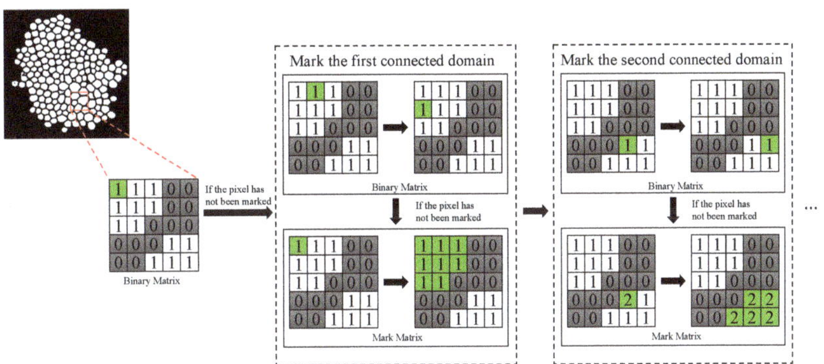

Figure 3. An example of the counting process based on the eight neighborhood search.

4. Experiments

4.1. Experimental Setting

4.1.1. Image Dataset

In our work, we use a yeast image dataset proposed in [77], containing 306 different images of yeast cells and their corresponding ground truth (GT) images. All images are resized to the resolution of 256 × 256 pixels, which are shown in Figure 4. Then, the original 306 images are rotated (0, 90, 180 and 270 degrees) and flipped (mirror), and thus the number of images in this dataset is augmented to eight times (2448 images).

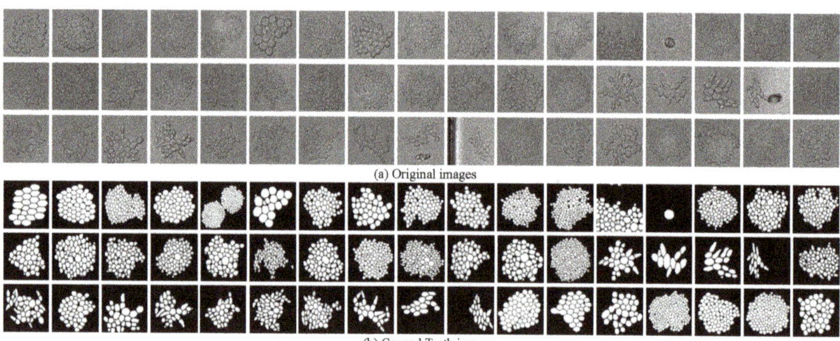

Figure 4. The images in yeast cell dataset. (**a**) The original yeast image and (**b**) the corresponding ground truth images.

4.1.2. Training, Validation and Test Data Setting

The original yeast image dataset was randomly divided into training, validation and test dataset with the ratio of 3:1:1, and then, each dataset was augmented eight times. Therefore, there 1470 images with their GT were applied as the training dataset, 489 images with their corresponding GT were applied for validation, and 489 original images were applied for testing.

4.1.3. Experimental Environment

The experiment was conducted by Python 3.8.10 in Windows 10 operating system. The experimental environment was based on Torch 1.9.0. The workstation was equipped with Intel(R) Core(TM) i7-8700 CPU with 3.20 GHz, 16 GB RAM and NVIDIA GEFORCE RTX 2080 8 GB.

4.1.4. Hyper Parameters

In the experiment of yeast cell counting, the purpose of image segmentation is to determine whether a pixel is a foreground (yeast cell) or background. The last part in the proposed PID-Net before the output is Softmax, which is applied to calculate the classification result of feature maps. The definition of Softmax is shown as Equation (1).

$$\text{Softmax}(z_i) = \frac{e^{z_i}}{\sum_{c=1}^{C} e^{z_c}}. \tag{1}$$

In Equation (1), z_i is the output in the ith node, and C is the number of output nodes, representing the number of classified categories. The classification prediction can be converted into the probabilities by using the Softmax function, which distributes in the range of [0, 1], and the sum of probability is 1. As the image segmentation for yeast counting is to distinguish the foreground and the background. Hence, it is a binary classification, Equation (1) can be rewritten as Equation (2).

$$\text{Softmax}(z_1) = \frac{e^{z_1}}{e^{z_1} + e^{z_2}} = \frac{1}{1 + e^{-(z_1 - z_2)}} = \text{Sigmoid}(\beta). \tag{2}$$

In Equation (2), β is $(z_1 - z_2)$, which means the Softmax function and the Sigmoid function are the same for binary classification (a little difference between them is, the number of the fully connected (FC) layer of Softmax is two to distinguish two different categories; however, the number of FC layer of Sigmoid is one, only to judge whether the single pixel is the object to be segmented).

The probability of the pixel to be classified as 1 is:

$$\hat{y} = P(y = 1|x). \tag{3}$$

Apparently, the probability of the pixel to be classified as 0 is:

$$1 - \hat{y} = P(y = 0|x). \tag{4}$$

According to the maximum likelihood formula, the joint probability can be expressed as:

$$P(y|x) = \hat{y}^y \cdot (1 - \hat{y})^{1-y}. \tag{5}$$

After that, *log* function is applied to remain the monotonicity invariance of the function:

$$logP(y|x) = log(\hat{y}^y \cdot (1 - \hat{y})^{1-y}) = ylog\hat{y} + (1 - y)log(1 - \hat{y}). \tag{6}$$

Hereto, the loss can be expressed as $-logP(y|x)$, and the loss function for multiple samples can be defined as the cross-entropy loss (N is the number of categories):

$$Loss = -[\sum_{i=1}^{N} y^{(i)}log\hat{y}^{(i)} + (1 - y^{(i)})log(1 - \hat{y}^{(i)})]. \tag{7}$$

In order to guarantee the stable and fast convergence of the proposed network, we deploy preliminary experiments to determine the choices of hyper parameters. Adaptive moment estimation (Adam) is compared with stochastic gradient descent (SGD) and natural gradient descent (NGD). Adam optimizer has the smoothest loss curves and stablest convergence, which performs best in microorganism-counting task. Adam optimizer is applied to minimize the loss function, which can adjust the learning rate automatically by considering the gradient momentum of the previous time steps [78].

The initial learning rate is set from 0.0001 to 0.01 in preliminary experiments. By observing the loss curves while training, the learning rate of 0.001 can balance the speed and stability of convergence. The batch size is set as 8 due to the limited memory size (8 GB). The selection of the hyper parameters above are optimal in preliminary experiments, and thus they are applied in our formal microorganism counting experiment. The epoch is set as 100 by considering the converge speed of experimental models, the example of loss and intersection over union (IoU) curves of models is shown in Figure 5.

Though there are 92,319,298 params to be trained in PID-Net; however, it can converge rapidly and smoothly without over fitting. There is a jump in loss and IoU plots for all three tested networks from 20 to 80 epochs, which is caused by the small batch size. Small batch size may lead to huge difference between each batch, and the loss and IoU curves may jump with convergence.

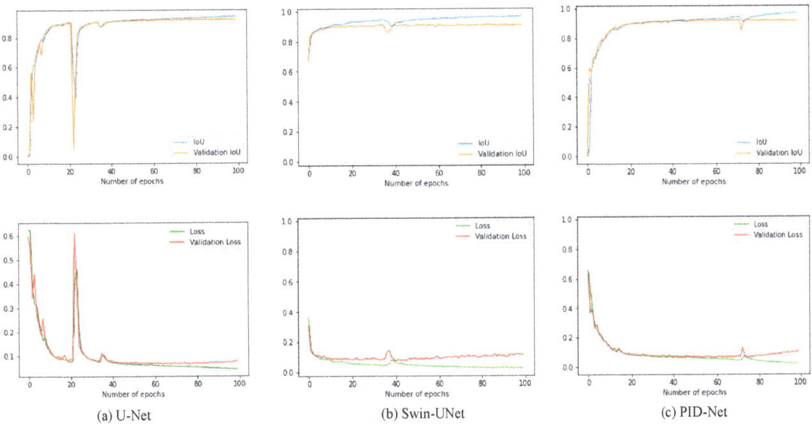

Figure 5. The IoU (**top row**) and loss (**bottom row**) curves in the training process.

4.2. Evaluation Metrics

In the task of dense tiny object counting, the evaluation of image segmentation is the most significant part. Hence, the widely applied segmentation evaluation metrics Accuracy, Dice, Jaccard and Precision are employed here to evaluate the performance of microorganism segmentation. Furthermore, the Hausdorff distance is applied to evaluate the shape similarity between the predicted image and GT. Finally, the counting accuracy is calculated to quantify the counting performance of the models.

Accuracy is applied to calculate the proportion of pixels that are correctly classified. The Dice coefficient [79] is applied to measure the similarity of the predicted image and GT. The similarity can be quantified range from 0 to 1 (1 means the predicted result coincides exactly with the GT). Jaccard [80], also named the intersection over union (IoU), is applied to compare the similarity and differences between the predicted image and GT image, focusing on whether the samples' common characteristics are consistent. Precision is defined as the proportion of positive pixels in the pixels, which are classified as positive.

The Hausdorff distance [81] is applied to measure the Euclidean distance between the predicted and GT images with the unit of pixels in per image. In contrast with Dice, the Hausdorff distance focuses on the boundary distance of two objects to measure the shape similarity; however, the Dice majors in the inner similarity. An example of the Hausdorff distance between GT and predicted image is shown in Figure 6. The Hausdorff is the maximum of the shortest distance between a pixel in a image and another image [82].

In the task of microorganism counting, the Hausdorff distance can be applied to measure the shape similarity between the GT and segmentation result, showing the performance of segmentation models. Finally, the performance of counting is measured using counting accuracy, which is defined as the proportion of the predicted number and GT number of yeast cell images.

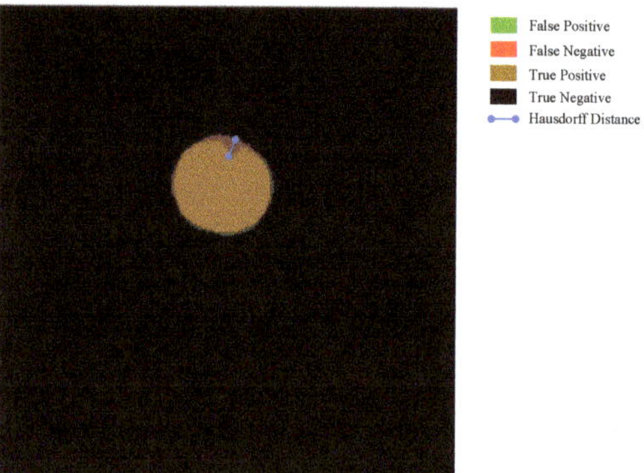

Figure 6. The visualization result of the Hausdorff distance between the GT and predicted image.

The definitions of the proposed evaluation metrics are summarized in Table 2. The TP (True Positive), TN (True Negative), FP (False Positive) and FN (False Negative) are basic evaluation metrics, which can be applied to measure the performance of segmentation in general. An example of a yeast cell image with its TP, TN, FP and FN is illustrated in Figure 7 for intuitive understanding. V_{pred} is the foreground after segmentation by using the model, V_{GT} is the foreground of the GT image. Furthermore, N_{pred} means the number of connected regions in the predicted image, N_{GT} means the number of connected regions in the GT image, which indicates the number of yeast cells. In the definition of the Hausdorff distance, sup is the supremum, and inf is the infimum.

Table 2. The definitions of evaluation metrics. CA and HD are abbreviations of the Counting Accuracy and Hausdorff Distance, respectively.

Metric	Definition	Metric	Definition
Accuracy	$\frac{TP+TN}{TP+TN+FP+FN}$	Dice	$\frac{2\times\lvert V_{pred}\cap V_{GT}\rvert}{\lvert V_{pred}\rvert+\lvert V_{GT}\rvert}$
Jaccard	$\frac{\lvert V_{pred}\cap V_{GT}\rvert}{\lvert V_{pred}\cup V_{GT}\rvert}$	Precision	$\frac{TP}{TP+FP}$
CA	$1-\frac{\lvert N_{pred}-N_{GT}\rvert}{N_{GT}}$	HD	$d_H(X,Y)=max(sup_{x\in X}inf_{y\in Y}d(x,y),sup_{y\in Y}inf_{x\in X}d(x,y))$

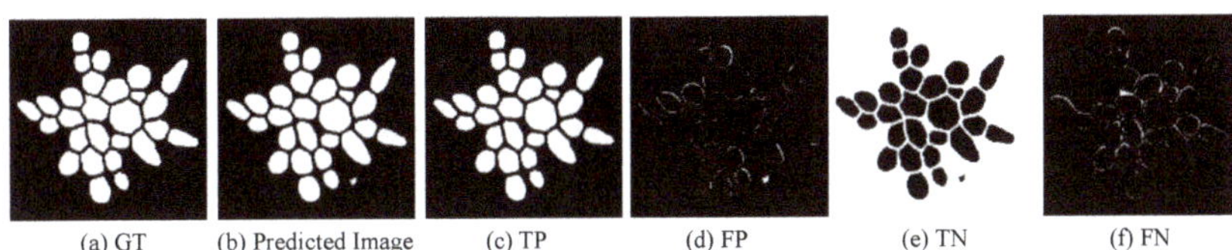

Figure 7. The illustration of TP, TN, FP and FN between the predicted image and GT image.

The proposed evaluation metrics, containing the Accuracy, Dice, Jaccard and Precision, are proportional to the segmentation performance of models. The Hausdorff distance has an inverse correlation with the segmentation performance. Counting accuracy can evaluate the final counting results of different models.

4.3. *Evaluation of Segmentation and Counting Performance*

To prove the satisfactory segmentation performance of the proposed PID-Net for dense tiny object counting, we compare different down-sampling methods to show the advancement of our proposed method. Furthermore, several state-of-the-art approaches are applied for comparative experiments. All of the experimental setting and evaluation indices are same for comparative experiment. Furthermore, the same dataset is applied for all comparative experiments, which is proposed in Section 4.1.1. The models are trained from scratch without pre-training and fine-tuning.

4.3.1. Comparison of Different Down-Sampling Methods

In this part, we compare the effect of different down-sampling and skip connection approaches for segmentation. In our proposed PID-Net, pixel interval down-sampling and max-pooling operations are concatenated to combine the dense and sparse feature maps after convolution operations. Then, in the process of hierarchy skip connection, max-pooling is applied to combine the high-level features and low-level features directly, which is beneficial to reduce the effect of resolution loss while up-sampling and help rebuild the segmentation result.

To show the effectiveness and reasonability of the proposed method, we change the approaches of down-sampling and hierarchy skip connection as PID-Net Modified-1 (PID-Net-M1) and PID-Net Modified-2 (PID-Net-M2). In PID-Net-M1, max-pooling operations are only applied in the process of hierarchy skip connection and not in down-sampling. The down-sampling block of PID-Net-M1 is illustrated in Figure 8. In PID-Net-M2, all down-sampling operations are realized using pixel interval down-sampling without max-pooling. The segmentation evaluations and counting performance of those approaches are shown in Table 3.

From Table 3, we find that the proposed PID-Net achieves the best counting performance. By comparing with the PID-Net-M1 and PID-Net-M2, the average accuracy is increased by 0.1% to 0.6%; the improvement of average Dice value is 0.1% to 1.1%; the average Jaccard is improved by around 0.3% to 1.6%. Furthermore, the mean Hausdorff distance of PID-Net is the shortest, which indicates the similarity between the predicted images and GT images is the highest. Finally, the counting accuracy achieved 96.97%, which shows the satisfactory counting performance of the PID-Net. Hence, the segmentation and counting performance of PID-Net is the best by referring to all evaluation metrics.

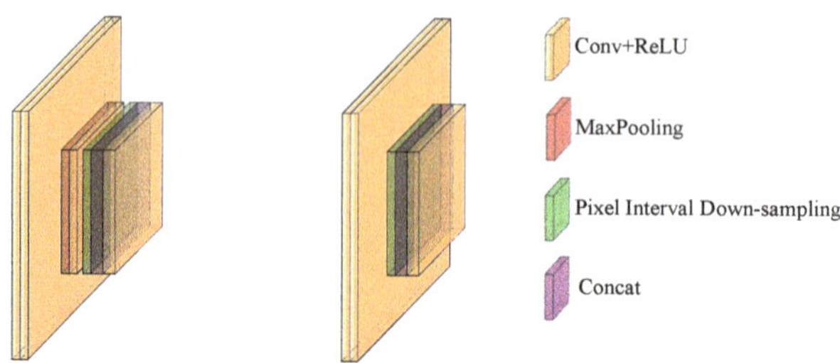

Figure 8. The down-sampling block of PID-Net and PID-Net-M1.

Table 3. The average segmentation evaluation indices of predicted images. A, D, J, P, C and H are abbreviations of the Accuracy, Dice, Jaccard, Precision, Counting Accuracy (in %) and Hausdorff Distance (in pixels/per image), respectively.

Methods	A	D	J	P	C	H
PID-Net	97.51	95.86	92.10	96.02	96.97	4.6272
PID-Net-M1	96.90	94.71	90.05	94.71	67.84	5.0110
PID-Net-M2	97.42	95.75	91.89	95.68	96.88	4.7204

4.3.2. Comparison with Other Methods

In this part, some comparative experiments are applied for the yeast cell counting task. Some classical methods proposed in Section 2 and deep-learning-based methods proposed in Section 3 are compared, consisting Hough transformation [83], Otsu thresholding, Watershed, SegNet and U-Net-based segmentation approaches. Furthermore, we conduct some extra experiments using state-of-the-art approaches, containing Attention U-Net [66], Trans U-Net [67] and Swin U-Net [68].

Due to the determination of k in clustering methods, such as k-means, is still an insoluble problem while counting; therefore, the clustering-based approaches cannot be applied here for dense tiny object counting. All comparative experiments have the same experimental setting, which can be referred to Section 4.1 for details. After image segmentation and object counting, the average evaluation indices are summarized in Table 4, and the example images of segmentation are shown in Figure 9.

Table 4. The average segmentation evaluation indices of predicted images. A, D, J, P, C and H are abbreviations of the Accuracy, Dice, Jaccard, Precision, Counting Accuracy (in %) and Hausdorff Distance (in pixels/per image), respectively.

Methods	A	D	J	P	C	H
PID-Net	97.51	95.86	92.10	96.02	96.97	4.6272
SegNet	94.69	90.34	84.02	88.50	68.82	6.3604
YeaZ (in [77])	-	94.00	-	-	-	-
U-Net	97.47	95.71	91.84	95.62	91.33	4.6666
Attention U-Net	96.62	93.36	88.96	92.67	83.44	5.1184
Trans U-Net	96.84	93.60	88.99	93.25	91.32	5.0715
Swin U-Net	96.47	92.99	88.32	92.43	91.95	5.3140
Hough	82.12	61.12	44.74	88.26	73.66	9.2486
Otsu	84.23	65.71	49.90	87.66	74.34	8.9165
Watershed	78.67	50.15	34.88	78.61	63.34	9.6873

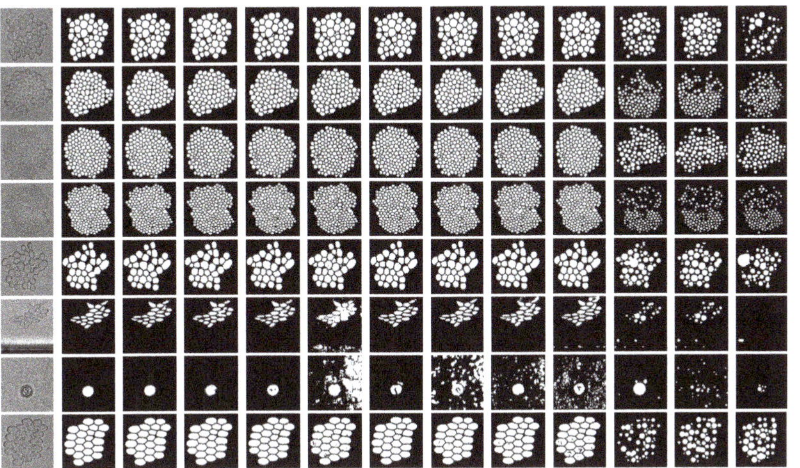

Figure 9. An example of segmentation images predicted by different models.

From the evaluation indices summarized in Table 4, we can find that the PID-Net has the highest Accuracy, Dice, Jaccard, Precision and Counting Accuracy and the lowest Hausdorff distance, which means the proposed model performs best in the task of dense tiny object counting by comparing with other models. Even more, the Jaccard of PID-Net is higher than the YeaZ who proposed this yeast cell dataset. In general, the approaches based on deep learning perform better than the classical approaches.

We find that the Counting Accuracy of several methods are very low abnormally, which may caused by the enormous difference between the GT and the predicted image. For instance, the single yeast cell image in Figure 9 performs unsatisfactory when the segmentation is not accurate. The segmentation results of SegNet and Attention-UNet have a large number of False Positive pixels, and the counting approach is based on the connected domain detection; hence, the value of $\frac{|N_{pred}-N_{GT}|}{N_{GT}}$ is much higher than normal.

From the best performance of the proposed PID-Net in the task of dense tiny object counting, we can infer that the down-sampling and skip connection part of PID-Net, which combines max-pooling and pixel interval down-sampling can obtain the feature maps of dense tiny objects and reconstruct the images better.

4.4. Repeatability Tests

Five additional experiments were repeated based on the original PID-Net model for repeatability tests. The evaluation indices are given in Table 5. From Table 5, we find that all evaluation indices of repeated PID-Nets are approximate, which shows satisfactory and stable counting performance for the dense tiny object counting task.

Table 5. The evaluation indices of Repeatability Tests. A, D, J, P, C and H are abbreviations of the Accuracy, Dice, Jaccard, Precision, Counting Accuracy (in %) and Hausdorff Distance (in pixels/per image), respectively.

Methods	A	D	J	P	C	H
PID-Net	97.51	95.86	92.10	96.02	96.97	4.6272
PID-Net (Re 1)	97.51	95.79	91.97	95.91	95.26	4.5865
PID-Net (Re 2)	97.33	95.59	91.62	95.70	96.25	4.7290
PID-Net (Re 3)	97.54	95.91	92.18	96.21	96.82	4.6023
PID-Net (Re 4)	97.37	95.64	91.70	95.70	95.51	4.7471
PID-Net (Re 5)	97.43	95.66	91.73	92.24	96.26	4.6395

4.5. Computational Time

The training time, mean training time, test time and mean test time are listed in Table 6. There are 1470 images in the training dataset and 489 images in the test dataset. The mean training time of PID-Net model is approximately 2.9 seconds higher than the time of U-Net, and the test time is about 0.4 seconds higher than U-Net. The memory cost of PID-Net is about 20MB, which is about 6 MB more than the cost of U-Net model, meanwhile, the PID-Net has better counting performance and lower memory cost than Swin-UNet (41 MB). The counting accuracy is increased about 6%; hence, the PID-Net has satisfactory counting performance and a tolerable computational time, which can be widely applied in accurate dense tiny object counting tasks.

Table 6. The summary of computational time (in seconds).

Model	Training Time	Mean Training Time	Test Time	Mean Test Time
PID-Net	10,438.86	7.10	454.68	0.93
U-Net	6198.00	4.21	257.64	0.53
Swin-UNet	7884.36	5.36	319.50	0.65
Att-UNet	6983.58	4.75	296.64	0.61

4.6. Discussion

Deep learning is essentially to build a probability distribution model driven by data. Therefore, as the deep-learning-network architecture becomes deeper, the quantity and quality of training data will have a greater impact on the performance of the model. However, in the imaging process of microorganism images, the amount of satisfactory data is relatively small due to some objective reasons, such as the impurities in the acquisition environment, uneven natural light and other adverse factors, which leads to insufficient training and poor performance in various tasks. Though the proposed PID-Net has excellent segmentation and counting performance for images with dense tiny objects, there still exists some mis-segmentation, causing the decrease of counting accuracy. Several incorrect segmentation results are shown in Figure 10.

There are three main problems for segmentation and counting, which are illustrated in Figure 10. The blue circle refers to the situation of under segmentation—that is, the neighbor yeast cells cannot be segmented, and the edges cannot be detected. Due to the counting method is based on the eight neighborhood search algorithm, the situation leads to under estimation of the real count of yeast cells. The green circle refers to a part of background is classified as yeast cells. As shown in Figure 10, most of the images are full of dense tiny yeast cells with irregular shapes, and the limitation of small dataset leads to inadequate training.

Therefore, the background between yeast cells with irregular shape is easily classified as a yeast cell, which results in over estimation of the real count of yeast cells. On the contrary, the red circle represents the part of yeast cell is classified as background. There are 1 to 256 yeast cells with different sizes in a single image in this dataset, and thus the shape and size of yeast cells have a great difference. Therefore, the tiny yeast cell between the larger cells has a great similarity with the background, which is difficult for models to discriminate especially in a small dataset. The situation leads to under estimation of the real count of yeast cells.

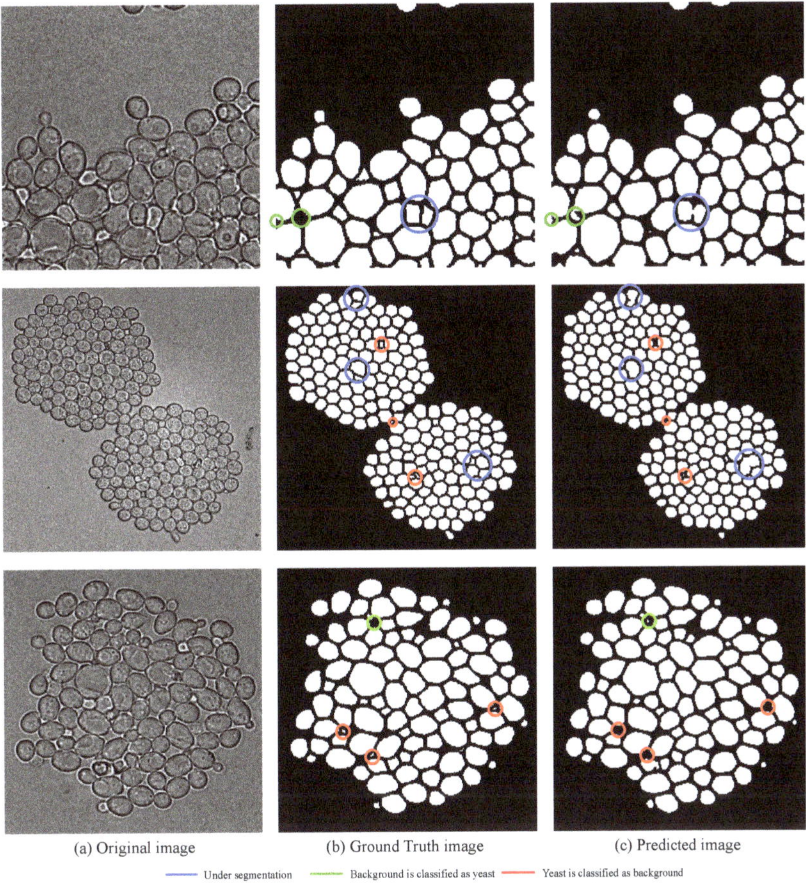

Figure 10. An example of incorrect segmentation results using the proposed PID-Net.

The situations of adherent yeast cells and mis-segmentation lead to counting error. Moreover, the training data is limited due to the small dataset, therefore, the models cannot be trained perfectly. The small dataset is a limitation of the yeast counting task. However, despite some cases of mis-segmentation, most of the yeast cells in the test dataset could be detected and segmented with other cells. The segmented region might be small but has little effect on the counting results calculated using the eight neighborhood search algorithm.

5. Conclusions and Future Work

In this paper, a CNN-based PID-Net was proposed for dense tiny objects (yeast) counting task. The PID-Net is an end-to-end model based on an encoder–decoder structure, and we proposed a new down-sampling model consisting of pixel interval down-sampling and max-pooling, which can serve to extract the dense and sparse features in the task of dense tiny object counting. By comparing with the proposed PID-Net and classical U-Net-based yeast counting results, the evaluation indices of Accuracy, Dice, Jaccard, Precision, Counting Accuracy and Hausdorff Distance of PID-Net were 97.51%, 95.86%, 92.10%, 96.02%, 96.97% and 4.6272, which are improved by 0.04%, 0.15%, 0.26%, 0.4% and 5.7%, respectively, and the Hausdorff Distance decreased by 0.0394.

Although the small image dataset resulted in some cases of mis-segmentation, the proposed PID-Net showed a more satisfactory segmentation performance than the other models in the task of dense tiny object counting on a small dataset.

In the future, we plan to apply PID-Net for more dense tiny object counting tasks, such as the *streptococcus* counting task and blood-cell-counting task. We will further optimize the PID-Net for better counting performance. For instance, object separation is one of the most significant parts in object counting; therefore, the Contour Loss [84] can be used by referring to our work to distinguish inner texture and contour boundaries for more accurate counting. We also consider using Knowledge Distillation [85] to reduce the memory cost of PID-Net, which can help to deploy the model on portable equipment.

Author Contributions: J.Z. (Jiawei Zhang): methodology, software, validation, formal analysis, data curation, writing—original draft preparation, review and editing and visualization; X.Z.: resources; T.J.: investigation and funding acquisition; M.M.R.: validation; Y.Y.: validation; Y.-H.L.: validation; J.Z. (Jinghua Zhang): formal analysis; A.P.: investigation; M.G.: investigation and C.L.: conceptualization, methodology, data curation, writing—original draft preparation, writing—review and editing, supervision and project administration. All authors have read and agreed to the published version of the manuscript.

Funding: This work was funded by the "National Natural Science Foundation of China" (No. 61806047) and the "Sichuan Science and Technology Plan" (No. 2021YFH0069, 2021YFQ0057, 2022YFS0565).

Data Availability Statement: Yeast dataset can be found at: https://www.epfl.ch/labs/lpbs/data-and-software/ (accessed on 16 January 2021).

Acknowledgments: We thank Zixian Li and Guoxian Li for their important discussion.

Conflicts of Interest: There is no conflict of interest in this paper.

Abbreviations

The following abbreviations are used in this manuscript:

PID-Net	pixel interval down-sampling network
CNN	convolutional neural network
EMs	environmental microorganisms
SVM	support vector machine
PCA	principal-component analysis
BPNN	back propagation neural network
ANN	artificial neural network
ReLU	rectified linear unit
Adam	adaptive moment estimation
SGD	stochastic gradient descent
NGD	natural gradient descent
GT	ground truth
FC	fully connected
IoU	intersection over union
TP	true positive
TN	true negative
FP	false positive
FN	false negative

Reference

1. Yu, M.; Wang, J.; Tang, L.; Feng, C.; Liu, H.; Zhang, H.; Peng, B.; Chen, Z.; Xie, Q. Intimate coupling of photocatalysis and biodegradation for wastewater treatment: Mechanisms, recent advances and environmental applications. *Water Res.* **2020**, *175*, 115673. [CrossRef]
2. Wang, Y.; Qiu, L.; Hu, M. Application of yeast in the wastewater treatment. *EDP Sci.* **2018**, *53*, 04025. [CrossRef]
3. You, L.; Zhao, D.; Zhou, R.; Tan, Y.; Wang, T.; Zheng, J. Distribution and Function of Dominant Yeast Species in the Fermentation of Strong-flavor Baijiu. *World J. Microbiol. Biotechnol.* **2021**, *37*, 26. [CrossRef]
4. Balestra, G.; Misaghi, I. Increasing the Efficiency of the Plate Counting Method for Estimating Bacterial Diversity. *J. Microbiol. Methods* **1997**, *30*, 111–117. [CrossRef]

5. Sambrook, J.; Russell, D.W. Estimation of Cell Number by Hemocytometry Counting. *Cold Spring Harb. Protoc.* **2006**, *2006*, pdb–prot4454. [CrossRef]
6. Gasol, J.M.; Del Giorgio, P.A. Using Flow Cytometry for Counting Natural Planktonic Bacteria and Understanding the Structure of Planktonic Bacterial Communities. *Sci. Mar.* **2000**, *64*, 197–224. [CrossRef]
7. Li, X.; Li, C.; Rahaman, M.M.; Sun, H.; Li, X.; Wu, J.; Yao, Y.; Grzegorzek, M. A comprehensive review of computer-aided whole-slide image analysis: from datasets to feature extraction, segmentation, classification and detection approaches. *Artif. Intell. Rev.* **2022**, *55*, 4809–4878. [CrossRef]
8. Li, Y.; Li, C.; Li, X.; Wang, K.; Rahaman, M.M.; Sun, C.; Chen, H.; Wu, X.; Zhang, H.; Wang, Q. A Comprehensive Review of Markov Random Field and Conditional Random Field Approaches in Pathology Image Analysis. *Arch. Comput. Methods Eng.* **2021**, *29*, 609–639. [CrossRef]
9. Zhou, X.; Li, C.; Rahaman, M.M.; Yao, Y.; Ai, S.; Sun, C.; Wang, Q.; Zhang, Y.; Li, M.; Li, X.; et al. A comprehensive review for breast histopathology image analysis using classical and deep neural networks. *IEEE Access* **2020**, *8*, 90931–90956. [CrossRef]
10. Li, C.; Chen, H.; Li, X.; Xu, N.; Hu, Z.; Xue, D.; Qi, S.; Ma, H.; Zhang, L.; Sun, H. A review for cervical histopathology image analysis using machine vision approaches. *Artif. Intell. Rev.* **2020**, *53*, 4821–4862. [CrossRef]
11. Liu, W.; Li, C.; Rahaman, M.M.; Jiang, T.; Sun, H.; Wu, X.; Hu, W.; Chen, H.; Sun, C.; Yao, Y.; et al. Is the aspect ratio of cells important in deep learning? A robust comparison of deep learning methods for multi-scale cytopathology cell image classification: From convolutional neural networks to visual transformers. *Comput. Biol. Med.* **2022**, *141*, 105026. [CrossRef]
12. Rahaman, M.M.; Li, C.; Yao, Y.; Kulwa, F.; Wu, X.; Li, X.; Wang, Q. DeepCervix: A deep learning-based framework for the classification of cervical cells using hybrid deep feature fusion techniques. *Comput. Biol. Med.* **2021**, *136*, 104649. [CrossRef]
13. Rahaman, M.M.; Li, C.; Wu, X.; Yao, Y.; Hu, Z.; Jiang, T.; Li, X.; Qi, S. A survey for cervical cytopathology image analysis using deep learning. *IEEE Access* **2020**, *8*, 61687–61710. [CrossRef]
14. Zou, S.; Li, C.; Sun, H.; Xu, P.; Zhang, J.; Ma, P.; Yao, Y.; Huang, X.; Grzegorzek, M. TOD-CNN: An effective convolutional neural network for tiny object detection in sperm videos. *Comput. Biol. Med.* **2022**, *146*, 105543. [CrossRef]
15. Chen, A.; Li, C.; Zou, S.; Rahaman, M.M.; Yao, Y.; Chen, H.; Yang, H.; Zhao, P.; Hu, W.; Liu, W.; et al. SVIA dataset: A new dataset of microscopic videos and images for computer-aided sperm analysis. *Biocybern. Biomed. Eng.* **2022**, *42*, 204–216. [CrossRef]
16. Ma, P.; Li, C.; Rahaman, M.M.; Yao, Y.; Zhang, J.; Zou, S.; Zhao, X.; Grzegorzek, M. A state-of-the-art survey of object detection techniques in microorganism image analysis: From classical methods to deep learning approaches. *Artif. Intell. Rev.* **2022**, 1–72. [CrossRef]
17. Zhao, P.; Li, C.; Rahaman, M.; Xu, H.; Yang, H.; Sun, H.; Jiang, T.; Grzegorzek, M. A Comparative Study of Deep Learning Classification Methods on a Small Environmental Microorganism Image Dataset (EMDS-6): From Convolutional Neural Networks to Visual Transformers. *Front. Microbiol.* **2022**, *13*, 792166. [CrossRef]
18. Kulwa, F.; Li, C.; Zhang, J.; Shirahama, K.; Kosov, S.; Zhao, X.; Jiang, T.; Grzegorzek, M. A new pairwise deep learning feature for environmental microorganism image analysis. *Environ. Sci. Pollut. Res.* **2022**, *29*, 51909–51926. [CrossRef]
19. Kosov, S.; Shirahama, K.; Li, C.; Grzegorzek, M. Environmental microorganism classification using conditional random fields and deep convolutional neural networks. *Pattern Recognit.* **2018**, *77*, 248–261. [CrossRef]
20. Li, C.; Shirahama, K.; Grzegorzek, M. Environmental microbiology aided by content-based image analysis. *Pattern Anal. Appl.* **2016**, *19*, 531–547. [CrossRef]
21. Li, C.; Shirahama, K.; Grzegorzek, M. Application of content-based image analysis to environmental microorganism classification. *Biocybern. Biomed. Eng.* **2015**, *35*, 10–21. [CrossRef]
22. Rahaman, M.M.; Li, C.; Yao, Y.; Kulwa, F.; Rahman, M.A.; Wang, Q.; Qi, S.; Kong, F.; Zhu, X.; Zhao, X. Identification of COVID-19 samples from chest X-Ray images using deep learning: A comparison of transfer learning approaches. *J. X-ray Sci. Technol.* **2020**, *28*, 821–839. [CrossRef]
23. Zhang, J.; Li, C.; Yin, Y.; Zhang, J.; Grzegorzek, M. Applications of artificial neural networks in microorganism image analysis: A comprehensive review from conventional multilayer perceptron to popular convolutional neural network and potential visual transformer. *Artif. Intell. Rev.* **2022**, 1–58. [CrossRef]
24. Kulwa, F.; Li, C.; Zhao, X.; Cai, B.; Xu, N.; Qi, S.; Chen, S.; Teng, Y. A state-of-the-art survey for microorganism image segmentation methods and future potential. *IEEE Access* **2019**, *7*, 100243–100269. [CrossRef]
25. Zhao, P.; Li, C.; Rahaman, M.M.; Xu, H.; Ma, P.; Yang, H.; Sun, H.; Jiang, T.; Xu, N.; Grzegorzek, M. EMDS-6: Environmental Microorganism Image Dataset Sixth Version for Image Denoising, Segmentation, Feature Extraction, Classification, and Detection Method Evaluation. *Front. Microbiol.* **2022**, *13*, 1334. [CrossRef] [PubMed]
26. Li, C.; Zhang, J.; Kulwa, F.; Qi, S.; Qi, Z. A SARS-CoV-2 Microscopic Image Dataset with Ground Truth Images and Visual Features. In *Chinese Conference on Pattern Recognition and Computer Vision (PRCV)*; Springer: Cham, Switzerland, 2020; pp. 244–255.
27. Zhang, J.; Xu, N.; Li, C.; Rahaman, M.M.; Yao, Y.D.; Lin, Y.H.; Zhang, J.; Jiang, T.; Qin, W.; Grzegorzek, M. An application of Pixel Interval Down-sampling (PID) for dense tiny microorganism counting on environmental microorganism images. *arXiv* **2022**, arXiv:2204.01341.
28. Zhang, J.; Li, C.; Rahaman, M.M.; Yao, Y.; Ma, P.; Zhang, J.; Zhao, X.; Jiang, T.; Grzegorzek, M. A Comprehensive Review of Image Analysis Methods for Microorganism Counting: From Classical Image Processing to Deep Learning Approaches. *Artif. Intell. Rev.* **2022**, *55*, 2875–2944. [CrossRef]

29. Chunhachart, O.; Suksawat, B. Construction and Validation of Economic Vision System for Bacterial Colony Count. In Proceedings of the 2016 International Computer Science and Engineering Conference (ICSEC), Chiang Mai, Thailand, 14–17 December 2016; pp. 1–5. [CrossRef]
30. Choudhry, P. High-throughput Method for Automated Colony and Cell Counting by Digital Image Analysis Based on Edge Detection. *PLoS ONE* **2016**, *11*, e0148469. [CrossRef]
31. Minoi, J.L.; Chiang, T.T.; Lim, T.; Yusoff, Z.; Karim, A.H.A.; Zulharnain, A. Mobile Vision-based Automatic Counting of Bacteria Colonies. In Proceedings of the 2016 International Conference on Information and Communication Technology (ICICTM), Kuala Lumpur, Malaysia, 16–17 May 2016; pp. 41–46. [CrossRef]
32. Li, C.; Wang, K.; Xu, N. A Survey for the Applications of Content-based Microscopic Image Analysis in Microorganism Classification Domains. *Artif. Intell. Rev.* **2019**, *51*, 577–646. [CrossRef]
33. Badrinarayanan, V.; Kendall, A.; Cipolla, R. SegNet: A Deep Convolutional Encoder-decoder Architecture for Image Segmentation. *IEEE Trans. Pattern Anal. Mach. Intell.* **2017**, *39*, 2481–2495. [CrossRef]
34. Ronneberger, O.; Fischer, P.; Brox, T. U-net: Convolutional Networks for Biomedical Image Segmentation. In *International Conference on Medical Image Computing and Computer-assisted Intervention*; Springer: Cham, Switzerland, 2015; pp. 234–241. [CrossRef]
35. Clarke, M.L.; Burton, R.L.; Hill, A.N.; Litorja, M.; Nahm, M.H.; Hwang, J. Low-cost, High-throughput, Automated Counting of Bacterial Colonies. *Cytom. Part A* **2010**, *77*, 790–797. [CrossRef] [PubMed]
36. Zhang, C.; Chen, W.B.; Liu, W.L.; Chen, C.B. An Automated Bacterial Colony Counting System. In Proceedings of the 2008 IEEE International Conference on Sensor Networks, Ubiquitous, and Trustworthy Computing (SUTC 2008), Taichung, Taiwan, 11–13 June 2008; pp. 233–240. [CrossRef]
37. Zhang, C.; Chen, W.B. An Effective and Robust Method for Automatic Bacterial Colony Enumeration. In Proceedings of the International Conference on Semantic Computing (ICSC 2007), Irvine, CA, USA, 17–19 September 2007; pp. 581–588. [CrossRef]
38. Garcia Arnal Barbedo, J. An Algorithm for Counting Microorganisms in Digital Images. *IEEE Lat. Am. Trans.* **2013**, *11*, 1353–1358. [CrossRef]
39. Yamaguchi, N.; Ichijo, T.; Ogawa, M.; Tani, K.; Nasu, M. Multicolor Excitation Direct Counting of Bacteria by Fluorescence Microscopy with the Automated Digital Image Analysis Software BACS II. *Bioimages* **2004**, *12*, 1–7. [CrossRef]
40. Ogawa, M.; Tani, K.; Yamaguchi, N.; Nasu, M. Development of Multicolour Digital Image Analysis System to Enumerate Actively Respiring Bacteria in Natural River Water. *J. Appl. Microbiol.* **2003**, *95*, 120–128. [CrossRef]
41. Ates, H.; Gerek, O.N. An Image-processing Based Automated Bacteria Colony Counter. In Proceedings of the 2009 24th International Symposium on Computer and Information Sciences, Guzelyurt, Turkey, 14–16 September 2009; pp. 18–23. [CrossRef]
42. Selinummi, J.; Seppälä, J.; Yli-Harja, O.; Puhakka, J.A. Software for Quantification of Labeled Bacteria from Digital Microscope Images by Automated Image Analysis. *Biotechniques* **2005**, *39*, 859–863. [CrossRef] [PubMed]
43. Brugger, S.D.; Baumberger, C.; Jost, M.; Jenni, W.; Brugger, U.; Mühlemann, K. Automated Counting of Bacterial Colony forming Units on Agar Plates. *PLoS ONE* **2012**, *7*, e33695. [CrossRef]
44. Masschelein, B.; Robles-Kelly, A.; Blanch, C.; Tack, N.; Simpson-Young, B.; Lambrechts, A. Towards a Colony Counting System using Hyperspectral Imaging. *Imaging Manip. Anal. Biomol. Cells Tissues X* **2012**, *8225*, 822510. [CrossRef]
45. Austerjost, J.; Marquard, D.; Raddatz, L.; Geier, D.; Becker, T.; Scheper, T.; Lindner, P.; Beutel, S. A Smart Device Application for the Automated Determination of E. coli colonies on Agar Plates. *Eng. Life Sci.* **2017**, *17*, 959–966. [CrossRef]
46. Boukouvalas, D.T.; Belan, P.; Leal, C.R.L.; Prates, R.A.; Araújo, S.A.D. Automated colony counter for single plate serial dilution spotting. In *Iberoamerican Congress on Pattern Recognition*; Springer: Cham, Switzerland, 2018; pp. 410–418.
47. Alves, G.M.; Cruvinel, P.E. Customized computer vision and sensor system for colony recognition and live bacteria counting in agriculture. *Sens. Transducers* **2016**, *201*, 65.
48. Yoon, S.C.; Lawrence, K.C.; Park, B. Automatic Counting and Classification of Bacterial Colonies using Hyperspectral Imaging. *Food Bioprocess Technol.* **2015**, *8*, 2047–2065. [CrossRef]
49. Zhang, R.; Zhao, S.; Jin, Z.; Yang, N.; Kang, H. Application of SVM in the Food Bacteria Image Recognition and Count. In Proceedings of the 2010 third International Congress on Image and Signal Processing, Yantai, China, 16–18 October 2010; Volume 4, pp. 1819–1823. [CrossRef]
50. Motta, M.d.; Pons, M.N.; Vivier, H.; Amaral, A.; Ferreira, E.; Roche, N.; Mota, M. The Study of Protozoa Population in Wastewater Treatment Plants by Image Analysis. *Braz. J. Chem. Eng.* **2001**, *18*, 103–111. [CrossRef]
51. Akiba, T.; Kakui, Y. Development of an in Situ Zooplankton Identification and Counting System Based on Local Auto-correlational Masks. In Proceedings of the Oceans' 97, MTS/IEEE Conference Proceedings, Halifax, NS, Canada, 6–9 October 1997; Volume 1, pp. 655–659. [CrossRef]
52. Blackburn, N.; Hagström, Å.; Wikner, J.; Cuadros-Hansson, R.; Bjørnsen, P.K. Rapid Determination of Bacterial Abundance, Biovolume, Morphology, and Growth by Neural Network-based Image Analysis. *Appl. Environ. Microbiol.* **1998**, *64*, 3246–3255. [CrossRef] [PubMed]
53. Ferrari, A.; Lombardi, S.; Signoroni, A. Bacterial Colony Counting by Convolutional Neural Networks. In Proceedings of the 2015 37th Annual International Conference of the IEEE Engineering in Medicine and Biology Society (EMBC), Milan, Italy, 25–29 August 2015; pp. 7458–7461. [CrossRef]

54. Ferrari, A.; Lombardi, S.; Signoroni, A. Bacterial Colony Counting with Convolutional Neural Networks in Digital Microbiology Imaging. *Pattern Recognit.* **2017**, *61*, 629–640. [CrossRef]
55. Tamiev, D.; Furman, P.E.; Reuel, N.F. Automated Classification of Bacterial Cell Sub-populations with Convolutional Neural Networks. *PLoS ONE* **2020**, *15*, e0241200. [CrossRef] [PubMed]
56. Perez, A.; Gonzalez, R.C. An Iterative Thresholding Algorithm for Image Segmentation. *IEEE Trans. Pattern Anal. Mach. Intell.* **1987**, *PAMI-9*, 742–751. [CrossRef]
57. Otsu, N. A threshold selection method from gray-level histograms. *IEEE Trans. Syst. Man, Cybern.* **1979**, *9*, 62–66. [CrossRef]
58. Magnier, B.; Abdulrahman, H.; Montesinos, P. A review of supervised edge detection evaluation methods and an objective comparison of filtering gradient computations using hysteresis thresholds. *J. Imaging* **2018**, *4*, 74. [CrossRef]
59. Levner, I.; Zhang, H. Classification-Driven Watershed Segmentation. *IEEE Trans. Image Process.* **2007**, *16*, 1437–1445. [CrossRef]
60. Yuen, H.; Princen, J.; Illingworth, J.; Kittler, J. Comparative study of Hough transform methods for circle finding. *Image Vis. Comput.* **1990**, *8*, 71–77. [CrossRef]
61. Jolliffe, I.T.; Cadima, J. Principal component analysis: A review and recent developments. *Philos. Trans. R. Soc. A Math. Phys. Eng. Sci.* **2016**, *374*, 20150202. [CrossRef] [PubMed]
62. Vishwanathan, S.; Narasimha Murty, M. SSVM: A Simple SVM Algorithm. In Proceedings of the 2002 International Joint Conference on Neural Networks, Honolulu, HI, USA, 12–17 May 2002; Volume 3, pp. 2393–2398. [CrossRef]
63. Li, Y.; Hao, Z.; Lei, H. Survey of Convolutional Neural Network. *J. Comput. Appl.* **2016**, *36*, 2508–2515. [CrossRef]
64. Dai, H.; MacBeth, C. Effects of learning parameters on learning procedure and performance of a BPNN. *Neural Netw.* **1997**, *10*, 1505–1521. [CrossRef]
65. Ghate, V.N.; Dudul, S.V. Optimal MLP Neural Network Classifier for Fault Detection of Three Phase Induction Motor. *Expert Syst. Appl.* **2010**, *37*, 3468–3481. [CrossRef]
66. Oktay, O.; Schlemper, J.; Folgoc, L.L.; Lee, M.; Heinrich, M.; Misawa, K.; Mori, K.; McDonagh, S.; Hammerla, N.Y.; Kainz, B.; et al. Attention U-net: Learning Where to Look for the Pancreas. *arXiv* **2018**, arXiv:1804.03999. [CrossRef]
67. Chen, J.; Lu, Y.; Yu, Q.; Luo, X.; Adeli, E.; Wang, Y.; Lu, L.; Yuille, A.L.; Zhou, Y. Transunet: Transformers Make Strong Encoders for Medical Image Segmentation. *arXiv* **2021**, arXiv:2102.04306. [CrossRef]
68. Cao, H.; Wang, Y.; Chen, J.; Jiang, D.; Zhang, X.; Tian, Q.; Wang, M. Swin-Unet: Unet-like Pure Transformer for Medical Image Segmentation. *arXiv* **2021**, arXiv:2105.05537. [CrossRef]
69. Simonyan, K.; Zisserman, A. Very Deep Convolutional Networks for Large-scale Image Recognition. *arXiv* **2014**, arXiv:1409.1556. [CrossRef]
70. Graham, B. Fractional Max-pooling. *arXiv* **2014**, arXiv:1412.6071. [CrossRef]
71. Zeiler, M.D.; Fergus, R. Stochastic Pooling for Regularization of Deep Convolutional Neural Networks. *arXiv* **2013**, arXiv:1301.3557. [CrossRef]
72. Gulcehre, C.; Cho, K.; Pascanu, R.; Bengio, Y. Learned-norm Pooling for Deep Feedforward and Recurrent Neural Networks. In *Joint European Conference on Machine Learning and Knowledge Discovery in Databases*; Springer: Berlin/Heidelberg, Germany, 2014; Volume 8724, pp. 530–546. [CrossRef]
73. Zeiler, M.D.; Taylor, G.W.; Fergus, R. Adaptive Deconvolutional Networks for Mid and High Level Feature Learning. In Proceedings of the 2011 International Conference on Computer Vision, Barcelona, Spain, 6–13 November 2011; pp. 2018–2025. [CrossRef]
74. Zeiler, M.D.; Krishnan, D.; Taylor, G.W.; Fergus, R. Deconvolutional Networks. In Proceedings of the 2010 IEEE Computer Society Conference on Computer Vision and Pattern Recognition, San Francisco, CA, USA, 13–18 June 2010; pp. 2528–2535. [CrossRef]
75. Boss, R.; Thangavel, K.; Daniel, D. Automatic Mammogram Image Breast Region Extraction and Removal of Pectoral Muscle. *arXiv* **2013**, arXiv:1307.7474. [CrossRef]
76. Wang, H. Nearest neighbors by neighborhood counting. *IEEE Trans. Pattern Anal. Mach. Intell.* **2006**, *28*, 942–953. [CrossRef]
77. Dietler, N.; Minder, M.; Gligorovski, V.; Economou, A.M.; Joly, D.A.H.L.; Sadeghi, A.; Chan, C.H.M.; Koziński, M.; Weigert, M.; Bitbol, A.F.; et al. A Convolutional Neural Network Segments Yeast Microscopy Images with High Accuracy. *Nat. Commun.* **2020**, *11*, 5723. [CrossRef]
78. Kingma, D.P.; Ba, J. Adam: A Method for Stochastic Optimization. *arXiv* **2014**, arXiv:1412.6980. [CrossRef]
79. Dice, L.R. Measures of the Amount of Ecologic Association between Species. *Ecology* **1945**, *26*, 297–302. [CrossRef]
80. Jaccard, P. The Distribution of the Flora in the Alpine Zone. 1. *New Phytol.* **1912**, *11*, 37–50. [CrossRef]
81. Huttenlocher, D.P.; Klanderman, G.A.; Rucklidge, W.J. Comparing Images using the Hausdorff Distance. *IEEE Trans. Pattern Anal. Mach. Intell.* **1993**, *15*, 850–863. [CrossRef]
82. Sim, D.G.; Kwon, O.K.; Park, R.H. Object matching algorithms using robust Hausdorff distance measures. *IEEE Trans. Image Process.* **1999**, *8*, 425–429.
83. Illingworth, J.; Kittler, J. A Survey of the Hough Transform. *Comput. Vision, Graph. Image Process.* **1988**, *44*, 87–116. [CrossRef]
84. Chen, Z.; Zhou, H.; Xie, X.; Lai, J. Contour Loss: Boundary-aware Learning for Salient Object Segmentation. *arXiv* **2019**, arXiv:1908.01975. [CrossRef]
85. Gou, J.; Yu, B.; Maybank, S.J.; Tao, D. Knowledge Distillation: A Survey. *Int. J. Comput. Vis.* **2021**, *129*, 1789–1819. [CrossRef]

MDPI
St. Alban-Anlage 66
4052 Basel
Switzerland
www.mdpi.com

Applied Sciences Editorial Office
E-mail: applsci@mdpi.com
www.mdpi.com/journal/applsci

www.ingramcontent.com/pod-product-compliance
Lightning Source LLC
LaVergne TN
LVHW071304170726
843515LV00006BA/1480

* 9 7 8 3 0 3 6 5 9 2 6 6 4 *